A History of British Actuarial Thought

Craig Turnbull

A History of British Actuarial Thought

palgrave
macmillan

Craig Turnbull
Actuary, Edinburgh,
United Kingdom

ISBN 978-3-319-33182-9 ISBN 978-3-319-33183-6 (eBook)
DOI 10.1007/978-3-319-33183-6

Library of Congress Control Number: 2016955056

Printed on acid-free paper

This Palgrave Macmillan imprint is published by Springer Nature
The registered company is Springer International Publishing AG
The registered company address is: Gewerbestrasse 11, 6330 Cham, Switzerland

To Lachlan and Leyla

Introduction

To paraphrase Winston Churchill, the title of this work doubtlessly requires the reader to indulge the author's terminological inexactitude. It certainly strives to tell a history, though it is told by an actuary rather than a historian. It focuses on a history that unfolded in Great Britain, yet the start of the history pre-dates the formation of that nation (and at times during the writing of this work it appeared as if its publication might post-date its demise).

It is intended to be a work on the development of actuarial *thought*—that is, the intellectual basis for good actuarial professional practice. The story cannot, of course, be told without reference to actuaries' professional bodies and the financial institutions in which they have worked, but this is not a history of those organisations. One of the central themes of how actuarial thought has developed through the centuries has been the way in which the profession has (or has not) utilised relevant ideas that have been developed from outside the profession. A proper examination of this theme requires some familiarity with this external work and an understanding of how it developed historically. Thus a substantial portion of this history of British actuarial thought is spent surveying the work of late-eighteenth-century French probabilists and American financial economists of the second half of the twentieth century.

This perhaps poses the question of whether a history of actuarial thought should be specific to any geographical region. This author would argue that the development of actuarial thought, as loosely defined above, generally bears little comparison with the progress of a universal, physical science. Good professional practices are shaped by the society in which the group of professionals practice, as well as by theoretical or conceptual technical developments: a society's laws, economic development, legal framework, tax regime and the form of financial risk-sharing between individuals, generations and the state that are acceptable to the society's social and political culture all help to define the context and parameters within which the actuarial profession must exercise its role. These factors mean that, for example, the intellectual research interests of British and continental European actuarial professions over the nineteenth and twentieth centuries have often been markedly different (though perhaps increasingly less so by the end of the period).

A history is about the past rather than the present. Our story therefore finishes some years before today—at around the start of the twenty-first century. In addition to providing an appropriate distance from contemporary actuarial activities, this also represents a natural end-point. Around this time, each of the key practice areas of the British actuarial profession—life, pensions and general insurance—had, after some traditionalist resistance, undergone intellectual revolutions of varying degrees that offered the prospect of better coming decades than those experienced by the British profession at the end of the twentieth century. The concluding section of the book, however, provides some final reflections on the possible implications of the tumultuous financial years since the global financial crisis of 2007/8 for actuarial thinking and practice.

This history attempts to span 350 years, the three core actuarial practice areas and the most relevant external developments in the fields of probability and statistics and financial economics. Inevitably, what has been striven for in breadth has necessitated compromises in depth. For the reader whose interests have not been satiated, the author can heartily vouch for the pleasures of reading some of the texts discussed in this work. Those with a deep interest in the history of probability and statistics should read Hacking, Stigler and Todhunter. For the history of financial economics, try Rubinstein. Walford and Ogborn will provide the interested reader with a fascinating history of eighteenth- and nineteenth-century British insurance. And for the most inspirational, original British actuarial thought, the reader will not be failed by the papers of Richard Price, John Finlaison, Frank Redington or Sidney Benjamin.

Acknowledgements

Over the course of my career to date, I have had the great fortune of working with many actuarial thought-leaders who have provoked, stimulated and inspired my own thinking. A few of these deserve particular credit for the direct support, input and advice they have provided throughout the development of this book. I would particularly like to acknowledge Seamus Creedon, Peter McDade and Colin Wilson in this regard.

The librarian team at the Institute and Faculty of Actuaries were unfailingly responsive to my seemingly endless stream of requests for obscure historical journal papers. The library is a remarkable asset of the profession that is easy to take for granted.

As an actuary whose only professional examination failure was in the written communications paper, I never presumed to write a book—it just sort of happened. It could not have happened without the patience, support, love and understanding of my family.

Contents

Author Biography

Craig Turnbull is an Edinburgh-based Fellow of the Institute and Faculty of Actuaries. He has published research papers in leading actuarial journals in the fields of life and pensions and was the recipient of the Institute's Peter Clark Prize for best research paper in 2013. He has served as a member of actuarial working parties at both the institute and the Actuarial Association of Europe. He is a regular speaker at actuarial conferences around Europe, and occasionally further afield, most frequently on the topics of investment, capital modelling and asset-liability management.

Craig was a Managing Director at Barrie & Hibbert, the leading financial risk modelling firm. In his role there he was based for several years in New York as well as in the UK. He is now an Investment Director at a major global asset management house. In his free time, he (clearly) likes reading history.

1

Probability and Life Contingencies, 1650–1750: The First One Hundred Years

Coming at the end of the Renaissance period and some 100 years after the Reformation, the mid-seventeenth century is generally regarded by historians as part of the early modern epoch. We should not, however, infer from this label that for Britain it was a time of stability in its political institutions or sophistication in its financial institutions. England spent the middle of the seventeenth century at war with the Netherlands, Scotland, Ireland and itself. The country recovered quickly from the tyranny of Cromwell's Commonwealth; the arrival of King William III from the Netherlands in 1689 brought peace with the Netherlands but still more wars in Europe, this time with France. In the second half of the seventeenth century, England lagged behind some of northern Europe, and most notably the Netherlands, in the sophistication of its financial systems. King William imported Dutch practices in the raising of long-term government funding, and his costly wars had much use for them.

Marine insurance was a well-established commercial activity in England and continental Europe by 1650. European trade and early colonisation created demand for the pooling or transfer of the risks associated with the transportation of valuable goods over long distances by sea. In contrast, very little life assurance business was conducted at this time. There were no significant financial institutions in England in 1650 that existed for the primary purpose of providing life assurance. The economy was predominantly agricultural. The industrial revolution and the professional middle class that it would foster were over a century away. Life assurance as insurance for a family's loss of a breadwinner's future salaried income was a concept whose time had not yet come.

C. Turnbull, *A History of British Actuarial Thought*,
DOI 10.1007/978-3-319-33183-6_1

This was also the midst of what is sometimes called the Scientific Revolution—the era of Galileo, Newton and Leibnitz. This title reflects the exceptional intellectual progress that occurred in the seventeenth century, which saw the development of the scientific method and the 'mathematisation of nature'. Intellectual progress in the fields of probability and life contingencies up to 1650 had been rudimentary. These disciplines, however, both now started to attract some of the greatest mathematical and scientific minds of this period of remarkable intellectual progress. The initial seeds of development in each of these fields were planted side by side and followed strikingly contemporaneous paths of growth over the next 100 years. The pricing of life contingencies emerged as one of the most relevant empirical applications of the evolving ideas of probability and statistics during this era and some of probability's leading figures actively applied these new concepts to the similarly nascent field of the pricing of life contingencies.

Probability and Life Contingencies in the Mid-seventeenth Century

Probability emerged as a discipline of applied mathematics in the second half of the seventeenth century. At this time, probability was focused on combinatorial problems and was usually applied to games of chance. The general objective was to derive probabilities for the possible outcomes of well-defined games where the underlying probability process was known (for example, the distribution of the sum of n fair dice). Statistical inference is the inverse problem—that is, the estimation of the characteristics of an unknown population based on a limited sample of n data points. Statistical inference followed as an application of probability but, as we shall see later, the transition from probability to statistics involved many mathematical and conceptual difficulties, and only fully emerged in second half of the eighteenth century.

Some antecedent discussions of probabilistic thought can be found in historical literature stretching back at least as far as the ancient Greeks, where gambling was an active pastime. There is some evidence of dice-based gambling games as far back as ancient Egypt. These games prompted some intellectuals to make probing observations in the field of randomness and frequency, but no probabilistic thinking from those days contains any significant mathematical content. There was nothing remotely close to a probabilistic equivalent of Euclid's *Elements*. The reasons for this are the subject of much historical and philosophical debate.[1]

[1] See Hacking (1975) Chapter 1 for a survey of ancient probabilistic thinking and a discussion of what may have limited it.

We have to go forward another 2,000 years to the thinkers of the Scientific Revolution of the sixteenth century before important precursors to modern probabilistic thinking can be identified. Cardano and Galileo provide two such examples. Gerolamo Cardano, a Dutch physician of high repute, wrote *Liber de ludo alae*, a significant book on the application of elementary combinatorial calculations to games of chance in the middle of the sixteenth century. The book analysed such problems as the probability distribution of the sum of two and three dice throws, enumerating the results of the various equally possible combinations of throws and hence outcomes for the sums. It was not published, however, until 1663, 100 years after it was written. Therefore, as its ideas had already been superseded by the time it was published, it was never influential in the early development of probabilistic thinking.

Galileo Galilei was the first of the great mathematicians and scientists who contributed to the development of probability, though his interest was no more than passing. Like Cardano, he considered the distribution of the totals of the throw of a number of dice. Apparently a friend or student noted that three dice can total each of nine or ten through six distinct combinations.[2] Yet gamblers viewed a total of ten as more likely than nine. Galileo tabulated all 216 possible results of the throwing of three dice and showed that a total of ten would be achieved in 27 of the cases whilst nine would arise in 25 cases. *QED*. This is an interesting anecdote in a couple of ways. Firstly, it highlights how tricky raw combinatorial logic can be to the human mind (the friend was correct to say each total could be achieved by six different distinct combinations of three dice scores, but he didn't allow for the fact that these six combinations were not equally likely to occur—a throw of 4, 3, 3 occurs in three different ways (i.e. 4, 3, 3; 3, 4, 3; 3, 3, 4), whereas a throw of 3, 3, 3 can only occur one way. Secondly, it highlights that small differences in probability could be discerned through empirical observation (27/216 versus 25/216), even though no systematic records of empirical frequencies are known to have been made.

Prior to the 1650s, probabilistic thinking was limited to some elementary combinatorial analysis of games of chance such as those of Cardano and Galileo—probability at this time might be considered 'pre-mathematical'. Conceptually, it was confined to calculating the ratio of favourable events to a total number of equally possible events in well-defined circumstances. So what of the practice of life contingencies during this period? There were two forms of life contingency in Renaissance Europe—short-term life insurance and life annuities—and their histories are quite different.

[2] A total of ten can be produced by six distinct combinations: 6, 3, 1; 6, 2, 2; 5, 3, 2; 5, 4, 1; 4, 4, 2; 4, 3, 3. A total of nine can be produced by the six distinct combinations: 6, 2, 1; 5, 3, 1; 5, 2, 2; 4, 4, 1; 4, 3, 2; 3, 3, 3.

As noted above, marine insurance and, to a lesser degree, fire insurance developed into significant commercial activities much earlier than life insurance. Marine insurance was established in major European ports such as Genoa, Barcelona and Amsterdam from as early as the fourteenth century. Some short-term life insurance policies were written in England from as early as the sixteenth century, but such life insurance writing was essentially an ad hoc offshoot of the core activities of the established marine insurance underwriters. No underwriters appear to have existed in this period for the primary purpose of providing life insurance products.

Approaches to the pricing of these life insurance contracts would have been similar to those used by the underwriters in other areas of their business. Even though significant volumes of marine insurance were being written in the sixteenth century and earlier, no attempt was made to record the statistical frequency of shipwrecks until the late seventeenth century. No intellectual apparatus for statistical inference existed at this time, and even if it had done, it would arguably have been of limited use: given the variability of the political and economic environments of the times, the past may simply not have provided much useful information about the future. Standardised tables of maritime insurance prices did arise in various ports—insurers were clearly risk-sensitive—but their methods were not derived from statistical analysis of empirical data. Similarly, life insurance at the time would have been priced based on rules of thumb and with regard to the specific known characteristics of the life insured.

One of the main uses of early life insurance business was to provide protection from the risk that significant creditors died before repaying their debts. It was also used to speculate on the potential impending deaths of famous people of the time, such as the pope or monarchs. Such was its association with gambling that life insurance was made illegal in most of continental Europe (but not in England) by the end of the seventeenth century and this remained the case until the nineteenth century.

Annuities were more established than short-term life assurance as a form of life contingency contract in Europe in the Renaissance period. Whilst the speculative abuse of life insurance resulted in it being outlawed, annuities, on the other hand, were a useful instrument for borrowing in a lawful way: the riskiness of the cashflows was a crucial characteristic in circumventing the usury laws of the time. As a result, many European cities of the fifteenth, sixteenth and seventeenth centuries issued annuities as a form of municipal funding. No systematic allowance for age was made in setting these annuity prices, though in some cases individual underwriting may have been used to set the annuity price.

Pascal and Fermat (1654)

Blaise Pascal and Pierre de Fermat's correspondence in a series of letters over the summer of 1654 represents the starting point for mathematical probability. The two men were the pre-eminent mathematicians of their generation (Newton and Leibnitz came a little later), and they achieved lasting fame for their work in the physical sciences and number theory respectively. They initiated a long line of seventeenth and eighteenth century mathematicians and physicists of the highest order who made fundamental contributions to probability and statistics whilst dedicating most of their time to other areas of scientific research. Halley, De Moivre, Gauss and Laplace certainly fall under this category. Leibnitz might also be added, though his contributions to probability were less direct.

Pascal and Fermat's letters discussed solutions to what was known as the 'problem of points'. Suppose we have a game of chance between two players where the winner is the first player to reach a specified number of points and the winner receives the two stakes of the players. The two players play consecutive rounds until one player reaches the required number of points; in each round one player wins one point, and we assume the players have equal probability of winning any round. If the players agree to stop playing part-way through their game, how should they split the stakes? To take the specific example used in the letters: if the stake to play the game is 32 pistoles, the winner is first to reach three points; the first player has two points and the second player has one point, so how should the total 64 pistoles be split between the two players?

The problem of points was not new—it was discussed at length in fifteenth century Italy and then again by Cardano in the sixteenth century—but a solution had continually eluded each period's best mathematicians. It may appear surprising that such a seemingly trivial problem caused the finest mathematical minds of the day such difficulty. After all, the first half of the seventeenth century saw complex mathematical developments being applied in fields such as astronomy with startling results. But the maths of the problem of points was not the hard part: the type of thinking required to solve it was *conceptually* new.

Pascal and Fermat both independently arrived at the correct solution to the problem of points, albeit in subtly different ways. Both approached the problem by defining the *mathematical expectation* of each player's pay-out from the game as the fair settlement. This mathematical formulation of expectation was the crucial conceptual breakthrough required for the problem of points and it represented new territory. The Pascal and Fermat letters provide the first recorded instance of mathematical expectation being explicitly calculated. But neither of them used the term 'expectation' in their correspondence. It was the Dutch mathematician, Christiaan Huygens, who first introduced the term 'expectatio' in the Dutch translation of his 1657 paper on why it was appropriate to use mathematical expectation to price claims on uncertain cashflows (he focused mainly on fair lotteries).[3] Pascal and Fermat's 1654 correspondence was not published until 1679, but Huygens visited Paris in 1655 and, whilst he did not meet Pascal or Fermat, he was introduced to their ideas by mutual acquaintances.

The idea of using mathematical expectations to value uncertain claims can be viewed as a more technical treatment of the already established legal doctrine of equity, and its application to the treatment of aleatory contracts.[4] The term 'aleatory' referred to contracts where there was a settlement of a fixed payment today in exchange for uncertain future cashflows. Insurance and annuities fell under this category, and so too did a wider array of contractual arrangements such as settlement of inheritance expectations or, more generally, risky business investments. Like all contract law, a principle of fair and equitable treatment for both parties existed. For aleatory contracts, a qualitative, heuristic notion of expectation was already part of established law in the mid-seventeenth century. This did not make explicit use of probabilities, but did recognise degree of likelihood as an important factor in determining equitable contract settlements. From here, the development of mathematical expectation can be viewed as a natural quantitative development.

Fermat's solution to the calculation of mathematical expectation was to tabulate all the possible (and equally likely) combinations of wins/losses that could occur over the remainder of the game. He then worked out the probability of each player winning by enumerating the winning combinations and dividing by the total possible number of combinations. The equitable split of the stakes was then found as the winning probability multiplied by the total stake, i.e. the expected pay-off from the game. This basic concept of aleatory probability as a ratio of the number of equally possible favourable events to the total number of equally possible events was not new—as we saw above, it was used by Cardano and Galileo a century earlier. The breakthrough was in the conception of mathematical expectation—the probability-weighted value—and its use as a measure of fair value. But from a mathematical perspective,

[3] Huygens (1657).

[4] See, for example, Section 1.3, Daston (1988) for a scholarly discussion of the legal doctrine of equity and its influence on the development of mathematical expectation.

Fermat's approach is unexciting. Furthermore, the process of tabulating each possible combination would be increasingly computationally cumbersome as the number of players or the number of points required to win the game became larger.

Pascal's proposed method was essentially mathematically identical to Fermat's, but had a subtle difference in implementation which was arguably quite profound. Pascal's solution used a method of backward recursion to calculate the mathematical expectation that applied at any point in the game. Todhunter's translation[5] of Pascal's explanation of his approach is worth recounting in full:

> Suppose that the first player has gained two points and the second player one point; they have now to play for a point on this condition, that if the first player gains he takes all the money which is at stake, namely 64 pistoles, and if the second player gains each player has two points, so that they are on terms of equality, and if they leave off playing each ought to take 32 pistoles. Thus, if the first player gains, 64 pistoles belong to him, and if he loses, 32 pistoles belong to him. If, then, the players do not wish to play this game, but to separate without playing it, the first player would say to the second 'I am certain of the 32 pistoles even if I lose this game, and as for the other 32 pistoles perhaps I shall have them and perhaps you will have them; the chances are equal. Let us then divide these 32 pistoles equally and give me also the 32 pistoles of which I am certain'. Thus the first player will have 48 pistoles and the second 16 pistoles.
>
> Next, suppose that the first player has gained two points and the second player none, and that they are about to play for a point; the condition then is that if the first player gains this point he secures the game and takes 64 pistoles, and if the second player gains this point the players will then be in the situation already examined, in which the first player is entitled to 48 pistoles, and the second to 16 pistoles. Thus if they do not wish to play, the first player would say to the second 'If I gain the point I gain 64 pistoles; if I lose it I am entitled to 48 pistoles. Give me then the 48 pistoles of which I am certain, and divide the other 16 equally, since our chances of gaining the point are equal'. Thus the first player will have 56 pistoles and the second player 8 pistoles.
>
> Finally, suppose that the first player has gained one point and the second player none. If they proceed to play for a point the condition is that if the first player gains it the players will be in the situation first examined, in which the first player is entitled to 56 pistoles; if the first player loses the point each player has then a point, and each is entitled to 32 pistoles. Thus if they do not wish to play, the first player would say to the second, give me the 32 pistoles of which I am certain and divide the remainder of the 56 pistoles equally, that is, divide 24

[5] Chapter II, Todhunter (1865).

pistoles equally. Thus the first player will have the sum of 32 and 12 pistoles, that is 44 pistoles, and consequently the second will have 20 pistoles.

This is the first record of a backward recursive method being used to evaluate expectations through a path of stochastic steps. Its appeal to Pascal lay in its relative computational efficiency and elegance, but, in retrospect, it offers a great insight into the behaviour of the prices of assets with claims on uncertain cashflows: his method provides insight not only into how to price the claim at any given point in time, but also on *how that price will change as uncertain events crystallise*. Those with a familiarity with standard option pricing theory will immediately recognise the similarity between Pascal's logic and the binomial tree approach to option pricing[6] (which is used both to intuitively illustrate the mathematics of option pricing to the uninitiated and to provide solutions to path-dependent option pricing problems that are too complex for analytical treatment). But the path behaviour of asset prices was not of interest to Pascal—when he realised that the arithmetical triangle could efficiently identify the binomial coefficients required by Fermat's 'brute force' method, he advocated using this approach to solving the problem of points and similar combinatorial problems.

In summary, Pascal and Fermat' s solutions to the problem of points weaved together a handful of concepts that were new or at best half-baked at the time of their writing:

- Mathematical expectation as the probability-weighted sum of uncertain outcomes, where the probability is calculated by defining the set of exhaustive and equiprobable outcomes and counting the relevant sub-set.
- The principle that an equitable claim (price) on an uncertain cashflow should be set by assessing the mathematical expectation of the cashflow. This can be viewed as an explicit probabilistic rendering of the legal concept of equity that had been an established part of the civil law relating to 'aleatory' contracts and which was applied using heuristic judgement.
- Backward recursion as an efficient way of calculating expectations (and hence prices) through stochastic paths.
- The arithmetic triangle as an efficient computational means of producing binomial coefficients. Pascal did not invent the arithmetic triangle (Cardano discussed it in the sixteenth century, and the Chinese and other ancient civilisations are also thought to have known of it). But Pascal was the first to identify its use in binomial expansions and how that could be applied to probability problems.

[6] Cox et al. (1979).

Together, these ideas and insights represent a fundamental breakthrough in thinking, both with respect to mathematical probability and to the related topic of valuing claims on uncertain cashflows.

John Graunt, Johan de Witt and Their Prototypical Mortality Tables (1662–1671)

Amongst the small group of men who made profound contributions to probability and actuarial thinking in the seventeenth century, John Graunt stands out for his relative mediocrity. He was not a world-leading mathematician of his generation (Pascal, Fermat), nor was he an Astronomer Royal (Halley), nor an international statesman of historical significance (de Witt). Graunt appears to us as a middling merchant, a son of a tradesman without formal academic training or any trace of earlier exposure to scientific analysis, armed only with a peculiar interest in the London Bills of Mortality.

The London Bills of Mortality recorded the numbers of deaths and christenings in the parishes of London. They were produced weekly and continuously from 29 December 1603. The Bills first appeared in 1592, but were shortly discontinued before being resumed in 1603. Both 1592 and 1603 were years of heavy mortality arising from the plague. Part of the government's motivation for the production of the Bills appears to be to ease the panic induced in the London population by showing that the impact of the plague on human mortality was not as severe as rumoured.[7]

Graunt's paper 'Natural and Political Observations made upon the Bills of Mortality' was published in 1662. It foreshadowed the use of statistical analysis of population data to steer government social policy that would emerge 150 years later with the development of government statistical offices in Britain, France, Germany and Prussia. For example, Graunt observed that the statistical rate at which beggars died from starvation was very low, and suggested that, as they were already living off the wealth of the nation, they might as well be paid a guaranteed wage to keep them off the streets. With the use of statistical observations to form such policy arguments, it is arguable that Graunt could be viewed more naturally as the antecedent of Condorcet, Quetelet and nineteenth-century 'social mathematics' rather than an as an early pioneer of actuarial science.

As a source of statistical data to estimate mortality rates, the Bills of the seventeenth century had serious limitations. The data provided no direct

[7] Francis (1853).

information on the distribution of the population by age (so there was no total 'exposed to risk' available from the data). Even more fundamentally, the information included in the records of deaths did not show the age of the deceased. Inevitably, any table of mortality rate as a function of age derived from this data would require some giant assumptions and extrapolations. Undaunted, Graunt made several such assumptions to produce what can be recognised as the first mortality table based on explicit statistical analysis.

Though the seventeenth-century London Bills of Mortality did not include the age of the deceased, in the 1620s some parish records did start to include a cause of death. The format of these records varied from parish to parish. Some simply recorded whether the deceased had died of plague or not. By 1632, the parish of Westminster used a total of 63 categories to attribute the causes of death (including, for example, 'bit with a mad dog'). The most common causes were 'Consumption' and 'Fever', and 'causes' that were actually references to the age of the deceased: 'Aged' and 'Chrisomes,[8] and Infants'. This data was recorded by 'Searchers', a form of local government official. They were generally not medically trained and, for social and cultural reasons, may not have been entirely unbiased in their reporting of the cause of death. Graunt felt compelled to make rough adjustments to their records—for example, he decides that 'we shall make it probable, that in years of plague, a quarter-part more dies of that disease than is set down'.

The cause of death data was the key to the deductions Graunt made in developing his mortality table. The table is not a major focus of his paper and Graunt is not explicit in his workings. He identified the causes of death that were most associated with death at a young age, and from this he estimated that slightly more than one third of deaths occured by the age of six. He made the assumption that, of every 100 births, there would be one survivor by age 76. In the intervening 70 years, he assumed there was an equal probability of dying in each decade. This implies an annualised mortality rate of around 8 % over the first six years of life and around 5 % for the next 70 years.[9]

Graunt's analysis was conjectural and heuristic rather than statistically rigorous. His focus was on providing sociological insight, not on making improvements in the pricing of life contingencies. Nonetheless, his work on the analysis of population mortality data broke new quantitative ground and provided a starting point for the actuarial thinking that would later be pursued by some of its greatest names. Whilst the publication of his paper had no

---

[8] Infants that have not yet been christened.

[9] The mortality rates implied by Graunt's survival table increase after age 56, but this arises due to rounding of the number of remaining survivors from a pool that begins with the arbitrary total of 100 ('for men do not die in exact proportions nor in fractions') rather than by explicit design.

discernible impact on what limited life insurance business was practiced at the time, it did highlight the latent potential in existing population records and the possibilities of statistical analysis. His paper was reviewed by William Petty in a Paris journal in 1666. In 1667, France started to collect statistical data of a similar form to that found in the London Bills. The influence and impact of Graunt's work was confirmed when he was made a Fellow of the Royal Society at the behest of King Charles II, who is said to have commended him to the sceptical and snobbish fellows with the remark 'that if they found any more such tradesman, they should admit them all'.[10]

Johan de Witt's life contrasts starkly with the humble background of John Graunt. De Witt was born in 1825 as the son of an influential politician, and by the age of 28 he had obtained the position of grand pensionary—roughly equivalent to prime minister—of the States of Holland, and held it for the following nineteen years. The Dutch Republic was one of the most significant European powers throughout this period, and de Witt established himself as one of Europe's pre-eminent statesmen of his time. His political career, and ultimately his life, was cut short in 1672 when 120,000 French troops of Louis XIV invaded Holland. De Witt resigned but was nonetheless assassinated, along with his brother, by a mob supportive of his political rival Prince William of Orange, the future King William III of England, Ireland and Scotland (William II in Scotland).

As a youngster, de Witt received tutelage from some of Holland's leading thinkers of the time, from which he obtained a lifelong interest in mathematics. Between plotting geopolitical strategic alliances and fighting wars with England and France, he also devoted some time to the theory of annuity pricing. His interest in actuarial science was not driven entirely by mere mathematical curiosity. The Netherlands was advanced in the sophistication of its approach to raising government funding and by the mid-seventeenth century it had a long-established practice of using perpetuities and life annuities to issue state debt. By contrast, England at the time had no facilities for raising funded government debt until the final years of the century (it had experimented with issuing life annuities in the first half of the sixteenth century, but never in significant volume).[11] De Witt was concerned that the state was paying too much for its life annuity funding relative to the cost of perpetual annuity funding, and he proceeded to produce the first rigorous analysis of annuity pricing to make his point.

[10] Francis (1853), Chapter 1, p. 11, Hacking (1975), Chapter 12, p. 106.

[11] See Homer and Sylla (1996), p. 112.

As a graduate of law and mathematics, de Witt was well equipped to apply mathematical expectation as a form of legal equity. In a similar vein to Graunt, de Witt made some sweeping assumptions about mortality rates. He weaved together Graunt's approach to basic mortality modelling with the mathematical expectations concept of valuing aleatory contracts pioneered by Pascal, Fermat and others such as Huygens, so as to develop prices for life annuities.

In 1671 de Witt wrote a series of letters to the Estates General (the body responsible for Dutch government funding) where he set out his analysis of annuity pricing. He argued that with perpetual debt achieving a price of 25 times its annual interest, life annuities should be priced higher than the fourteen times annual income at which they were being sold at the time of writing.

He first suggested that an uncertain cashflow should be valued at its expectation: 'The value of several equal expectations or chances, a certain sum of money or other objects of value pertaining to each chance, is found to be exactly determinable by adding the money or other objects of value represented by the chances, and by then dividing the sum of this addition by the number of chances: the quotient or result indicates with precision the value of all these chances.'[12] His demonstration of the reasonableness of this assumption borrows liberally from his compatriot Huygens and showed how it was consistent with the result obtained from a series of equitable exchanges of aleatory contracts.

In a similar heuristic style to Graunt, de Witt assumed that mortality rates were uniform between the ages of three and 53. An important clarification is required when describing what de Witt meant by a uniform mortality rate assumption. He meant that, if we consider a starting pool of people of a given age, a constant number from that pool would die every future year. The size of the pool shrinks as people die, and so the mortality rate at which the survivors consequently die is always increasing. Thus, de Witt supposed that, if we start with a pool of 128 lives aged three, two would die every year over the next 50 years, leaving 28 remaining alive aged 53. This implies that the mortality rate for a 53 year-old is more than four times higher than for a three year-old.[13] Such a progression in mortality rates would seem intuitively desirable to modern eyes, but it does not occur in de Witt's table by explicit design. De Witt did not appear to appreciate the increase in probability of death that the specification of a constant number of deaths from a fixed starting pool implied—the modern conception of a mortality rate had simply not yet been realised. In contrast to Graunt, de Witt was happy to work with fractional lives in his projected pool: he assumed that between ages 53 and 63 people die from the remaining pool (of 28) at a rate of 4/3 per year; between 63 and 73

[12] De Witt (1671).

[13] $1 - 28/30 = 6.7\ \%$; and $1 - 126/128 = 1.6\ \%$.

they die at a rate of one per year; and between 73 and 80 they die at a rate of 2/3 per year. All 128 lives are therefore extinguished by age 80.

Unlike Graunt, de Witt offered no empirical basis for these mortality assumptions. De Witt had some correspondence on these assumptions with Johannes Hudde, a contemporary mathematically minded Dutch politician who was mayor of Amsterdam during the period. Hudde developed an empirical mortality analysis from the records of the annuities sold by the Dutch government between 1586 and 1590. This qualifies as the first mortality experience analysis of annuity business. He grouped the data by age of the annuitant when they purchased the annuity and tabulated the number of years for which each annuitant received their annuity payment. De Witt was satisfied that this data was consistent with his assumptions, noting that the data suggested that the proportion of 50 year-olds dying by age 55 was 1/6, and of 55 year-olds dying by age 60 was 1/5 (although de Witt's assumptions implied both these proportions would be very close to ¼).

Armed with the above mortality decrements and the 4 % interest rate derived from the perpetual annuity price, de Witt then undertook the arithmetic computation of mathematical expectations discounted by the time value of money in order to calculate the price of an annuity for someone aged three on purchase. He found that the annuity price was sixteen, thus supporting his initial statement that the price of fourteen, at which the annuities were currently being sold, was too low. De Witt then went on to argue that the selection effect 'of choosing a life, or person in full health, and with a manifest likelihood of prolonged existence' should significantly increase the price of the annuity further. This is notable as perhaps the first published discussion of the impact of selection on the price of life contingencies.

De Witt's work can be viewed as a synthesis of the probabilistic and valuation thinking developed by Pascal, Fermat and Huygens together with the pragmatic mortality modelling introduced by Graunt. Its originality lay in demonstrating how these emerging ideas could be used to rationally obtain fair prices for annuities, and how such pricing must consider practical effects such as anti-selection. The application of the then-recent developments in probabilistic thinking to the valuation of life contingencies had begun.

Edmond Halley's Breslau Table (1693)

Some forty years after the ground-breaking correspondence of Pascal and Fermat, another excellent mathematician dabbled in the emerging subjects of probability and the valuation of life-contingent claims. Edmond Halley, his interest piqued by the British government's plans for a significant issuance of life annuities, presented a paper to the Royal Society in 1693 on mortality modelling and annuity pricing.[14] It is little more than a footnote in the published work of the famed Astronomer Royal, but the paper fundamentally advanced the methods of mortality modelling and what we might now call actuarial statistics.

Over the preceding 40 years, several seeds of actuarial thought had been planted. Probability as a discipline of applied mathematics had been tentatively established and had attracted some of the greatest mathematical minds of the generation. This calculus of probabilities had been synthesised with contemporary legal doctrine to provide an intellectual basis for the use of mathematical expectations in valuing future uncertain cashflows, such as those arising from life annuities. Graunt's analysis of the empirical data provided by the London Bills of Mortality illuminated the possibilities of what might be inferred from the analysis of existing datasets. De Witt had taken similar mortality assumptions to those of Graunt and had applied them to develop prices for life annuities.

Up until this time, however, empirical mortality data had not been used to provide rigorous or granular insights into how mortality rates varied by age. Graunt and de Witt's mortality tables had relied on sweeping assumptions and simplifications and were not based on what would be recognised today as statistical fitting to a dataset. This was largely because of the limitations in the empirical data. In particular, the lack of data on both age at death and the numbers alive at each age severely limited the opportunity to conduct a straightforward statistical analysis of probability of death by age.

In his Royal Society paper of 1693, Halley first identified the shortcomings in the London Bills of Mortality data used by John Graunt. He noted that the population numbers of each age were unknown; that the ages of the people dying were unknown; and that the population size was unstable due to immigration flows. Halley, apparently with the assistance of Leibniz,[15] located bills of mortality for the German town of Breslau for the years of 1687–1691. The Breslau bills were the only such records known of at the time that recorded age at death. Halley also argued that the town of Breslau was a small, sleepy place that did not experience much immigration or emigration compared to the growing metropolis of London. Thus, two of three defects in Graunt's London data were arguably addressed by the Breslau bills.

Halley used this superior mortality data to produce the first mortality table with a granular description of how mortality rates varied as a function of age.

[14] Halley (1693).

[15] Hacking (1975), p. 113.

However, his paper is silent on the quantitative steps he took to transform the five years of raw data into this mortality table. Specifically, he does not explicitly describe what steps he took to work around the remaining defect of the data—the unknown total number alive at each age. Halley states that he assumed a stable population. He most likely used this assumption to infer the numbers alive at each age from the numbers that died: if the population is stable, the numbers alive at any given age must be equal to the sum of all those dying at that age and older. This observation allows the mortality rates to be inferred without any explicit data on the numbers alive. The approach is well-explained by Richard Price[16] some 80 years after Halley's paper:

> In every place that just supports itself in the number of its inhabitants, without any recruits from other places; or where, for a course of years, there has been no increase or decrease, the number of persons dying every year at any particular age, and above it, must be equal to the number of living at that age.... From this observation it follows, that in a town or country, where there is no increase or decrease, bills of mortality which give the ages at which all die, will show the exact number of inhabitants; and also the exact law, according to which human life wastes in that town or country.

Figure 1.1 compares the three seventeenth-century mortality tables discussed so far. Halley's table has a regularity that is absent from that of both Graunt and de Witt, whose mortality rates are subject to jumps that result from arbitrary assumptions. Halley's access to improved data permitted an advancement in methodology that resulted in a mortality table recognisable to modern eyes. To the extent there was such a thing, it remained the standard reference mortality table until the second half of the eighteenth century.

Halley's paper went on to apply his mortality table to the valuation of life annuities for both single and joint lives (and, in keeping with the time, these include novel geometric representations!). In a similar vein to De Witt, Halley concluded that government annuity prices appeared cheap relative to long-term government bonds. Halley noted that with the long-term interest rate at 6 %, his table implied a fair annuity rate of less than 10 % for a 40 year-old, yet the government's annuity pricing would offer an annuity income of 14 %.

Halley's motivation for his mortality and annuity investigation arose from the English government's plans to imitate the Dutch and make more use of life annuities as a means of government borrowing. We now turn to the English government annuity issuance that occurred in the year of the publication of

[16] Price (1772), Essay IV.

Fig. 1.1 Three seventeenth-century mortality tables

Halley's paper, and consider it alongside the other practices that were taking place in the life contingencies business at the end of the seventeenth century.

Life Contingencies at the End of the Seventeenth Century

The sale of life annuities is known to have occurred since late Roman times and perhaps even earlier. The Roman juror Ulpian produced a table of annuity prices in the third century AD[17]: a life annuity would cost a twenty year-old 30 times its annual income, whilst a 60 year-old's annuity would cost seven. The need for these annuity prices was not driven, however, by an active market in annuities. Rather, it was common for bequests to take the form of life incomes, and a capital value for these income streams was required to assess estate duty.[18] Further, there is nothing to suggest that a statistical approach to mortality modelling was used in the derivation of these prices. In Tudor England, government life annuities were sold at £7 for £1 of annual income (an annuity rate of around 14 %), irrespective of age. Interest rates at the time were around 12 %,[19] suggesting this annuity pricing was not particularly generous.

[17] Hacking (1975), Chapter 13, p. 111.

[18] Ogborn (1953), Phillips in Discussion, p. 196.

[19] Chapter IX, Homer and Sylla (1996).

Following the glorious revolution of 1688, King William III was in much need of funding for his European wars, and the English government looked to raise £1 million through the issuance of life annuities in 1692. Interest rates had fallen significantly since the Tudor annuity issuance: between the sixteenth and late seventeenth centuries, English interest rates had dropped from around 12 % to 6 %. Despite this substantial fall in interest rates, the 1692 annuity issuance was offered at exactly the same price as the one of the 1540s! Once again, the annuities were priced at a multiple of seven times annual income with no variation by age or sex of the nominated life.

Halley's Breslau table implied that the government's annuity pricing was extremely generous. He included a table of life annuity prices using an interest rate of 6 %, a reasonable assumption for the long-term risk-free interest rate prevailing at the time.[20] The government also sold 99-year fixed-term annuities at a price of 15.5 × annual income,[21] again consistent with the 6 % interest rate assumption. Using his table and this interest rate assumption, Halley produced an annuity price for a ten year-old of 13.4 × annual income; for a 40 year-old a price of 10.6; and for a 60 year-old a price of 7.6. Despite the apparent generosity of the government pricing, the life annuities sold poorly—less than £110,000 of the targeted £1 million was raised from the sale of annuities on 1,002 lives.[22] But many of the investors who did participate were well aware of the opportunity created by the age-insensitive annuity pricing: more than half of the nominated lives were under eleven years of age, and the annuities were therefore obtained at almost half of the fair price implied by Halley's Breslau table.

Full records of the mortality experience of these annuitants have not been found, but it is known that 503 of the 1,002 lives were alive in 1730; 175 lives were still alive in 1749 and that the last survivor died in 1783.[23] This mortality experience is quite similar to the ten year-old in Halley's table. The government borrowed at a realised interest rate of 12 % when they could have borrowed at a rate of 6 % by issuing long-dated gilts.

The historical literature offers different interpretations of the influence that Halley had on the government's annuity pricing policy. Some modern historians have suggested that his annuity pricing recommendations were rejected by the government.[24] But Walford[25] noted that the government subsequently

[20] Homer and Sylla (1996) p. 127.

[21] Francis (1853), Chapter 3, p. 55.

[22] Leeson (1968), p. 1.

[23] Leeson (1968), p. 1.

[24] Daston (1988), p. 139.

[25] Chapter IV, Walford (1868).

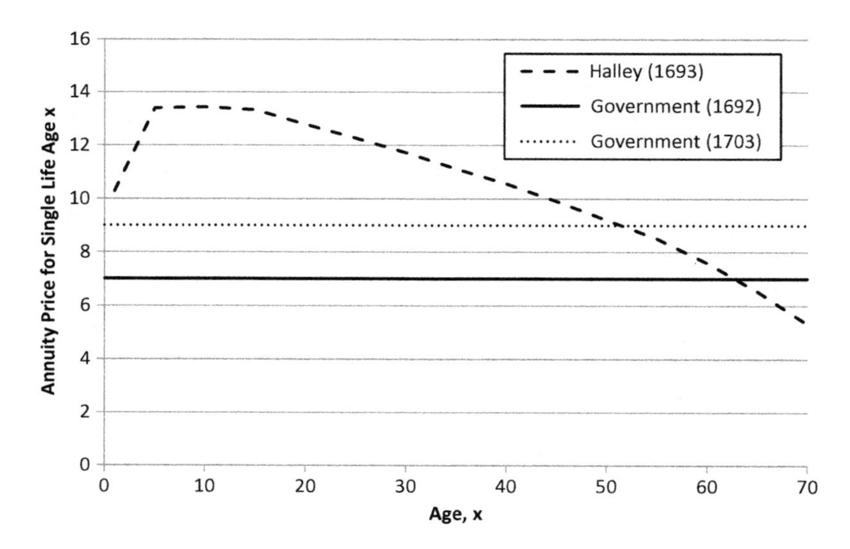

Fig. 1.2 Government annuity prices and Halley

increased its single life annuity prices in a 1703 Act of Parliament from a multiple of annual income of seven to nine, and Francis attributes this revision to the influence of Halley's paper.[26] The conventional contemporary historical position is that the early mortality tables had no discernible impact on the practice of insurers or governments in their pricing of life contingencies.[27] Figure 1.2 compares Halley's annuity prices with the government prices of 1692 and 1703.

Irrespective of its direct impact on contemporary government annuity pricing policy, Halley's paper introduced a new standard of method for mortality modelling and the Breslau table remained the benchmark mortality table until well into the second half of the eighteenth century.

We saw above how, despite the emergence of an actuarial approach to annuity pricing, at the end of the seventeenth century the English government did not feel compelled to move to an annuity pricing approach that priced annuities as a function of age (or indeed any other variable). Virtually no life insurance business existed in the private sector at this time. The growth of this industry would gather pace in the following decades; but in the final decade of the seventeenth century, there were only one or two examples of businesses that were transacting life contingencies.

[26] Francis (1853), p. 59.

[27] For example, see Daston (1988) Section 3.4, Hacking (1975), p. 113.

The Mercers' Company, established in London, was one such example. It launched a form of reversionary annuity product in 1698. This product paid a widow an annual income of £30 for the remainder of her life following the death of her husband for every £100 of single premium. This annuity rate did not vary as a function of the age of either the husband or the wife. Mercers was ultimately unable to pay the annuities promised by this product. It reduced the size of the promised annuity payments by one third in 1738 and finally required an Act of Parliament in 1747 to facilitate a partial government bail-out.

Halley's Breslau table was published several years prior to the launch of the Mercer reversionary annuity product. It is natural to consider what annuity rates were implied by the Breslau table and how they compared to the practice of the Mercers' Company. The Breslau table implies that the fair annuity rate varies substantially as a function of the age of husband and wife. For a husband and wife both aged 40, it implies that the Mercer product's promised annuity income was actually lower than the fair value—at 6 % interest and before allowance for expenses, the Breslau table implies an annuity income of around £45 per £100 of single premium. But for a husband aged 60 and wife aged 30, an annuity income of only £20 can be provided for according to the table. So the general level of the pricing is not obviously inconsistent with Halley's table. But the fair pricing of the product is highly sensitive to the ages of the husband and wife (and, in particular, to the difference in their ages). The decision not to vary the pricing by age of husband or wife therefore heavily exposed the company to selection effects.

Today we can only speculate as to whether it was this selection risk that was the root cause of the business's failure. The experience with the 1693 government annuity issuance, where most of the nominated lives were under eleven years of age, suggests enough of an understanding of annuity pricing existed for investors to take advantage of this selection opportunity. Many decades later, in 1772, Richard Price, of whom we will hear much more later, wrote of the fall of the Mercers' Company: 'The rapid fall of interest of money; their admitting purchasers [i.e. husbands] at too advanced ages; and, particularly, their paying no regard to the difference of age between husband and their wives, must have contributed much to hurt them.'[28]

The Mercers' Company also offered one-year life assurance policies that paid a lump sum on death. The lump sum, however, was not a fixed amount. The total premiums (plus income less expenses) received from the cohort of lives purchasing insurance for the year were divided at the end of the year

[28] Price (1772), p. 105.

amongst the policies of those who had died. No mortality risk was transferred to the insurer, the company simply acted as an administrator of the direct pooling of mortality risk amongst the policyholders.

In summary, at the start of the eighteenth century, the theoretical probabilistic and actuarial developments of the preceding 50 years had not yet significantly impacted on the life contingencies pricing practices of insurers or governments. Relatively little mortality risk transfer took place in this era, and the examples we do have are mainly of life annuities which resulted in significant losses or failures for the recipients of the risk (whether government or private sector company).

Jacob Bernoulli and the First Law of Large Numbers (1692/1713)

Jacob Bernoulli (1654–1705) was the first in a veritable dynasty of Bernoullis who made significant contributions to mathematics and physics in the eighteenth century. At least five of the clan published on probability, but none with greater influence than Jacob. His book, *Ars Conjectandi*, is a seminal piece in the history of probability and statistics. It likely was written over the 1680s and 1690s—and the most important part, the first limit theorem in probability, is believed to have been written in 1692—but it was never published in Bernoulli's lifetime. His nephew Nicolas, another famous mathematical Bernoulli (of whom more will be said later), edited and finally published the book in 1713, several years after Jacob's death.

The first part of *Ars Conjectandi* considers combinatorial problems and games of chance in a similar way to those dealt with by Pascal, Fermat and Huygens over the preceding decades. This section of the book is most notable for containing the first derivation of the binomial probability distribution. But the most important section of the book is in Chap. 4, where Bernoulli derives the first limit theorem in probability: Bernoulli's Theorem, or what is often now referred to as the *weak law of large numbers*. The theorem is the first to consider the asymptotic behaviour of a random sample from a population with known characteristics.

Bernoulli uses the analogy of an urn filled with black and white balls, where the proportions of black and white balls in the urn are known—say the proportion of all balls that are black is p. n balls are randomly selected, with replacement, from the urn. The number of black balls, m, in the n randomly sampled balls is a random variable, as is the sampled proportion of black balls, m/n. At the time of writing, the intuition that a larger sample size n would result in m/n being a more stable and reliable estimate of p already existed. Bernoulli writes: 'For even the most stupid of men, by some instinct of nature, by himself and without any instruction (which is a remarkable thing), is convinced that the more observations have been made, the less danger there is of wandering from one's goal.'

But no quantification of how the accuracy of a sample increased with sample size had been developed. Even more vitally, it had yet to be determined whether there was some fundamental limit to the amount of certainty that a random sample could provide, even for very large sample sizes. Bernoulli's treatment of these questions arguably signifies the moment where probability emerges as a fully formed branch of applied mathematics.

Mathematically, Bernoulli's Theorem, or the weak law of large numbers, states:

$$\lim_{n \to \infty} \Pr\left(\left|p - m / n\right| < \varepsilon\right) = 1$$

where p, m and n are defined as above, and ε is an arbitrarily small number.

The proof of the theorem was mainly a matter of combinatorial algebra. Bernoulli recognised that m would have the binomial probability distribution that he developed earlier in the book. He then used algebra's Binomial Theorem to expand the combinatorial terms arising in the probability distribution and show how they behaved in the limit for n.

Bernoulli's Theorem showed that a random sample of independent trials, each with a known and constant probability of 'success', would provide an unbiased estimate of the population 'success' probability. Furthermore, it showed that the noise in the sample probability asymptotically reduced to zero as the sample size increased. This work also provided, again for the first time, quantification of the sampling error associated with samples of a given finite size n (under the assumption that the population probability, p, is known). Bernoulli developed some quantitative examples. He considered the case where $p = 0.6$, and he asked how big n must be in order for the sample probability, m/n, to be between 31/50 and 29/50 with a probability of 9999/10,000 (a probability level Bernoulli referred to as moral certainty). He calculated that n needed to be at least 25,550. In his *History of Statistics*,[29] Stigler noted that *Ars Conjectandi* abruptly ends with this calculation, and he speculates that the magnitude of this sample size would have been disheartening at a time when examples of large-scale empirical sampling had yet to emerge.

---

[29] Stigler (1986).

Bernoulli's Theorem fundamentally requires the population probability, p, to be known. The theorem tells us about the behaviour of a sample when the characteristics of the population are already given. The inverse problem of statistical inference—of developing estimates of the characteristics of a population given the observed sample—is not advanced by the theorem. It does tell us that an infinite sample will provide an unbiased estimate of the population probability, but it does not say anything about the behaviour of finite samples from an unknown population—the crux of statistical inference. For example, if we do not know p, and we observe $m = 4$ and from a sample of size $n = 10$, the theorem does not tell us anything about the interval we can infer for p with some specified probability from these observation, or even that 0.4 is the 'best estimate' of p. Bernoulli's writing seemed aware of this limitation, but he also understood that a solution to the statistical inference problem was the greater prize and he tried hard to find applications of his theorem to it.

Bernoulli's efforts to wring some application to statistical inference out of his theorem led to a series of letters between Leibniz and Bernoulli during 1703. Leibniz disputed that Bernoulli's Theorem could be meaningfully applied to statistical inference. Bernoulli claims in a letter: 'Had I observed it to have happened that a young man outlived his respective old man in 1,000 cases, for example, and to have happened otherwise only 500 cases, I could safely enough conclude that is twice as probable that a young man outlives an old man as it is that the latter outlives the former'. But he was not able to mathematically define 'safely enough', nor was he able to define in what sense the 2–1 probability estimate in his example was the 'best' estimate. Leibniz used an analogy: for any finite number of sample points, an infinite number of curves can be made to pass exactly through all of them, and there was no means of establishing which one is best. Similarly, for a finite sample that gives a particular sample probability m/n, there are many values of p that feasibly could have created such an outcome, and so the choice of value remains arbitrary. Concepts like least squares and maximum likelihood—foundations of statistical inference—would take at least another 100 years before they started to appear.

De Moivre's Contributions to Probability and Life Contingencies (1718–1724)

Abraham de Moivre is another mathematician of the highest calibre who applied himself to probability whilst dedicating most of his energy to the more conventional mathematical fields of his time. In de Moivre's case, he achieved particular renown in astronomy and complex analysis. He is perhaps most well-known for his eponymous formula on complex numbers and trigonometric functions, which is a precursor to the famed Euler formula (which frequently tops surveys of the most beautiful formulae in mathematics). But his contribution to probability is also highly regarded. Isaac Todhunter, that most authoritative of nineteenth-century mathematical historians, writes of de Moivre, 'it will not be doubted that the Theory of Probability owes more to him than to any other mathematician, with the sole exception of Laplace'.[30]

Born in France in 1667 to a Huguenot (Protestant) family, de Moivre moved to London in 1687 to escape religious persecution. He lived in London for the remainder of his life, and died in 1754. Despite being a recognised leading mathematician and Fellow of the Royal Society, he was never able to obtain a major position at one of the universities. He spent his entire life in relative poverty and supported himself by working as a mathematical tutor.

De Moivre is notable in the history of actuarial thought for the publication of two books that span the related fields of probability and life contingencies: *Doctrine of Chances*,[31] which was first published in 1718; and *Annuities on Lives*,[32] first published in 1724. The latter is the main focus of our interest, but *Doctrine of Chances* is remarkable for including the first published derivation of the normal probability distribution. De Moivre considered Jacob Bernoulli's binomial probability distribution and, like Bernoulli, he considered the properties of samples from the distribution as the sample size increased to very large numbers. The factorial calculations entailed by the binomial distribution for large n were extremely unwieldy given the computational limitations of the time. De Moivre was able use factorial algebra to derive the normal distribution as a good approximation to the binomial distribution for large n. (This calculation was further refined by James Stirling, and now bears Stirling's name.)

At the time of publication of *Doctrine of Chances*, the wider implications of the normal distribution were not yet understood. De Moivre's work can now be seen as a special case of the Central Limit Theorem that was to emerge at the start of the nineteenth century. But it was still a very useful result at the time of publication: it substantially reduced the computational burden of calculating the probabilities associated with large sample sizes. It also provided new conceptual insights. In particular, de Moivre's work highlighted that the variability of a sample probability calculated from a known population will

[30] Todhunter (1865).

[31] De Moivre (1718).

[32] De Moivre (1724).

decrease in proportion to the square root of the sample size. The primitive statistical practices of the time, such that they existed in areas like the treatment of discordant astronomical observations, had assumed that the sample variability decreased in direct proportion to the sample size, rather than its square root.[33]

The section of *Doctrine of Chances* containing the derivation of the normal distribution is entitled 'A Method of approximating the Sum of Terms of the Binomial $(a + b)^n$ expanded into a Series, from whence are deduced some practical rules to estimate the degree of assent which is to be given to experiments'. De Moivre believed that his approximation result supported the inversion of Bernoulli's Theorem for use in statistical inference or induction—that is, to make statements about the likely behaviour of a population based on the behaviour of a sample, as opposed to making statements about the behaviour of a sample, based on knowledge of a population's characteristics. De Moivre argued from Bernoulli's Theorem that if a large sample could be observed to 'converge' on a given probability, then that must provide the unknown population probability. This idea of convergence recognised that the sample ratio is an unbiased estimator of the population probability, but de Moivre did not suggest any means of determining how large the sample must be to obtain convergence, or how much uncertainty is in a given sample when the population probability is not known. De Moivre's uncovering of the normal distribution is arguably a profound moment in probability. The mathematics of statistical inference, however, was not directly furthered by it.

De Moivre's *Annuities on Lives*, published in 1724, was the most significant work on life contingencies since Halley's seminal Breslau paper. De Moivre did not seek to improve on Halley's mortality table. Rather, his objective was to show how Halley's work could be more readily applied to the valuation of annuities by reducing the computational burden associated with calculating annuity prices from age-specific mortality rates. Halley himself had pointed out in his Breslau paper that the calculation of annuity prices from the mortality rates of his table 'will without doubt appear to be the most laborious calculation' and noted that the production of the annuity table in his paper was 'the short result of a not ordinary number of arithmetical operations'. De Moivre specified a simple mathematical form for the behaviour of mortality as a function of age that permitted a straightforward annuity pricing formula to be obtained which required a much smaller set of arithmetic operations:

> Consulting Dr Halley's Table of Observations, I found that the Decrements of Life, for considerable intervals of time, were in arithmetic progression; for instance, out of 646 persons of twelve years of age, there remain 640 after one

[33] Stigler (1986), p. 84.

year; 634 after two years; 628, 622, 616, 610, 604, 598, 592, 586, after 3, 4, 5, 6, 7, 8, 9, 10 years respectively, the common difference of those whole numbers being 6.[34]

De Moivre derived a relatively simply pricing formula for a single life annuity when the decrements of life followed this arithmetic progression:

$$Annuity\ Price = \frac{1 - \frac{(1+r)P}{n}}{r}$$

where r is the annually compounded rate of interest, n is the number of years until the number of lives is reduced to zero by the arithmetic progression, and P is the price of an annuity certain payable for n years. To set the n parameter, de Moivre assumed that the population would be reduced to zero by age 86, and so $n = 86 - x$, where x is the current age of the annuitant.

Of course, the whole of Halley's mortality table did not conform to de Moivre's assumption of a constant arithmetic progression of the decrements of life. Figure 1.3 shows the actual decrements produced by Halley's table.

This highlights that de Moivre was being somewhat selective in his quoting of the stability of the decrement of life in Halley's table. It is indeed constant for the ten years from age ten, as de Moivre points out (and also for fifteen years from age 54), but there inevitably must be significant variation—first upwards as mortality rates rise with older age; and then downwards as the population reduces to very small numbers.

How do the limitations of de Moivre's approximation of Halley's mortality assumptions impact on annuity pricing? Consider the price of an annuity on a 30-year-old life. De Moivre's approximation assumes that 9.5 lives of a starting number of 530 die every year for the next 56 years. Figure 1.3 shows that the actual decrement in Halley's table was eight per year at age 30, rising to ten or eleven over the next few decades, before falling sharply from age 75 onwards. So de Moivre's assumption is sometimes higher and sometimes lower than the 'true' values of Halley's table. The resultant approximate annuity price is surprisingly accurate—de Moivre's price is 11.6, and the exact price from Halley's table is 11.7 (at 6 % interest). Figure 1.4 shows how de Moivre's approximation compares with the exact prices from Halley's table for various ages of annuitants. It is less accurate for young ages, but impressively accurate for annuitants of age 30 or older.

[34] De Moivre (1724).

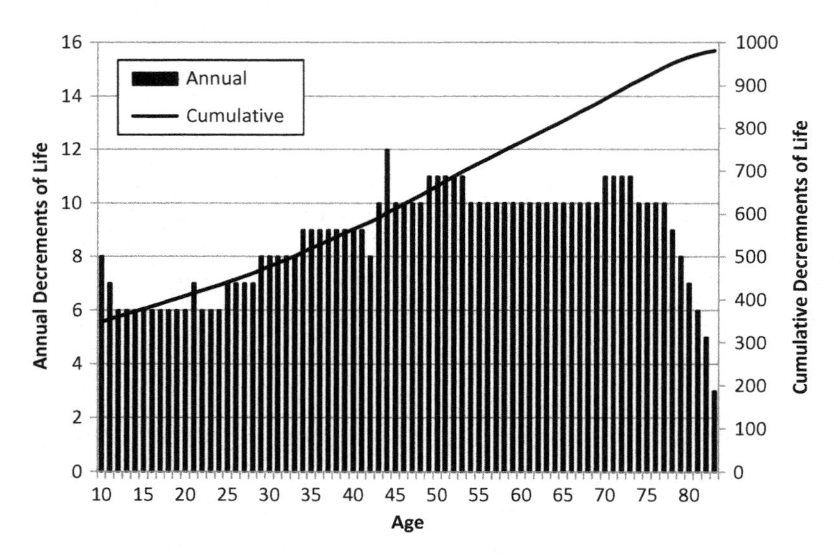

Fig. 1.3 Decrements of life from ages 10 to 82 (from 1,000 births); Halley's Breslau table

Fig. 1.4 A comparison of de Moivre's annuity price approximation with exact values calculated from Halley's Breslau table (at 6 % interest)

The remainder of *Annuities on Lives* is dedicated to the pricing of annuities on multiple lives. He first focused on finding pricing formulae for joint life annuities on two and three lives (where the annuity payments cease on the first death). He again focused on finding computationally tractable formulae by specifying an assumed mathematical form for the behaviour of mortality.

By doing so, he was able to find joint life annuity prices that were simple functions of single life annuities such as:

$$\text{Joint Life Annuity Price} = \frac{M.P}{M + P - rM.P}$$

where M is value of the single life annuity on the first life; P is the value of the single life annuity on the second life; and r the rate of interest.

The derivation of the above joint life annuity price relies on a different assumption for the mathematical model of mortality than the one he used in developing the single life annuity pricing approximation. Instead of assuming that the decrements of life follow an *arithmetic* progression, he here assumes that they follow a *geometric* progression. Expressed in modern actuarial terminology, he assumes q is constant across age (though it does not need to be the same q for each of the lives).

The assumptions used to derive M and P above, and the assumptions used to derive the above joint life formula are not only different, they are inconsistent and mutually incompatible. But this does not necessarily matter—de Moivre was not deducing a theorem, he was attempting to find a computational shortcut that produced reasonable estimates. So how effective was this formula against this criteria? Suppose we obtain the single life annuity prices M and P above using de Moivre's single life formula, and then plug these prices into his joint life formula—how does such a joint life annuity price compare with the exact value implied by Halley's table? Inevitably, the additional layer of approximation amplifies the errors in the valuation. For example, consider the joint life annuity price on two 30 year-olds. The exact price according to Halley's table (and at 6 % interest) is 9.5. And whilst de Moivre's single life annuity price provides a very close approximation to the exact single life price (11.6 versus 11.7), his joint life annuity price is 8.9, whereas the exact price 9.5. Nonetheless, in the great scheme of approximations involved in obtaining an early eighteenth-century annuity price, perhaps 5 % or so is not too bad.

Life Contingencies in the First Half of the Eighteenth Century (1700–1750)

We now turn again to developments in the British life assurance industry and its institutions. At the start of the eighteenth century, insurance was underwritten by private individuals, partnerships and mutual associations. The vast bulk of insurance activity at this time was in marine insurance. Life assurance was essentially a spin-off activity of the individuals and businesses that were focused on other forms of insurance.

In the early years of the eighteenth century, a number of new insurance ventures emerged. Foreshadowing a trend that was to continue until the final quarter of the century, many of these businesses were ostensibly life assurers, but actually provided a vehicle for speculation on a wide range of contingencies that could occur to third party individuals. The breadth of possible speculations was constrained by an Act of Parliament of 1711 that made insurance of events such as marriages, births and christenings illegal. But the legal concept of insurable interest was not introduced until later in the century and, in the meantime, life assurers continued to provide a vehicle for gambling and speculation on the health or otherwise of (usually famous) other people. In his review of life assurance in the early eighteenth century, Francis notes:

> From 1720 much of the legitimate business had been usurped by it, policies being opened on the lives of public men, with a recklessness at once disgraceful and injurious to the morals of the country. That of Sir Robert Walpole was assured for many thousands; and at particular portions of his career, when his person seemed endangered by popular tumults, as at the Excise Bill; or by party hate, as at the time of threatened impeachment; the premium was proportionately enlarged. When George II fought at Dettingen, 25 per cent was paid against his return.[35]

The most notable new insurer to emerge in the period of 1700–1720 was the Amicable Society for a Perpetual Assurance Office (generally known simply as the Amicable), which was founded in 1706. Like the other businesses that emerged around this time and prior to it, the Amicable's life assurance policies were essentially a form of mutual benefaction rather than insurance. That is, rather than providing a fixed sum assured, the premiums received from all policyholders in a given year were distributed amongst the arising claimants in that year.

The joint-stock company emerged as an increasingly important form of business corporation during this era. Such companies had been formed in Britain in the previous century, mainly for the purposes of funding colonial expansion. The East India Company is perhaps the most famous example. At this time, joint-stock companies could only be established by a Royal Charter of incorporation, and permission for such a charter would usually be granted

[35] Chapter IX, Francis (1853).

by an Act of Parliament. Typically, the Royal Charter would grant the joint-stock company some exclusive privileges in exchange for the company making a form of contribution to the state. For example, the East India Company was granted exclusive rights to trade with India in exchange for agreeing to loan the government £1,662,000.

In 1720, an Act of Parliament gave permission for two new joint-stock companies to be created for the specific purpose of writing insurance. Royal Exchange Assurance and London Assurance were thus founded. The Act of Parliament precluded any other joint-stock company from writing marine insurance. Whilst this seemingly granted the two companies significant market power, it did not prevent private individuals from continuing to write marine insurance. As this was the long-standing practice in the marine insurance market, it did not directly impact on the established marine underwriters' ability to compete with the new companies. But in return for the exclusive right to be the only joint-stock companies permitted to conduct insurance, the two companies agreed to pay £300,000 each to the Exchequer.[36] 1720 was also the year of the South Sea bubble. The first half of the year saw much activity in the forming of petitions for various new joint-stock companies amid an atmosphere of stock market speculation. Somewhat ironically, the Act of Parliament that provided the permission for the Royal Exchange Assurance and London Assurance charters was primarily focused on restraining the explosion of new joint-stock companies (it is often referred to as the 'Bubble Act'). The act went some way to pricking the South Sea bubble, and, as a result, the new insurance corporations were founded at the very peak of a stock market boom.

The total British life assurance market of this time is estimated to have been no greater than a few thousand policyholders. Life assurance was not included in the two new insurance companies' charters as one of the forms of insurance that the executive privilege applied to: initially, only marine insurance was covered by the charter. In the months following the granting of the charter in June 1720, the two companies' major preoccupation was with the turbulent stock market and the need to meet scheduled payments of the £300,000 government donation (Royal Exchange also committed, in addition, to buying £156,000 of government stock). Following some aborted equity capital issues in the autumn of that year, Royal Exchange failed to meet the scheduled payment dates for their outstanding sum of £200,000. The companies argued with the Treasury that their ability to pay their government obligations would

[36] Part A of Supple (1970) provides an excellent account of the establishment and early growth of Royal Exchange Assurance.

be materially improved by extending their charter to include fire insurance. Life assurance was then tacked on to this application for an additional charter, and this was granted one year later.

The life assurance business written by the two corporations over the following years was small. Between 1721 and 1783, Royal Exchange Assurance generated an average annual premium income from life business of £550. In contrast, the fire insurance business generated premiums of £6,000 in 1730 and £25,000 in 1780. The life assurance product offering was limited to a one-year term assurance. Though a simple product, it did represent a significant shift in practice from earlier times: the product provided genuine insurance in the sense that a specified fixed sum assured was payable on death—the insurance company now bore the mortality risk. The Royal Exchange Assurance's pricing approach was very straightforward: a premium rate of 5 % was charged, irrespective of age. This appears to have been the norm of the era. There is no evidence of any market participant (government, corporation, mutual association or individual) directly using Halley's or anyone else's mortality tables to price any life assurance or annuity product in the first half of the eighteenth century.

There has been much scholarly debate as to why life insurance practice in the first half of the eighteenth century chose to ignore the developments in mortality statistics of Halley and others that had emerged in the preceding years. There are many practical factors that may have contributed to this:

- The demand for the only life insurance product in existence at the time (one-year term assurance) appears to have been so small that it failed to provide any incentive to invest significant effort in refining its pricing.
- In this era, age may not have been the most important factor in determining a fair one-year term assurance price. Individual risk factors such as class, career, location and health may all have been considered by insurers when deciding whether or not to write a life policy.
- There would have been an intuitive understanding that for mortality statistics to be useful as a set of forward-looking estimates, a temporal stability in mortality behaviour was required that did not match the reality: the era was one where mortality rates could vary significantly from year to year as outbreaks of virulent diseases such as the plague came and went.
- Even if such stability were present, it should be remembered that the lack of an inferential statistical science meant that there was no understanding of how much confidence could be placed in the estimates derived from limited sample sizes. No one knew how to ask what the standard error was in Halley's q(60), never mind how to answer it.

Some historians have also argued that the era's predominant use of life assurance as a form of gambling and speculation was another important factor that delayed the application of actuarial techniques to life insurance business.[37] It is near impossible to apply conventional statistical methods to the analysis of highly specific contingencies such as whether a prime minister's proposed tax increases will result in him being the victim of a rioting mob, or whether the king will lose his life on a foreign battlefield. Yet it may still appear odd that no attempt was made to distinguish between the typical mortality rate of a twenty year-old and 50 year-old in standard term assurance pricing. We should not, however, lose sight of the fact that, at this time, quantitative statistical data had never actually been used to do anything. The first applications of statistics to scientific problems such as astronomical observation—which arguably were more amenable to quantitative analysis as they had more statistical regularity than social and demographic behaviour—did not fully emerge until the nineteenth century.

Daniel Bernoulli: The Beginnings of Utility Theory and Risk-Adjusted Valuation (1713/1738)

We read earlier how, in the second half of the seventeenth century, Pascal, Fermat, de Witt and Huygens developed and applied the concept of pricing uncertain cashflows by assessing their mathematical expectation. The discounting of expected cashflows to allow for the impact of the time value of money was understood and well demonstrated by the annuity pricing work of de Witt and Halley; in both cases, they explicitly calculated the mathematical expectations of the future cashflows and then discounted them into a present value using a government bond yield. Up until this time, this strand of thinking had not considered whether the valuation should make any allowance for the riskiness of the future cashflows (except insofar as it impacts on the size of the expected cashflows).

In 1713, a seemingly trivial valuation problem was introduced in a letter from Nicolas Bernoulli (who we briefly met earlier as the editor of his Uncle Jacob's *Ars Conjectandi*) to another eminent French mathematician Pierre Montmort. This valuation riddle highlighted the possible need for a more refined treatment of risk in the valuation of variable future cashflows. The valuation problem—known as the St Petersburg Paradox—can be described as follows. Suppose we have a fair coin. The coin is tossed and if it

[37] See, for example, Section 3.4.2, Daston (1988).

lands heads, player A pays player B £1. If it lands tails, then the coin is tossed again. If it lands heads on this turn, then player A pays player B £2. If it lands tails, then the coin is tossed again. If it lands heads on this turn, then player A pays player B £4. And so on, with the amount player A must pay player B when heads lands doubling each time the coin lands tails. An interesting difficulty arises when mathematical expectation is used to determine how much player B should pay player A to play this game: the mathematical expectation of this cashflow is infinite. This was viewed as a paradox because it was thought that no reasonable person would pay more than, say, £10 or £20 as player B in the game, even though he would receive a pay-out that had an infinite expectation.

The problem posed by Nicolas was eventually tackled by another Bernoulli—Daniel, another nephew of Jacob Bernoulli. Daniel was born in 1700 and established himself as another Bernoulli mathematical genius, making lasting contributions to mathematics and physics, most notably in fields such as fluid mechanics. In 1738, Daniel published a solution to the paradox in the annals of the University of St Petersburg (from which the paradox derives its name).[38] Bernoulli's solution proposed the use of a new form of expectation. Instead of pricing a contract using mathematical expectation (i.e. by considering the product of the possible cashflows and their respective probabilities), he proposed using what he termed *moral expectation*, which he defined as the product of the *utility* the recipient gained from each possible cashflow and their respective probabilities. This was the first time a quantitative concept of expected utility was considered in the context of valuation of risky cashflows.

Bernoulli went on to show that an intuitive price could be obtained for the St Petersburg game when moral, rather than mathematical, expectation was used. To do this, he supposed that increases in wealth were directly proportional to increases in utility (that is, a logarithmic utility function). He noted that the risk-aversion embedded in this non-linear utility function could rationalise demand for insurance and diversification practises (such as dividing cargo amongst several ships) as well as help to solve the St Petersburg Paradox.

To find the maximum price that someone would pay as player B in the game, Bernoulli argued we should find the price that equates his expected utility after playing the game with the utility he obtains from his wealth without playing the game. If the utility function is linear, then the expected utility from playing the game is infinite for any finite price. The non-linearity of the logarithmic function places less value on the extremely unlikely upside pay-

[38] Bernoulli (1738).

offs of the game. As a result, a finite value for the game is obtained. The risk-aversion embedded in the utility function also implied that the maximum value a player would rationally pay to play the game would be a function of his current level of wealth—the smaller his current wealth, the less he would be prepared to pay to play, even though the expected pay-out from playing is technically infinite.

Interestingly, Nicolas Bernoulli, who originally considered the paradox 25 years earlier, did not accept Daniel's concept of moral expectation. Nicolas was by this time a professor of law at the University of Basel. He maintained that for a contract to be legally equitable it should be priced such that the buyer and seller have an equal chance of winning or losing. Daniel's more sophisticated treatment had moved the pricing of uncertain claims from the realm of jurisprudence to that of economics.

His cousin was not the only one to object to Daniel Bernoulli's conception of risk-adjusted valuation of stochastic cashflows. Jean D'Alembert, another contemporary French mathematician, argued that the introduction of moral expectation was an ad hoc solution to the St Petersburg Paradox. He suggested a simpler way of reducing the value that the mathematical expectation attached to the extremely unlikely but extremely large pay-offs that arose in the game: he argued that beyond a particular probability level, an event is not physically possible, and it should make no contribution to the valuation. But the choice of such a probability level is similarly ad hoc and arbitrary.

One of the implications of Bernoulli's utility treatment is that the price a player would be willing to pay to play the game increases as his starting wealth goes up relative to the stake of the game. This implication is consistent with modern economic thinking on pricing of risky cashflows. The pay-out from the game is, in economic terms, a diversifiable risk. If the game can be played an increasingly large number of times for increasingly small stakes (relative to starting wealth), the utility function and starting wealth of the player becomes increasingly irrelevant, and the value of the game will tend to its mathematical expectation. So Bernoulli's solution to the paradox arguably only applies when it is assumed that the game must be played for stakes that are a material portion of the player's current wealth. This line of argument was pursued by the respected English mathematician Augustus De Morgan in the early nineteenth century. In his discussion of the St Petersburg Paradox he concludes: 'The results of all which precedes shows us that great risks should not be run, unless for sums so small that the venture can afford to repeat them often enough to secure an average.'[39]

[39] De Morgan (1838), Chapter V, p. 101.

The ultimate modern solution to the paradox, as argued by Samuelson,[40] is that no rational person would accept playing as player A for any finite price, and hence the game cannot be played and no paradox can arise. But this doesn't really address the original point of the paradox, which was that, if the game was offered, no reasonable person would intuitively be prepared to pay more than ten or twenty times the pay-off from the first coin toss, even though the total expected pay-off is infinite.

Concluding Thoughts

The 100 years between 1650 and 1750 saw remarkable progress across the broad spectrum of human endeavour and societal development. The Scientific Revolution was in full flow, with fundamental breakthroughs occurring in mathematics, physics, astronomy and other branches of science, as well as in the development of the scientific method itself. In Britain, the Royal Society, founded during Charles II's reign, had established itself as a significant arbiter of intellectual endeavour by the early eighteenth century. The Enlightenment and the age of reason brought important developments in philosophy and economics that would further influence and challenge society.

From the perspective of British finance, London emerged particularly strongly from this period. The City of London transformed itself from being a laggard of financial sophistication relative to practices in cities such as Amsterdam, Antwerp and Florence into a leading banking and insurance centre, with a respected Bank of England.

The developments in probability and life contingencies over the period can be viewed as intrinsic parts of these broader scientific, economic and industrial developments. Mathematical probability fully emerged as a branch of applied mathematics over this period, and its early development fully represented its holistic nature, involving contributions from disciplines as varied as jurisprudence and number theory. The use of statistical population data to model mortality and analyse its implications for the pricing of life contingencies developed contemporaneously. These early forms of statistical analysis pre-dated a rigorous theory of statistical inference, and did not directly rely on the theoretical developments in mathematical probability that were occurring at the same time. The two disciplines overlapped, but were generally each developed by a different cast of characters. Occasionally, however, titans such as de Moivre straddled both these disciplines and made substantial and lasting contributions to both.

---

[40] Samuelson (1960).

Up until 1750, the developments in thinking in probability and life contingencies were pregnant with possibility, but were still awaiting application to real life. The tiny level of demand for short-term life assurance that existed during the first half of the eighteenth century limited the incentive to develop practical applications of the emerging actuarial science. The application of this new discipline would require a societal change in demand for mortality risk transfer, and a form of life contingency product that could meet the needs of the growing, prudent and increasingly wealthy Victorian middle class of the industrial revolution.

In the field of probability, there was an awareness of the need for a theoretical breakthrough to make it fully applicable to scientific endeavour. Mathematical probability had been limited to finding probability distributions of random samples of populations with known properties. The new scientific method would increasingly produce quantitative observational data from which inferences would be made: modern thinking increasingly called for inductive, rather than deductive, reasoning. The inversion of mathematical probability into statistical inference—to make rigorous statements about populations based on sample data observations, rather than to make statements about sample probabilities based on known or assumed population characteristics—was a problem whose solution had so far evaded all the impressive group of thinkers who had tackled probability.

2

Revolutionary Developments Between 1750 and 1810

The second half of the eighteenth century was a remarkable period of development for both probability (and statistics) and actuarial thought (and practice) more broadly. Two fundamentally important fields of statistical and actuarial thought emerged over the period that can each now be seen as a natural flowering of the seeds planted over the previous 100 years. First, a number of major theoretical breakthroughs were made that created the permanent foundations for the inversion of mathematical probability into inferential statistics. Second, the whole-of-life with-profit policy was conceived and successfully brought to market. This transformed life assurance from a short-term insurance contract into a long-term savings vehicle that strongly resonated with the emerging professional classes of late-Georgian Britain.

Over this period these two fields of mathematical statistics and actuarial science followed increasingly distinct paths of development: the primary application that drove statistical thinking over this period was found in the physical sciences, and especially astronomy, rather than finance; similarly, the success of new and more complex life assurance products meant that actuarial thinking and practice had to wrestle with a broader set of challenges than 'just' making good statistical estimates of mortality rates. One man, Richard Price, notably transcended both fields during this era, marking him as one of the most important historical figures in early actuarial history.

C. Turnbull, *A History of British Actuarial Thought*,
DOI 10.1007/978-3-319-33183-6_2

From Probability to Statistics, 1764–1810

Bayes and His Application of His Theorem (1764)

Jacob Bernoulli's binomial analysis of *Ars Conjectandi* had been driven by the aim of making a breakthrough in the theory of statistical inference. He ultimately failed in this objective, although he established some other fundamental probability laws along the way. Thomas Bayes built on Bernoulli's binomial analysis, and, with a novel alternative physical analogy of a table instead of an urn, he introduced a new conceptual framework for inferential methods.

Bayes was a Fellow of the Royal Society but by all accounts led a quiet life as an English clergyman, largely detached from the intellectual fervour of the period. His now famous paper 'An essay towards solving a problem in the doctrine of chances' was unpublished during his lifetime.[1] It, along with his other papers and possessions, was bequeathed to a fellow clergyman, Richard Price. Price was a significant mathematician and philosopher and he, like Bayes, was also a Fellow of the Royal Society. As a thinker aware of the promise of statistical inference for empirical science, Price immediately recognised the value of Bayes' paper. He edited it and presented it to the Royal Society for publication in its journal a year or so after Bayes' death in 1762.

Bayes' paper was both ahead of its time in concept and yet presented in an anachronistic geometric style. Both these properties led to the paper being somewhat impenetrable. The nineteenth-century mathematical historian Todhunter wrote that the paper's introductory discussion of the general laws of probability was 'excessively obscure, and contrasts most unfavourably with the treatment of the same subject by De Moivre'.[2] But the paper's statement of intent was quite clear:

> *Given* the number of times in which an unknown event has happened and failed. *Required* the chance that the probability of it happening in a single trial lies somewhere between any two degrees of probability that can be named.

In essence, how can the population probability distribution be inferred from the observation of a limited sample of outcomes from that population? Richard Price viewed statistical inference as the rational solution to David Hume's sceptical problem of induction, in which it was argued that no

[1] Bayes (1764).

[2] Todhunter (1865).

amount of empirical observation of the past could provide knowledge of what would happen in the future.[3] In Price's introduction to the paper, he argued that the problem stated above by Bayes is 'necessary to be solved in order to assure foundation for all our reasoning concerning past facts, and what is likely to be hereafter ... it is necessary to be considered by anyone who would give a clear account of the strength of analogical or inductive reasoning'.[4] He believed that Bayes' new approach to statistical inference provided the fundamental solution to this problem of inductive reasoning.

The paper is perhaps most well-known for its eponymous Bayes' Theorem:

$$P(A \mid B) = \frac{P(B \mid A).P(A)}{P(B)}$$

This theorem, when considered separately from its application to statistical inference, is a straightforward conditional probability statement that is uncontroversial and unambiguously derived from the basic axioms of mathematical probability. In fact, it is so rudimentary and universal that scholars are unsure whether Bayes was in fact the first to use it.[5] But the structure of the equation—where probabilities of B conditional on A can be transformed into probabilities of A conditional on B—opens the door to what was then often referred to as 'inverse probabilities'. Mathematical probability was about making statements about the probabilities of observing specific outcomes, given known population characteristics. Bayes' Theorem alluded to a potential means of inverting these probability statements into ones that infer something about the probability characteristics of the population, given some observed sample outcomes.

Bayes used a physical analogy to develop his application of this probability theorem to statistical inference. Where Jacob Bernoulli had used an urn filled with black and white balls, Bayes used a table and balls. Whilst his table was two dimensional, his analogy only considers how the balls were distributed along one dimension. The use of a table rather than a one-dimensional line facilitated a geometric interpretation of his probability calculations.

He supposed a first ball was randomly thrown onto the table, and that this was followed by a further n balls that were independently thrown randomly onto the table. He considered how knowledge of whether each of the sub-

[3] Hume (1739).

[4] Bayes (1764).

[5] Hacking (1965), p. 190.

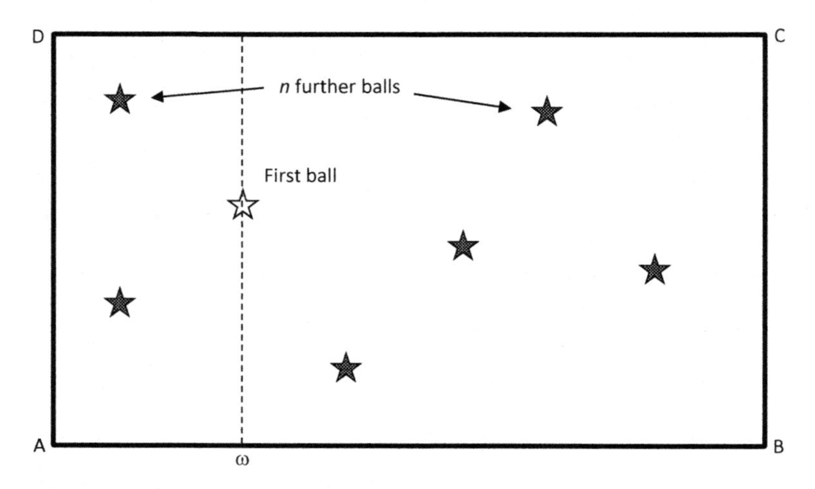

Fig. 2.1 Bayes' table

sequent n balls landed to the right or left of the first ball's position could be used to make a statement about the probability distribution of ω, the position along the 'x-axis' of the table of the first ball. This set-up can be summarised simply in the Fig. 2.1 above.

Bayes wished to infer the probability distribution of the position, ω, of the first ball along the line AB, given only the number of times that the n subsequent random balls landed to the right or left of the first ball. For any *given* value of ω, the probability θ of a subsequent ball landing to the right of the first ball is simply the ratio of the line distances ωB/AB (as the ball's position on AB is uniformly distributed). For a given ω, the number of times, M, that the n subsequent balls land to the right of the first ball is therefore binomially distributed with parameters n and θ. So, we can write:

$$Prob\left(M = x \middle| \frac{\omega B}{AB} = \theta \right) = \binom{n}{x} \theta^x (1-\theta)^{(n-x)}$$

So far, this looks extremely similar to Jacob Bernoulli's work on the distribution of the number of black balls chosen from an urn that has a known proportion of black and white balls. But Bayes had merely laid the preparations for his breakthrough step. He then noted that his theorem allowed him to find an expression for the probability that ω is between two points b and f on the line AB:

$$Prob\left(b<\omega<f|M=x\right)=\frac{Prob\left(M=x\mid\omega\right)Prob\left(b<\omega<f\right)}{Prob\left(M=x\right)}$$

By construction of his physical example, where each ball is independently and randomly thrown onto the table, the position of every ball on the line AB is uniformly distributed. This means that the unconditional probability of M = x is $1/(n+1)$ for all x: the number of balls out of a total of n that lands to the right of the first ball can take a total of $(n+1)$ different values (0, 1, 2, ... n), and if the balls are uniformly distributed then each of these outcomes are equally likely. With this and the above equation, Bayes obtained:

$$Prob\left(b<\omega<f|M=x\right)=\frac{(n+1)!}{x!(n-x)!}\int_b^f \theta^x\left(1-\theta\right)^{(n-x)}d\theta$$

Bayes had provided a rigorous mathematical solution for the probability distribution of the location, ω, of the first ball along the line AB, conditional on the observed results of how many of the subsequent n randomly distributed balls were positioned to the left or right of the first ball. Today, this distribution is known as the beta distribution. The integral is intractable for large values of n and $(n-x)$, but is very straightforward for relatively small values. For example, consider the case of $n = 6$ given in Fig. 2.1 above. Without any knowledge of the distribution of the subsequent balls, the probability of ω being in the right-hand half of the line AB is clearly simply 1/2. If three of the six balls were observed to land to the right of ω, the above distribution implies that the probability would, intuitively enough, remain 1/2. However, if, instead, four of the balls were found to have landed to the right of ω, the probability of ω being in the right-hand half of AB would be reduced to 29/128 (slightly less than a quarter).

Thus far, Bayes' work is a novel but still uncontroversial application of mathematical probability—within the physical framework set out, his results are unambiguously rigorous mathematics. In Bayes' physical model of the table with randomly thrown balls, a 'uniform prior distribution' for ω arises by construction. It has an unambiguous physical interpretation that derives from the set-up that the balls are randomly thrown onto the table.

Bayes argued that the 'inversion' logic of his theorem was naturally applicable to statistical inference. That is, to making conditional statements about the properties of unknown populations (analogous to ω on his table) based on observed sample data (the position of the n balls). The more general application of this approach to the problem of statistical inference required treating the unknown parameters of the population that were being estimated as random variables (θ in the above example). The *prior* probability distribution would be a description of the information (or lack thereof) known about the parameter before the statistical observations were considered. The *posterior* distribution was the parameter's probability distribution after being updated to reflect the information in the statistical observations. In his physical analogy, Bayes' set-up had implied a uniform prior and he had derived a beta posterior distribution.

This 'Bayesian' approach to statistical inference involves two fundamental steps that have invoked much philosophical and mathematical controversy, particularly in the twentieth century when it was criticised by leading statistical thinkers such as Sir Ronald Fisher. The first issue is how to interpret the idea of modelling the unknown (but fixed constant) population parameter as a random variable. This made sense in Bayes' analogy when the position ω was physically distributed with a given probability distribution. But the standard deviation, say, of a normal distribution is not distributed in this way—it is simply an unknown fixed quantity. The use of a probability distribution to represent the degree of knowledge or prior information available about the parameter is a conceptual leap. To move from Bayes' physical analogy to the application of Bayes' Theorem in more general statistical inference, the physical randomness of Bayes' ball must be translated into an epistemological statement of ignorance. This translation is in some sense philosophical rather than purely mathematical, and as such its viability is open to some debate.

The second issue closely follows: in the absence of *any* knowledge or information about the parameter, is it appropriate to assume the prior distribution is *uniform*? Bayes appears to have thought deeply about this question and he tackled it head on in the paper. He argued that for events where 'we absolutely know nothing antecedently to any trials made concerning it' then 'concerning such an event, I have no reason to think that, in a certain number of trials, it should rather happen any one possible number of times than another'. This assertion of equiprobability straightforwardly translates into the assumption of a uniform probability distribution.

In modern times, this is sometimes referred to as the Principle of Insufficient Reason and it may still be invoked as the logical basis for the uniform prior distribution. It has been speculated that Bayes' lack of confidence in the soundness of this argument restrained him from publishing his paper during his lifetime, but this is largely conjecture. It is equally possible that his inability to find a workable mathematical approximation for his beta posterior distribution was the main reason for his reluctance to publish. In any case, this argument attracted little dissent until around 100 years after the publication of the paper. As we shall see in the next section, the Bayesian uniform prior was adopted whole-heartedly by no lesser mathematicians than Laplace and Gauss in the decades following Bayes' paper. In 1854, George Boole, an English mathematician and philosopher (and the father of Boolean logic) was the first to publish a significant critique of the logic of Bayesian statistical inference, arguing:

> It has been said that the principle involved ... is that of the equal distribution of our knowledge, or rather of our ignorance—the assigning to different states of things of which we know nothing, and upon the very ground that we know nothing, equal degrees of probability. I apprehend, however, that this is an arbitrary method of procedure.[6]

The philosophical argument that the assumption of a uniform prior distribution was in fact an arbitrary and therefore illegitimate general approach to statistical inference gained further traction through the latter part of the nineteenth century and the first half of the twentieth century. Sir Ronald Fisher, who was one of the most important statistical thinkers of the twentieth century, famously argued that, in the absence of any prior knowledge, the assumption that the prior distribution was a squared sine function was as logical a candidate for the prior distribution as the uniform distribution.[7] In twentieth-century statistical practice, Bayesian statistical inference started to be superseded by other methodologies such as maximum likelihood that were argued to be less subjective or arbitrary. But the controversy reigns on, and the Bayesian method has become increasingly popular once again in the twenty-first century (so far!).

Laplace's Bayesian Analysis (1774–1781)

Pierre-Simon Laplace was the pre-eminent mathematician of the second half of the eighteenth century. He made particularly important contributions in the fields of astronomy and celestial mechanics. Over his lifetime, he also sporadically committed significant time and energy to the study of probability and statistics. He did more than anyone else of his time to develop the mathematics of Bayesian statistical inference (including Bayes). Laplace's employment of the Bayesian approach and the use of the uniform prior distribution doubtless did much to add credence to the logic of the Bayesian approach. Writing in the 1950s, Sir Ronald Fisher noted:

[6] Boole (1854), p. 370.

[7] Fisher (1956), p. 16.

> The superb pre-eminence of Laplace as a mathematical analyst undoubtedly inclined mathematicians for nearly fifty years to the view that the logical approach adopted by him had removed all doubts as to the applicability in practice of Bayes' theorem ... In spite of the high prestige of all that flowed from Laplace's pen, and the great ability and industry of his expositors, it is yet surprising that the doubts which such a process of reasoning from ignorance must engender should begin to find explicit expression only in the second half of the nineteenth century, and then with caution.[8]

Laplace produced two important papers on Bayesian statistical inference in the two decades following the publication of Bayes' seminal paper: in 1774, 'Memoir on the Probability of Causes of Events'[9] was published in the journals of the French Royal Academy of Sciences; and 'Memoir on Probabilities'[10] was presented to the Academy in 1780 and published in 1781. Bayes' paper did not become known in continental Europe until around 1780—the first Laplace paper was published without any knowledge of Bayes' developments. But Laplace developed a remarkably similar line of attack to the problem of statistical inference. He introduced his 1774 paper with the following statement:

> If an event can be produced by a number *n* of different causes, then the probabilities of these causes given the event are to each other as the probabilities of the event given the causes, and the probability of the existence of each of these is equal to the probability of the event given that cause, divided by the sum of all the probabilities of the event given each of these causes.

The above statement is Bayes' Theorem with a uniform prior distribution![11] Laplace had independently developed the Bayesian framework for 'inverting' probability statements about the behaviour of a sample given a population into statements about a population given sample data via the assumption of a uniform prior distribution. Unlike Bayes however, who presented a proof that was detailed to the point of philosophical obscurity, Laplace presents this principle without any proof or even justification. Bayes had thought deeply and philosophically about the fundamentals of probability and inference. Laplace, on the other hand, was a mathematician whose primary interest was in applying his renowned analytical capabilities to the mathematical problems that probability presented.

---

[8] Fisher (1956), p. 20.
[9] Laplace (1774).
[10] Laplace (1781).
[11] Stigler (1986), Chapter 3, p. 103.

Laplace considered the same binomial problem as Bayes. That is, given a sample of n binary observations of which x are 'successes', what is the probability distribution of the population probability of success, ω? He obtained the same result as Bayes for the posterior distribution of ω. We noted above that Bayes was unable to find an analytical solution, or indeed a good approximation, for the integral when n and $(n\text{-}x)$ were large. Laplace, however, used his superior mathematical skills to manipulate the integral so as to provide a much more accurate numerical solution than the crude limits that Bayes (and Price) had been able to find. The posterior probability distribution that could be inferred from a finite sample of binomial observations could now be quantified quickly and accurately.

In his papers of 1774 and 1781, Laplace took a number of further important steps beyond Bayes. Bayes' paper had presented how to generate a posterior probability distribution for the probability of the 'success' of a binary event, based on a sample of n observations and the assumption of a uniform prior distribution. But he had not explicitly considered how to use this posterior distribution to infer a single 'best estimate' for the success probability. Some measure of central tendency of the posterior distribution would be a natural candidate for the estimate, but Bayes never explicitly discussed this topic. Laplace addressed this directly, and suggested two possible approaches for determining the best estimate: the value that makes it equally likely that the 'true' value will be larger or smaller according to the posterior distribution, i.e. the posterior median; and the value that minimises the probability-weighted sum of the absolute differences between the observed value and the best estimate, where the probability weights were obtained from the posterior distribution. Laplace then proved that these two approaches were mathematically identical and hence would always produce the same value for the best estimate.

Laplace also moved beyond Bayes in terms of to what he applied the Bayesian prior/posterior framework. Bayes' paper had focused on binary events (success or failure; a ball landing to the right or left of another). Laplace moved onto variables that could take a continuum of sizes. His main motivation for this was found in astronomical observation. Astronomers of the time found that observations of an object in the sky, such as the planets Jupiter or Saturn, were not perfectly consistent with each other—some empirical observation error would inevitably arise from the physical, manual process of observing planetary positions in the sky. How should these observations be combined to find the best estimate of the position of the planet? This problem had bedevilled some of the greatest mathematicians of the age, including Euler, who could not find a satisfactory way of 'solving' a system of inconsistent linear equations.[12]

[12] Euler (1749). See Chapter 1 of Stigler (1986) for further discussion of Euler and the treatment of discordant astronomical observations.

Laplace wished to find a best estimate for some continuous variable *V*, based on a limited sample of observations, say, $v_1, \ldots v_n$. He tackled this using the Bayesian framework—that is, by obtaining the posterior distribution produced by assuming a uniform prior distribution for *V*, and then integrating the conditional probabilities of observing $v_1, \ldots v_n$, over all possible values of *V*. Laplace's best estimate would then be the median of this posterior distribution. In Bayes' specification of the problem, the conditional probabilities that he had to sum across were already defined: because he was considering binary events, the sum of observations had a binomial distribution. Laplace's more general problem meant that he had to specify the conditional probability distribution for observing $v_1, \ldots v_n$ for a given value of *V*. But how could he determine a universally useful form of distribution for these conditional probabilities?

In order to tackle this generic specification of the sampling distribution, he made a subtle change of tack. Instead of considering the conditional probabilities for $v_1, \ldots v_n$, he considered the probabilities of the *differences* or observation *errors* between v_i and *V*: $e_i = v_i - V$. This simple change immediately allowed some intuitive general criteria to be established for the shape of the sampling probability distribution. In his 1774 paper, Laplace specified three criteria: the error probability distribution should be symmetric around zero (as the errors should be as likely above as below); the error probability should tend to zero as the error tends to infinity in both directions (small errors should be more probable than large errors); and, of course, the area under the error probability distribution must integrate to one. This still left an unlimited number of possible functions that could be specified for the error probability distribution and at the time no good reason existed to choose a particular one amongst them. He considered some specific error distribution choices, but was unable to find approaches that were amenable to analytical solution for large sample sizes and that did not involve the introduction of additional arbitrary parameters. A 'best estimate' of a population parameter as a function of sample data could not be mathematically defined without an explicit error distribution. Laplace's journey to a general solution to statistical inference had reached a dead end.

Least Squares, Errors and the Central Limit Theorem (1805–1810)

New conceptual breakthroughs were required in order to make further progress in the general framework of statistical inference. Relatively little progress was made in the twenty years following Laplace's 1781 paper. Laplace himself appears to have put probability to one side and refocused on mathematical astronomy, where he made terrific contributions. Two vital breakthroughs were then delivered in succession in 1805 and 1809 by another two pre-eminent European mathematicians—Legendre and Gauss. Laplace then returned once again to the subject of probability and delivered the final fork of this trident in 1810. The synthesis of the three new concepts that were introduced over this five-year period represented a fundamental and permanent development in mathematical statistics.

Like so many of his peers, Adrien-Marie Legendre's primary focus during his remarkable career had been in pure mathematics and astronomy. His interest in probability and statistics arose through the contemporary problem that had similarly attracted Laplace: finding a 'best estimate' from a number of discordant astronomical observations. In 1805 he published a short book on modelling comets, *Nouvelles methods pour la determination des orbites des cometes.*[13] The book included an appendix entitled '*Sur la method des moindres quarres*'. In this appendix, he advocated choosing a parameter estimate by minimising the squared errors that it produces for a given set of observations. That is, using the notation developed above, if we have v_1, ... v_n observations and we wish to make a best estimate of V, we would write $e_i = v_i - V$, and find the estimator that minimises the sum of e_1^2, ... e_n^2. He noted that, in this form of example, the estimator produced by least squares would be the arithmetic mean of the sample (and this applied irrespective of the specific form of the error probability distribution). He did not attach any profound importance to this observation, but viewed it as a desirable property.

Legendre's suggestion of the use of the sample arithmetic mean was clearly easy to practically implement, and had a beguiling mathematical elegance. But he appeared to have plucked the least squares fitting rationale out of thin air. He did not provide any fundamental mathematical or metaphysical rationale for why the least squares estimator property of the sample arithmetic mean made it, in some deeper sense, 'best'. Moreover, Laplace had already established that, for an arbitrary choice of error probability distribution, the arithmetic mean of the sample would not necessarily be best (in the sense of being the median of the posterior distribution). Legendre had worked out that the arithmetic mean of the sample was the least squares estimator of the population mean, but for this observation to progress from being a curious nicety to a pillar of a rigorous probabilistic approach to statistical estimation, more was required.

The great Carl Gauss then took an important step towards an explanation of *why* the least squares estimator could be mathematically described

[13] Legendre (1805).

as a 'best' estimate. Like Laplace and Legendre, he focused the bulk of his energy and talents on pure mathematics and astronomical problems. In 1809, he published a major paper on the mathematics of planetary orbits, '*Theoria motus corporum celestium*'.[14] Like Legendre's paper on the orbit of comets, he included as an appendix a piece on statistical inference. Gauss's statistical piece started where Laplace had finished. That is, he considered the Bayesian approach to developing a posterior distribution based on an assumption of a uniform prior distribution and the conditional distributions of the observed errors. He showed that the choice of parameter value which was implied by the median of the posterior distribution in this case was also the choice of parameter value which maximised the value of the joint conditional probability distribution of the errors. So, the parameter value could therefore be found by differentiating the joint error probability distribution and setting it equal to zero. Interestingly, this was very close to the maximum likelihood concept that Fisher developed in the early twentieth century. However, without defining the form of the error probability distribution, Gauss had not practically advanced beyond where Laplace had reached twenty years earlier.

To break the impasse, Gauss did something rather ad hoc. Instead of specifying a form of error distribution and then deriving the optimal parameter estimate that it generated, he specified a particular form of parameter estimate (the arithmetic mean of the sample), and then derived the form of error distribution that implied this form of parameter estimate was optimal. He found that the arithmetic mean of the sample maximises the joint distribution of the errors only when the error distribution has the form:

$$z(x) = \frac{h}{\sqrt{\pi}} e^{-h^2 x^2}$$

This is the normal distribution, though at this point in time no particular importance was attached to it. A probabilistic rationale for Legendre's least squares method had now been produced: Legendre had shown that the arithmetic mean was the least squares estimator, and Gauss had now shown that the least squares estimator was the optimal estimator when the error distribution was normal. However, Gauss had not provided any good reason why the normal distribution was the best choice of error distribution. For as long as the normal distribution was just an arbitrary assumption for the error distribution, a significant 'so what' would hang over Gauss's intriguing finding.

[14] Gauss (1809).

This 'so what' did not hang in the air for long. Laplace had returned to the subject of mathematical statistics during the first decade of the nineteenth century, and, without any knowledge of the work of Gauss, published in 1810 his '*Memoire sur les approximations des formules qui sont fonctions de tres grand nombres et sur leur application aux probabilities*'.[15] This paper contained Laplace's most important contribution to probability and statistics: the *Central Limit Theorem*. Speaking loosely, this said that the sum of any independent variables would be approximately normally distributed when the number of terms in the sum is large.

This suddenly propelled the normal distribution from obscurity to centre stage. De Moivre's derivation of the normal distribution as the limiting case of the binomial distribution almost 100 years earlier could now be seen as just one example of an all-pervasive phenomenon. In an instant, Gauss's work on normal error distributions went from being an interesting piece of ad hoc analysis to being the defining statement of statistical inference. By Laplace's Central Limit Theorem, error distributions could be assumed to be normal (for large samples). In which case, Gauss's optimal estimator for the population mean was the arithmetic mean, which Legendre had shown was the least squares estimator. An elegant and profound synthesis had been achieved that would form part of the permanent foundation of inferential statistics.

The Emergence of with-Profits, 1756–1782

The previous section discussed how the work of Bayes, Laplace and others during the decades following 1760 provided a fundamental breakthrough in mathematical statistics. Contemporaneously, another small group of men were revolutionising what life assurance could be and how it could be delivered in a sustainable and equitable way. Richard Price uniquely spanned both these sets of men and their ideas, and for this he must be regarded as one of the most important people in the history of actuarial thought. But before Price, first came James Dodson.

James Dodson and Whole-of-Life Assurance (1756–1772)

James Dodson, born in England in 1710, was a friend and pupil of Abraham de Moivre. It is reasonable to assume that his interest in life contingencies

[15] Laplace (1810).

arose from de Moivre's influence. Dodson published on mathematics, including topics in annuity valuation, at a level that saw him inducted as a Fellow of the Royal Society in 1755. He worked as a schoolmaster in London and supplemented his income by consulting on financial matters relating to annuities. The story, perhaps apocryphal, goes that he was refused a life assurance policy by the Amicable in 1755 due to their then-policy not accept lives over age 45 (a policy that was necessitated by their use of a flat 5 % premium rate for annual life assurance for all permitted ages at the time). This motivated Dodson to advocate developing rational age-dependent life assurance pricing. In 1756, he published a treatise, *First Lectures on Insurance*, espousing his ideas and he initiated a project to create a life assurer that would implement them. This project ultimately led to the foundation of Equitable Life, though he did not live to see it as he died prematurely in 1757, aged only 47 (perhaps somewhat ironically, Amicable therefore did well to avoid insuring his life!).

Though Dodson was clearly an able mathematician, his permanent contribution to actuarial science is not in mortality modelling or statistics, but in his conception of a new life assurance product that could be appealing to the emerging salaried and prudent middle class. Up until this time, the only form of life assurance written was short-term (usually one-year) assurance. As noted in Chap. 1, the demand for this business was very small, and was largely limited to insuring the lives of business creditors and to speculative gambling activities related to the mortality prospects of the celebrities of the age. Dodson proposed a new form of life assurance product: instead of buying life assurance for one year, the policy would offer insurance over the entire remainder of the policyholder's life. Instead of a short-term insurance policy that would only pay out in circumstances of extreme misfortune, Dodson envisioned a policy that would pay out the fixed sum assured *with certainty*. The only variable was the *timing* of the payment. In his *Lectures*, he went on to show how this whole-of-life policy could be paid for by regular even premiums. As mortality rates tend to increase with age, the regular premium in the early years of the contract would be greater than the regular premium payable for a single-year term assurance contract. This excess funded a reserve, which was essentially a long-term savings element of the contract. The reserve would be used to partly fund the claims that would arise many years later, when the regular premiums payable in those years would not be sufficient to fully meet the cost of that year's cover. This conception of a whole-of-life policy transformed life assurance into a form of long-term savings vehicle.

By paying for the contract with regular premiums, the policyholder was providing a life annuity to the insurer. Dodson, as an expert in annuity pricing, showed how the regular premium could be set such that their present value equated to the present value of the sum assured benefit for a policyholder of a given age, according to a given mortality table and interest rate. Dodson developed some example premium calculations using a mortality basis derived from the London Bills of Mortality from 1728 to 1750. In 1728, the London Bills had started to include age of death in its data (though even then it was recorded only by decade rather than year of age).

Significantly, Dodson argued that the pricing basis for setting the whole-of-life premium rate should be set by referring to the *worst* mortality that was experienced in any given year in the data sample, rather than the *average*. Thus he derived two sets of mortality rates: a 'mean deaths' table and a 'greatest deaths' table. Dodson used the bills to show that the average number of deaths over the sample period was 26,207, whilst the worst year experienced 32,169 deaths (this occurred in 1741, where severe weather resulted in high food prices and near famine in parts of England). Thus the 'greatest deaths' basis had mortality rates that were almost ¼ higher than the 'mean deaths' basis. Dodson also noted that even the 'mean deaths' basis would likely provide some margin over the experience of a well-run life assurer:

> As the Bills of Mortality contain the deaths of all kinds of people healthy and unhealthy and as care will be taken not to insure those lives which are likely to be soon extinct therefore in all probability fewer of the persons insured will die in proportion to their number than those who are not insured, which will also contribute to the gain of the corporation since the premiums are proportioned to the Bills.[16]

The use in the pricing basis of the 'greatest deaths' rates experienced in any one year of a 22-year period may appear today as a somewhat ad hoc and noisy way of setting a pricing margin. But he provided an explicit rationale for this margin, based on a distinction between those policyholders who would bear the risks and those who would not. In his conception of a mutual life assurer, there would be two classes of policies: one that would underwrite the guarantees and participate in the profits of the business; and another that would not participate in the risks or profits of the business. Furthermore, at this time limited liability was only applicable by exception to companies that were granted it by royal charter. So the participating policyholders would have unlimited liability, and in the event that claims were due which could not be met by the assets of the corporation, they would face a call on their personal assets.

Dodson's view was that a non-participating policyholder should be charged *more* than a participating policyholder for the same sum assured, so that the

[16] Dodson (1756), Chapter 1.

participating policyholder was offered a compensatory expected return for the risks they were underwriting. His proposal was therefore that the participating business be priced using 'mean deaths' whereas the non-participating premium basis would use the 'greatest deaths' table: 'if the persons who shall desired to be insured without being liable to such a call should be rated in proportion to the greatest number of deaths that have happened within our knowledge as I think they ought, then the latter ought to pay near ¼ part in the premium more than the former'.[17] For those aged 40–50, Dodson's 'greatest deaths' table would result in a non-participating pricing basis that was quite similar to the 5 % assumption used by the Amicable.

In his *Lectures* Dodson produced tabulations of multi-year projections of accumulated assets, premium payments and claim outflows to illustrate the workings of his regular premium whole-of-life policy. Figure 2.2 charts one of these tables. This is a projection of a non-participating product. The projection is made on the 'mean deaths' basis but the premium basis used is the 'greatest deaths' table.

The regular premium rate for a 40 year-old was calculated to be 4.625 %. The starting annual regular premium income for the cohort of 8,165 lives

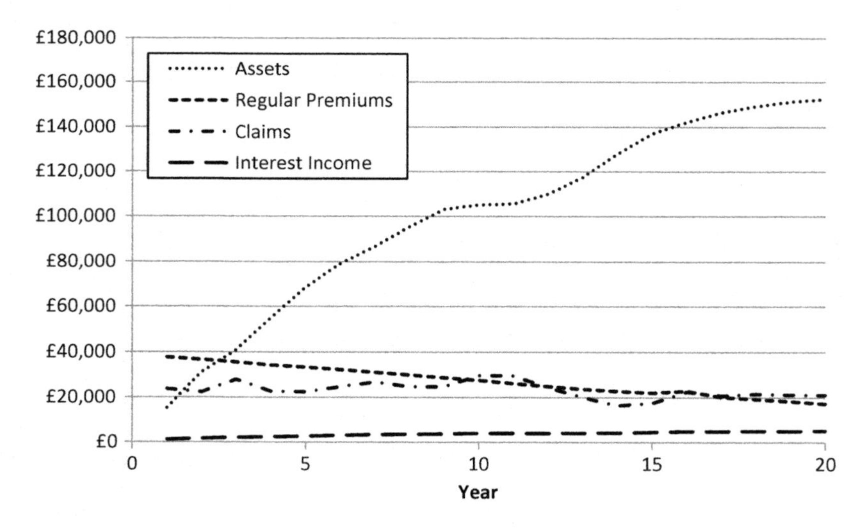

Fig. 2.2 Dodson's twenty-year projection of a regular premium whole-of-life policy: Cohort of 8,165 40 year-olds, sum assured of £100

[17] Dodson (1756), Chapter 1.

with a sum assured of £100 each is therefore £37,763. By the end of Dodson's twenty-year projection, 4,647 of the 8,165 lives in the cohort have died, leaving a total of 3,518 alive (and hence the total sum assured outstanding is £351,800), while assets of £152,392 have been accumulated. Not all of these assets represent a surplus—some of it is required as a reserve for the heavier mortality experience that is expected in the later years of the cohort's lives. Dodson calculated that, at the end of year twenty, the present value (on the 'mean deaths' basis) of the future claims of the remaining 3,518 lives was £265,917; and that the regular premium annuity stream (again on the 'mean deaths' basis) from the 3,518 remaining 60 year-olds had a present value of £134,412. Thus, a surplus of assets over liability reserve of £152,392 - (£265, 917 - £134,412) = £20,887 had, so far, been generated by the pricing margin implied by using the 'greatest deaths' table in pricing.

Interestingly (in the context of actuarial debates to come much later), Dodson presented this prototypical liability reserving calculation by explicitly appealing to the logic of valuing the liability with reference to the cost of transferring the liabilities to a third party:

> Now let us suppose that the corporation will contract with some other body, take the insuring of these off their hands by a payment of a sum in hand, being allowed discounts for the money so advanced them by computing the sum for which 100 may be insured for a life of 60 on the manner already planned.[18]

Having established the principle that non-participating business should be priced to generate a surplus for participating policyholders, the question arose of how to distribute this surplus. The equitable distribution of the surplus amongst the participating policyholders is discussed in Dodson's *Lectures*. There he argued that the profit should only be distributed to participating policyholders who have already themselves generated a profit for the assurer by paying a total amount of regular premium that exceeded their sum assured, and then in proportion to that excess:

> It seems, therefore, unreasonable to divide any part of the profit [from non-participating business] among those who have not paid as much as they receive because they have evidently a profit by being paid their claims which profit is very great in first years ... it seems reasonable therefore to divide the profit amongst the persons to have paid more than claims and to do this in proportion to the sums so overpaid.[19]

[18] Dodson (1756), Chapter 3.

[19] Dodson (1756), Chapter 4.

He suggested this surplus be distributed to the participating policyholders in the form of a cash dividend. He did not consider the possibility of distributing the surplus in the form of an increase in their sums assured.

Dodson argued that it would be unnecessary for the business to hold any initial cash reserve as a contingency for adverse experience. This may seem a peculiar position for a prototypical actuary who understands risk well enough to advocate charging on the basis of the worst single year in 22 years of experience. But he argued that the presence of participating policyholders' unlimited liability meant it was superfluous: even if a funded reserve was established, it would not absolve the participating policyholders from potential calls on their personal assets, and the security of the non-participating business would still at least partly rely on such calls being made good. It should also be noted that it was typical practice of joint-stock companies of the time to only require a small portion of equity capital to be 'paid-up', with the remainder again being funded by future calls on shareholders when the company required the cash. So, in the broader context of the historical corporate practice, this position was reasonably natural.

Dodson's work was less concerned with technical innovations, such as those developed by Halley and de Moivre over the preceding 60 years. He was motivated by a vision of how such techniques could be applied to the provision of a new form of life assurance that was really quite revolutionary in the context of the life assurance practices of the time. The summary features of his system as set out in *First Lectures in Insurance* were:

- The use of mortality tables to produce age-specific premium rates for life assurance. At the time, a flat non-age-specific premium basis was standard market practice.
- The development of a whole-of-life assurance policy, available in single and regular premium forms. At the time, only short-term (usually one-year term) assurance was sold.
- A with-profit system where participating and non-participating policies were simultaneously provided. Non-participating policies would be more expensive and would be priced to make it unlikely (relative to experience) that losses would be suffered. Participating policyholders would nonetheless have an unlimited liability exposure to calls on their personal assets in the event of losses exceeding the assets in the fund.
- The surplus on non-participating business would be distributed equitably amongst participating policyholders, in proportion to how much profit their own policies had generated for the corporation by virtue of their longevity.

- Contingency capital would be funded by calls on shareholders (the long-term policyholders in a mutual scheme) rather than by cash funding of a contingency reserve.

Everything advocated by Dodson as described above was implemented by Equitable Life within 25 years of the publication of *Lectures*. The consequent success of the Equitable meant that it would come to be regarded as the model of life assurance practice. But this revolution would occur after Dodson's death and the significant contributions of some talented others would be required to bring his vision to fruition.

The Beginnings of the Equitable (1757–1772)

> It has been frequently said that the history of the Equitable Office is the history of life assurance in England. If this be not literally true, it certainly is true that the readiness with which the conductors of this Office have ever listened to and embraced all real improvements, not only led it to the high position which it so speedily attained, but has tended to advance both the theory and practice of Life Assurance in this Kingdom.[20]

As *Lectures* went to press in 1756, Dodson organised a series of City of London meetings to secure the financial and political support he would require to bring his conception to fruition. Dodson secured significant backing from influential City and political figures to create a new life assurer. His City backers whole-heartedly bought into Dodson's vision with one important exception: Dodson had envisioned the life assurer being established as a mutual partnership with unlimited liability, but his backers believed that the venture would be more marketable as a corporation that provided limited liability to its participating policyholders. The share capital would still be largely unfunded and so participating policyholders would still be 'on call' to inject capital if required, but this would be subject to a maximum amount. Their first objective was therefore to obtain a royal charter for incorporation. They presented a petition of application to the Privy Council in 1757.

The three corporations writing life assurance in Britain at the time (the Amicable, Royal Exchange and London Assurance) lobbied for the petition to be rejected. These firms argued that the size of the life assurance market was so small that it could not support another corporation. They also

[20] Walford (1868), Chapter 4.

argued that it would be unsafe for the new insurer to be established without the backing of a capital fund. The petition was rejected by government in 1760, and in the intervening time since the application, Dodson had died. A report was published by the Attorney General and Solicitor General in 1761 explaining the reasons for the rejection of the petition. A few quotes from the report can give a sense of the nervousness about the revolutionary life assurance design being proposed by the petitioners.[21]

On the funding of risk capital through capital calls rather than a 'paid-up' fund:

> Because it appears to us altogether uncertain whether this project will or can succeed in the manner in which it is proposed, and if the success is uncertain, the fund for supporting it, which is to arise from the profits of the undertaking, will be precarious. This last objection is, in our opinion, a fatal objection to the scheme; for, though an undertaking plainly calculated for the benefit of the public may, in some instances, deserve encouragement, even where success is dubious, yet, in such cases, the projectors alone ought generally to abide the peril of the miscarriage. In the present proposal, therefore, whatever else may be hazardous, *the capital or fund to answer the losses ought to be certain and liable to no casualty*.

On the use of mortality tables to set premiums by age:

> The success of this scheme must depend upon the truth of certain calculations taken upon tables of life and death, whereby the chance of mortality is attempted to be reduced to a certain standard: this is a mere speculation, *never yet tried in practice*, and consequently subject, like all other experiments, to various chances in the execution.

On the use of population data to set life insurance premiums:

> The register of life and death ought to be confined, if possible, for the sake of exactness, to such persons only as are the objects of insurance. Whereas the calculations offered embrace the chances of life in general, the healthy as well as unhealthy parts thereof, which, together with the nature of such persons' occupations, are unknown numbers.

[21] The following quotations from the report of the Attorney General and Solicitor General are obtained from Ogborn (1962), Chapter II.

On the prospect of the business being profitable:

> We are more apt to doubt the event, because it has been represented to us ... that all profit that has been received by the Royal Exchange Assurance Office, from the time of its commencement to the present time (40 years) amounts only to a sum of £2,651, the difference between £10,915 paid in premiums and the sum of £8,263 disturbed in losses ... If then, this corporation, who are charged with taking big unreasonable premiums have reaped no greater profit, we can hardly expect a more considerable capital to arise from lower premiums, and the hazard of loss will be increased in proportion as the dealing will be more extensive.

And their final advice:

> If the petitioners, then, are so sure of success, there is an easy method of making the experiment, by entering into voluntary partnership, of which there are several instances now subsisting in the business of insuring; and, if upon a trial these calculations are found to stand the test of practical experiment, the petitioners will then apply with a much better grace for a charter than they can at present, whilst the scheme is built only upon speculative calculations.

The promoters immediately took the advice of the Attorney General and Solicitor General and produced a draft deed of settlement that would create a new life assurer as an unincorporated partnership of participating members (just as Dodson had always advocated). The Society for Equitable Assurances on Lives and Survivorships was thus established in September 1762.

For the next ten years, the Equitable developed as an established life assurer in a small life assurance market. Like most businesses, it had its fair share of corporate politics and power struggles, but it stayed largely true to Dodson's vision. By 1772, it had accumulated assets of around £30,000, with an annual premium income of £7,558 and total sums assured of £174,282 on 567 assurance policies. The policies were a mix of participating and non-participating contracts, and whole-of-life and term assurance policies. There were also 20 annuitants receiving a total of £694 per annum. The further development of Dodson's vision and indeed the next developments in actuarial thought required impetus from a new character.

Richard Price and His *Observations* (1772)

Richard Price is known for two particular achievements that made permanent contributions to the history of actuarial thought and beyond. As we saw above, he presented Bayes' revolutionary work on statistical inference to the

Royal Society. Secondly, Price produced the Northampton mortality table, which was the market standard for life assurance pricing in Great Britain and North America for nearly a century—over which period the industry finally and substantially blossomed.

His contributions to actuarial thought, however, are even wider than these two singular achievements. From 1768 through to the 1780s, the Equitable consulted him as an actuarial trouble-shooter whenever challenging technical issues arose. When his advice was given, it appears to have been consistently taken and acted upon. More generally, his actuarial writing reveals a deep understanding of both theoretical and practical issues in pricing and reserving for life assurance. His work provides perhaps the earliest examples of a rounded actuarial analysis—quantitative and technical, but only as a means to an end, and with the clear objective of developing pragmatic solutions to real and complex financial problems, implemented with judgement and an awareness of the limitations and complications of business reality and human behaviour.

Like Thomas Bayes, Richard Price was a clergyman, a mathematician, a philosopher and a Fellow of the Royal Society. Price was also actively involved in political debate, arguing passionately in favour of freedom of thought and political reform in areas such as the extension of the voting franchise, the reduction of the national debt and in favour of the revolution in America. Politically, he was a prototypical liberal Victorian. Intellectually, he was a classic late Enlightenment figure: an intense man who was passionate about rational thinking and reasoned debate, and who obtained the highest regard of influential men across a diversity of intellectual circles. Price died in 1791, aged 68.

The philosopher and historian Ian Hacking ranks Price's Northampton tables highly amongst the historical mortality tables. They were 'perhaps the first statistical results to be taken seriously ... The Institute of Actuaries did not do anything much better until 1869.'[22] The first version of the Northampton table was published by Price in 1772 (along with tables based on population data for Norwich and London) in the second edition of his *Observations on Reversionary Payments*.[23] This book addressed a wide selection of topics relating to life contingencies (and a number of financial topics outside the field of life contingencies). The construction of mortality tables is addressed in Essay IV of the book, which, in keeping with the style of the times, had an exhaustively descriptive title: 'On the Proper Method of constructing Tables for determining the rate of human mortality, the number of inhabitants, and the values of Lives in any Town or District, from the Bills of Mortality in which are given the numbers dying annually at all ages'.

[22] Hacking (1975), p. 113.

[23] Price (1772).

Halley's Breslau table had remained the standard reference table from its publication in the 1690s until the arrival of Price's tables. No practical use was made of Halley's table by life assurers, but it was used by other writers such as de Moivre in their research on life contingencies. Halley's paper did not include an explicit description of his assumptions and methodology. Price was more forthcoming and provided a detailed discussion of exactly what steps he believed should be taken in transforming raw data on deaths into a finished mortality table.

In the earlier discussion of Halley's tables, we noted that Price had specified how mortality rates could be derived from ages at death when the underlying population is at a stable level and there is no immigration or emigration (recall that the technical challenge here arose from the lack of information about the number *alive* at each age, and this therefore needed to be inferred from the data on numbers of deaths). In his essay, Price showed how mortality rates could be derived when there is a stated level of net immigration or emigration, setting down the following general rule:

> From the sum of all that die annually, after any given age, subtract the number of annual settlers after that age; and the remainder will be the number of the living at the given age.[24]

He noted that the effects of immigration would be particularly significant in London, and he analysed its impact on estimated mortality rates by calculating rates from the London Bills of Mortality 1759–1768 data with and without his immigration adjustment. Naturally, the modelled impact of immigration must be a function of assumptions about the ages at which immigration occurs. Price noted that deaths significantly exceeded reported births in London, even though London was generally reckoned to have increased in population over the period. The available birth data of the period was somewhat unreliable and he expected it would underestimate the true levels of births, but he still conjectured that around one quarter of the annual deaths were of immigrants to the city. He assumed that all immigrants would be aged 20 at the time they entered.

Based on these assumptions, the mortality rates for ages less than 20 would be *understated*, as the no-immigration assumption would result in an *overestimation* of the numbers alive at ages up to 20 (recall that the estimation of the numbers alive at age x is based on the total numbers annually dying aged more than x). With the assumption that all immigration occurs at age 20

[24] Price (1772), p. 246.

and none occurs beyond that age, no adjustment for immigration would be implied for mortality rates from age 20 onwards.

Figure 2.3 compares the no-adjustment and with-adjustment results obtained by Price for ages up to 20.

The chart shows that mortality rates at very young ages had, until this time, been significantly understated by calculations using the Bills of Mortality (such as those produced by Dodson for use by the Equitable). This did not have much direct consequence for the pricing of life contingencies as most policies were written on lives older than twenty years of age. It did, however, have implications for the assessment of life expectancy at birth and for government health policies. Great Britain underwent an unprecedented period of population growth in the second half of the eighteenth century. The population is estimated to have grown from around 6.3 million to 9.2 million between 1751 and 1801, and most of the growth over this period is believed to have occurred form 1780 onwards.[25] A sharp fall in infant mortality rates is generally believed to have been a significant factor in this population growth. Price's estimates certainly highlight how shockingly high infant mortality rates were at the start of this period—according to his calculations, most of London's newborn babies of this time would not survive to see their fifth birthday.

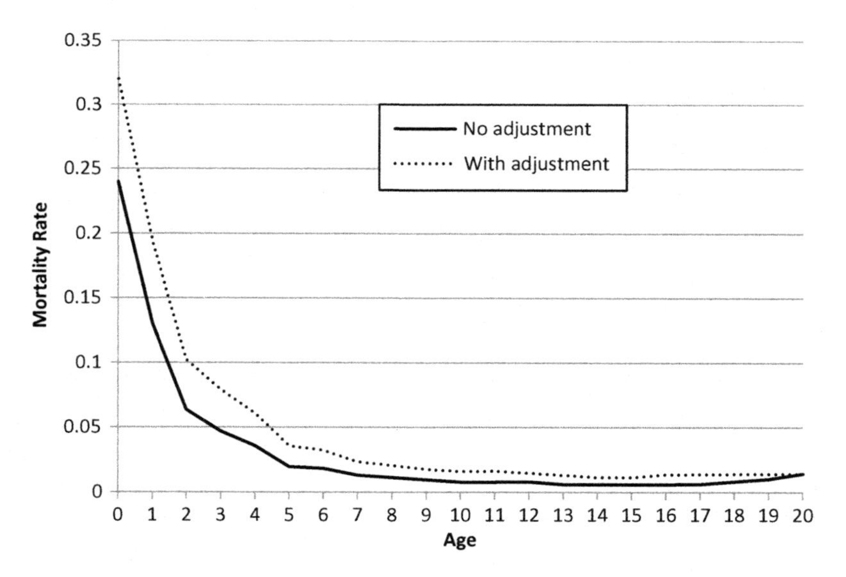

Fig. 2.3 Price's mortality rates derived from London Bills of Mortality 1759–1768 data, with and without adjustment for immigration

[25] Deane and Cole (1967), Chapter 1, Section 2.

Price then identified Northampton and Norwich as towns that had kept Bills of Mortality which included ages at death for many decades. He applied his methodology, including his adjustment for immigration, to both these datasets to produce tables for each town. He also compared these results with Halley's Breslau table. The mortality rates of these two tables, together with Price's London table and Halley's Breslau table, are compared below in Fig. 2.4.

The consistency of these tables is quite striking, and Price noted, with an unbridled satisfaction, 'there is a striking conformity between all the three Tables [Norwich, Northampton, Breslau], which gives them great weight and authority'. The London table produced noticeably higher mortality rates than the other three tables and this was unsurprising: John Graunt had noted a century earlier the relatively lower mortality rates associated with country living.

As we might expect from Price given his history of involvement with statistical inference, he gave some consideration to the potential sampling error in his data. He did not attempt to calculate an explicit standard error for his mortality rates (no such concept existed at this time). Instead, he used an empirical approach noting that whilst 30 years of data were used in each of the Northampton and Norwich tables, the results for each were very similar when any ten-year period within the 30 years was used instead. Based on this analysis, he concluded: 'These Tables, therefore, are founded on a sufficient

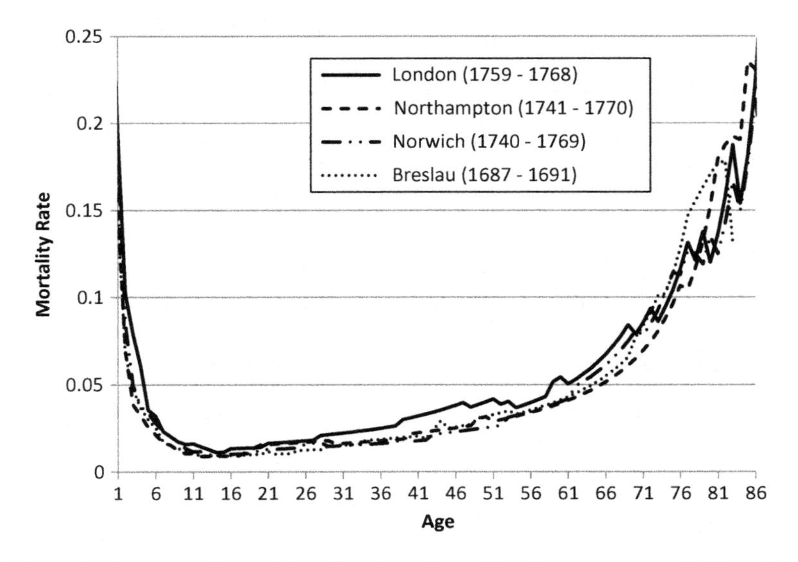

Fig. 2.4 Price's mortality tables for London, Northampton and Norwich and Halley's Breslau table

number of observations; and it appears, that there is an invariable law which governs the waste of human life in these towns.'[26]

Price's *Observations* is most renowned for these mortality tables. But there was much else in the book that had an important intellectual impact on the developing actuarial thinking of the time. *Observations* was based on Price's experiences as an advisor to a number of embryonic or would-be insurance companies that were considering how to price a variety of forms of annuity contract. The book begins with an exhaustive discussion of his experiences in this area, giving annuity pricing examples for a number of variations of reversionary and deferred forms of annuity in single premium and regular premium forms. In these examples, he computed exact valuations rather than relying on the sort of approximations that de Moivre had developed. In the intervening 40 years since de Moivre's *Annuities on Lives*, methods of arithmetic computation had become somewhat more efficient, making the exact calculations more feasible, if still highly laborious. Price included an essay on de Moivre's annuity pricing approximations that warned of the limited accuracy of his joint life pricing formula. In particular, Price showed how this could lead to substantial proportional errors when valuing a reversionary annuity (which is valued as the *difference* between the single life annuity and the joint life annuity).

Observations also included discussion of a number of annuity products that had been launched by life and annuity businesses since the 1750s. In these discussions, he repeatedly found that the products were substantially underpriced and concluded it unlikely that policyholders would receive their full promised benefits. This treatment was applied to the reversionary annuity products of the London Annuity Society and the Laudable Society, and more widely: 'There are in this kingdom several institutions for the benefit of widows, besides the two on which I have now remarked; and in general, as far as I have had any information concerning them, they are founded on plans equally inadequate.'

The actuarial management of the life assurance companies was considered too. Whilst the Royal Exchange and London Assurance were long-established writers of life assurance with guaranteed sums assured, Price focused on the Amicable, which wrote significantly higher volumes of business, though, as was discussed earlier, without offering guaranteed sums assured. Price strongly criticised this mutual benefaction model of life assurance that Amicable provided: 'It is obvious that regulating the dividends among the nominees by the number of members who die every year is not *equitable*; because it makes

[26] Price (1772), pp. 257–258.

the benefit which a member is to receive to depend, not on the value of his contribution, but on a *contingency*; that is, the number of members that shall happen to die in the same year with him.'[27] He also noted that charging all members the same premiums irrespective of their age was not equitable. He went on to contrast these imperfections with the Society for Equitable Assurances on Lives and Survivorships (of which he was a paid consultant) and provided some public advice to its managers:

- First, on the need for rigorous underwriting to avoid anti-selection: 'I have already more than once observed that those persons will be most for flying to these establishments, who have feeble constitutions, or are subject to distempers, which they know render their lives particularly precarious; and it is to be feared that no caution will be sufficient to prevent all danger from hence.'[28]
- Then on adverse variation in experience: 'The calculations only determine probabilities; and agreeably to these, it may be depended on that events will happen on the whole. But at particular periods, and in particular instances, great deviations will often happen; and these deviations, at the commencement of a scheme, must prove either very favourable, or very unfavourable.'[29]
- And finally, with impressive prescience given the tumultuous debates that would embroil the Equitable in the coming decades, on the prudent distribution of surplus: 'the encouragement arising from the possession of a large surplus (should) be led to check or stop the increase of its stock by enlarging its dividends too soon, (otherwise) the consequences might prove pernicious.'[30]

These pieces of advice highlight the completeness of Price as an actuarial thinker. He strongly advocated a reasoned approach to pricing based on the rigorous analysis of statistical data, and he furnished the profession with much by way of methods and results of such analysis. But he also understood the practical aspects that must attend the successful and sustainable actuarial management of a life assurance business, and he anticipated some that had not yet arisen at the time of his writing. He used both these technical and practical insights to actively direct the development of new actuarial tech-

[27] Price (1772), p. 124.
[28] Price (1772), p. 129.
[29] Price (1772), p. 129.
[30] Price, *Observations*, p. 131.

niques at the Equitable, and these would be a critical part of that business's huge success in the following decades. Added to these contributions is his vital role as the midwife of Bayes' revolutionary work on statistical inference. All this taken together, Richard Price stands as a titan of actuarial thought. If an actuary is someone who provides rational and rigorous advice on the sustainable long-term financial management of life contingencies business, then Price was arguably the first actuary, and inarguably one of the most important and influential actuaries in the profession's history.

William Morgan and the Actuarial Management of with-Profit Business (1775–82)

Several developments important to the nascent life assurance industry occurred over the dozen years following the Equitable's founding in 1762. Price's *Observations*, with its new mortality tables and sound actuarial insights, brought actuarial thinking on the pricing of life contingencies to a new level of competence. The Equitable had established itself and its whole-of-life assurance product as successful features of the British life assurance industry. It had become the leading provider of life assurance, but the size of this market remained modest relative to other forms of insurance business. The Life Insurance Act of 1774 further improved the conditions for future growth in the industry. This act introduced the legal concept of insurable interest. Henceforth, life assurance could not be used as a means of speculative gambling, but only as a form of risk mitigation where the buyer had a real and legitimate exposure. This focused life assurance business on products that were generally most amenable to actuarial and statistical analysis. Thus the scene was set for the next step in the development of life assurance business and the actuarial practices required to make it sustainable.

William Morgan, nephew of Richard Price, and the actuary of the Equitable for a full 55 years from 1775 to 1830, played a most significant role in shaping this next stage of actuarial practice. Morgan was appointed assistant actuary in 1774, serving under John Pocock, who was elected actuary a month earlier following the death of John Edwards, the previous holder of this post. John Pocock had been in office less than a year when he died unexpectedly in early 1775. Morgan had distinguished himself sufficiently well during his short term as assistant actuary for him to be elected actuary in an uncontested election at a general court of the society in 1775. Morgan did not attend university but did receive some medical training as an apprentice at an apothecary prior to joining the Equitable. He was tutored in life assurance mathematics by Price in 1773 before being appointed assistant actuary. At this time, the actuary's role in a life assurance business was not yet well defined. Edwards and Pocock had been book-keepers rather than statisticians. It would not have been viewed as essential for Morgan to demonstrate advanced mathematical learning, and the tutelage under Price would likely have carried some weight with the directors of the business.

At the start of Morgan's tenure as actuary, there was an emerging interest in the distribution of the capital that was evidently accumulating in the business. A satisfactory way of assessing the appropriate level of reserves for in-force business had yet to be fully implemented (actuarial methodology and the computational burden of policy-by-policy valuation both had to be addressed). A rigorous and accepted means of determining how much of the visible asset accumulation could be considered as surplus available for distribution was therefore wanting. With Morgan still establishing himself in his new role, the directors looked to Price for guidance and he duly provided them with a number of recommendations. Price suggested the society should pursue three approaches to analysing the current adequacy of the accumulated capital fund: a comparison of the actual mortality experience to date with the mortality assumptions used in the premium basis; a comparison of the amount of premiums being received with the claims being paid (he suggested, as a rule of thumb, that around two thirds of premiums received in the early years of a whole-of-life policy would be required to adequately build up the reserve for the claims that would arise in its latter years); and a comparison of the accumulated assets of the society with a valuation of the liabilities.

In anticipation of a surplus being identified by the above methods, Price also provided some advice on the *form* of distribution of surplus to the participating policyholders. In particular, he advised against the payment of any cash dividends. Instead, he argued, if any distribution of surplus was deemed appropriate, it should be made by either increasing the sums assured or reducing the future regular premiums of the participating policies. Price's advice was acted upon by Morgan and, like virtually everything else Price did in the field of actuarial science, his recommendations became part of standard actuarial practice for decades (or, as in this case, centuries).

The market value of the assets of the society at the end of 1773 amounted to £33,000. Price surmised that a valuation of the liabilities might be around £22,000, implying a surplus of £11,000. It was Morgan's job to perform the first rigorous valuation of the society's liabilities and hence determine a reliable measure of the surplus. He and his small team of assistants valued each of the 922 policies in-force individually. Four months and several thousand manual calculations later the full results were ready to report. The results were even better than anticipated by Price. At the start of 1776, the assets were valued at £41,928 (their market value) and the liabilities were valued (using the premium basis) at only £16,786, implying a surplus of £25,142. This startling result raised two fundamental questions: how did such a large surplus arise? And what should be done with it?

The surplus primarily arose from the much lighter mortality experienced to date than that assumed in the premium basis. This was especially true in the non-profit business, where, as discussed earlier, the premium basis was more expensive than the premium basis applied to with-profit policies. But even the with-profit experience was very substantially better than assumed in its premium basis. Morgan himself noted that the 1776 investigations had concluded 'the probabilities of life in the Society had been higher than those in Mr Dodson's Table, from which its premiums were computed, in the proportion of three to two'.[31]

This substantial mortality margin had several contributory factors. The premium basis used Dodson's recommended London Bills of Mortality basis. This was based on mortality data from 50 years prior at a time when life expectancies were improving significantly due to medical developments and improvements in economic and social conditions. The period on which the bills were based included 1740, a year of weather-induced food shortages that resulted in mortality almost as severe had been experienced in bouts of the plague. The bills were based on the experience of London, which was probably the least healthy place in the country; and they were based on the full breadth of the London population, not the emerging middle class that was buying the Equitable's whole-of-life assurance policies. Furthermore, the business had an active approach to underwriting, with every new life assured being reviewed and approved by a director.

There were other factors beyond the mortality basis that also contributed significantly to this accumulation of surplus. A number of somewhat arbitrary additional loadings were applied on top of the premiums implied by Dodson's tables. These included loadings for larger assurances, young lives and females (sex-specific mortality tables were not in use at this time). And then a final 6 % margin was additionally applied to all premiums. Furthermore, wars in Europe and America had resulted in significant increases in government bond yields, and much new money was therefore invested at yields well in excess of the 3 % assumed in the premium basis.

These profits were amplified by the gearing effect whereby the surplus generated by non-participating short-term business was left to be shared amongst the remaining participating policyholders. A decade later, in an analysis of the

[31] Morgan (1829), p. 20.

surplus that had emerged between 1768 and 1787, Morgan concluded that more than half was due to profits made on temporary (and non-participating) assurances.

Finally, there was another source of surplus that was important during this period. Of the whole-of-life assurances written during the society's first twenty years, less than half ultimately resulted in claims. The vast majority of those policies that did not result in claims were forfeited without any compensation. The society did have a policy of offering fair surrender values, but it may have been that many policyholders never actually applied to surrender their policy but merely forfeited them when they chose not to make any further premium payments. Furthermore, the surrender basis used by the Equitable was based on a form of net premium valuation rather than an asset share-based calculation. This prospective approach to assessing the surrender value was highly sensitive to the mortality basis used in the valuation. When the Equitable moved its premium basis to the Northampton table in 1780 (discussed further below), it meant that many of the regular premium whole-of-life contracts written on the old premium basis were worth very little on this prospective basis, or even had a negative value to the policyholder![32] So the surrender values were much lower than the accumulated value of the assets that the surrendered policies had contributed (even after deducting the cost of the life assurance that had been provided over the duration of the policy to date).

What to do with the surplus? A series of meetings of the general court of the society in 1777 determined that a 10 % reduction in all future premiums should be effected. Moreover, the in-force with-profit policyholders would have this reduction back-dated: it would be calculated for all the premiums they had paid to date, and this would be offset against their next regular premium. To the extent that the refund was greater than the next regular premium, the remainder would be paid to the policyholder in cash. So the back-dated element was essentially a one-off cash dividend.

The premium reduction and refund of 1777 was a temporary stop-gap measure that alleviated the building pressure from participating policyholders. But a more fundamental review of pricing and surplus distribution was called for. As usual, the intellectual impetus was provided by Richard Price. In 1780, he advocated a change in the underlying mortality tables that were used in the premium basis. Dodson's London Bills of Mortality premium basis was still in use, and this was clearly generating premiums that were much larger than was reasonable. Indeed, it is interesting that the business could continue to generate significant new business at these rates. This perhaps

[32] Morgan (1829), p. 38.

reflects the lack of (credit-worthy) competition in the life assurance market of the time. It may also suggest that policyholders increasingly viewed the whole-of-life with-profit policy not merely as an insurance product, but as an investment product (with likely future participation somehow in the surplus). Price recommended using a mortality basis derived from the Northampton or Norwich tables that he published in *Observations*. The directors, as usual, followed this recommendation and decided upon the Northampton table. Price updated this table to include Northampton Bills of Mortality data up to 1780 (so the table was based on data from 1735 to 1780). The directors opted to add a 15 % loading to the premiums implied by this table. Even Price, whose advice to the society tended to be universally prudent and frugal, referred to this loading as 'exorbitant'.

This new premium basis would be applied to new business. It did not address if and how to distribute the surplus that had accrued to date to existing policyholders. How much of the surplus should be distributed to the existing with-profit policyholders, and what form should this distribution of surplus take? The directors recommended that an addition should be made to with-profit policyholders' sums assured of 1.5 % of the sum assured for each annual regular premium paid (hence a policy that had been in-force for the full 19 years since the founding of the society would obtain a 28.5 % proportional increase in their sum assured). The proposal was approved by the members, and the first with-profit reversionary bonus was thus paid in 1781.

No explicit historical explanation is available, but it is likely that the size of the reversionary bonus was chosen to distribute a targeted portion of the surplus. What drove Price's preference for distributing surplus through increases in sums assured rather than cash dividends? It was perhaps because retaining the assets within the society for as long as possible fitted with his vision of a long-term growing business of increasing influence and scale. In Ogborn's outstanding history of the Equitable,[33] he suggested that the increase in government bond yields that occurred in the period of 1770–1780 may also have played a part. The American War of Independence had placed considerable strain on the British Government's finances, and long-term government bond yields had increased from around 3 % in 1770 to around 6 % by 1780.[34] Most of the assets of the society were invested in long-term government bonds, and there may have been a reluctance to be forced to sell at these deflated price levels to fund a cash distribution (the assets, however, were valued at these market values in Morgan's surplus analysis). The society had also

[33] Ogborn (1962), p. 110.

[34] Homer and Sylla (1996), pp. 156–157.

recently started to invest a significant portion of its assets in mortgages. The illiquidity of these assets might also have tempered the directors' enthusiasm to distribute significant amounts of cash (particularly if it created the expectation that more might be follow).

A further reversionary bonus of the same form was made in 1792, this time of 2 % of the sum assured for each regular premium paid. This same level of bonus was added again in 1800 when the surplus was calculated to be £484,000 (and the bonus was calculated to reduce this surplus to £225,000). The bonus system established by Morgan proved highly attractive to prospective policyholders. Furthermore, it was not a system that could be quickly replicated by competitors.

Despite spending much of the period at war, between 1781 and 1800 the British economy grew rapidly as the early stages of the industrial revolution gathered momentum. Increases in overseas trade and substantial population growth fuelled impressive economic growth.[35] For the emerging salaried middle class, the whole-of-life with-profit policy was a highly compelling composite of a short-term life assurance and a long-term savings vehicle. The Equitable was perfectly positioned to take advantage of this demand. The business of the Equitable exploded in scale over this period: between 1776 and 1800, the assets of the Equitable grew from £42,000 to over £1 million.[36]

The greatest challenges facing Morgan in the following decades were related to managing the consequences of this remarkable success. In particular, he was occupied with determining how to limit the number of new policyholders being admitted to the society so as to avoid 'carpet-bagging'; ensuring an equitable distribution of surplus between the 'old' and 'new' members; and, more generally, managing the expectations of policyholders as to the amount of surplus that it was reasonable to distribute over time. In 1800, the society introduced two safeguards to over-distribution of surplus: bonuses would only be distributed once every ten years; and the present value of the size of the bonus should not exceed two thirds of the surplus assessed at that point in time. A further, rather more controversial safeguard was put in place in 1816: bonuses would only ever be applied to the oldest 5,000 policies of the society at the date of the bonus distribution. At the time of its implementation there were around 9,500 policies in-force.[37] This created a 'waiting period' for new policyholders before they were entitled to participate in any bonus distribution. Morgan expected at the time of the rule's implementation that, based

[35] Deane and Cole (1967), Chapters I and II.

[36] Ogborn (1962), pp. 104 and 123.

[37] Ogborn (1962), p. 155.

on historical discontinuance rates, this waiting period would be around five years. Whilst controversial at the time of its inception, this rule remained in place until 1893.

Thus, by the start of the nineteenth century, much of the actuarial *ethos* of with-profit management had fully emerged: Dodson, Price and Morgan had envisioned, analysed and implemented a financial product—the whole-of-life with-profit policy—that had radically reshaped life assurance as a product, business and important element of the financial sector. With William Morgan as the first role model, the actuary had emerged as a person of considerable power in a life assurance company: his considered judgement and detachment from short-term commercial pressure was required to determine equitable and prudent distributions of the surplus of a with-profit fund. Morgan was almost a caricature of this role: in later years, he would face much criticism for his near pathological prudence, resisting pressure from managers and members to reduce premiums or increase surplus distributions. Indeed, he felt compelled to defend his actions in a book that he published a year before retiring from his position as actuary in 1830, explaining in its Introduction:

> It is indeed much against my inclination that I have engaged in the present work; but my anxious desire to justify the conduct of a Society for whose welfare I have laboured so many years, has impelled me to notice some of the misrepresentations with which it has been assailed, and, as the best refutation of them, to give a short history of the Society.[38]

Morgan's time was many years before the introduction of financial regulators and statutory capital standards. There was perhaps less need for them when men like him were in the chair.

[38] Morgan (1829), Introduction.

3

Life from the Napoleonic Wars to the Second World War

> When the romantic formative period of a science is over, it gradually settles down into one of ordered progress, which while supremely useful has less about it to attract the imagination.[1]

So said Samuel George Warner in his 1917 presidential address to the Institute of Actuaries. And yet, with the benefit of 200 years of hindsight, this statement might be more applicable to the beginning of the nineteenth century. In Warner's era, actuarial thought and practice was largely reactive in the context of a series of unprecedented events and circumstances: global war, high inflation, interminable economic depression, low interest rates, more global war, government-induced ultra-low interest rates. At the start of the nineteenth century, life assurance practices still had to navigate the political and economic challenges of their time, but the actuaries of the nascent life assurance industry were arguably freer to plot the course of their profession and its practices. The trail had been blazed in the final half of the eighteenth century and a revolutionary vision of life assurance as a long-term savings vehicle for the middle classes had been realised with the remarkable success of the Equitable in the final two decades of the century. Many new life offices would follow in Equitable's wake. The pioneering work of Dodson, Price and Morgan now had to be systematically developed into robust best practices for an embryonic profession that was invested with substantial powers and discretion in a fast-growing and increasingly significant element of the financial sector.

---

[1] Warner (1918), p. 16.

C. Turnbull, *A History of British Actuarial Thought*,
DOI 10.1007/978-3-319-33183-6_3

The modelling of mortality and its implications for the pricing of life contingencies continued to dominate actuarial thought in the first half of the nineteenth century. Whilst important, even fundamental, developments in mortality modelling continued through to the end of the nineteenth century, a sense emerged by the middle of the century that mortality was increasingly well-understood. Likewise, it was recognised that a greater amount of professional energy should be invested in the further development of thought in other important aspects of actuarial responsibility within a life assurer such as reserving methods, surplus distribution approaches and the assurer's investment policy. The forming of the Institute of Actuaries in 1848 (and Faculty of Actuaries in 1856), and the arrival of their professional journals and sessional meetings, was a catalyst for more collaborative and co-ordinated thought-leadership amongst actuaries. It is perhaps no coincidence that the 25 years following the formation of the institute was a period that saw the crystallisation of many important ideas and principles that acquired permanence in actuarial thought.

The technical considerations involved in the valuation of liabilities, and hence surplus, emerged as a major field of actuarial research in the mid-nineteenth century, together with the related topic of determining how the surplus should be distributed amongst the with-profit policyholders in a way that was just and equitable. To a lesser degree, the second half of the eighteenth century also saw actuaries give greater consideration to the asset side of the balance sheet, and the first actuarial papers on investment strategy for life assurance liabilities were produced. But just a handful of investment papers emerged in the nineteenth century, and this field was only to become an area of real focus for the profession in the twentieth century.

The economic turmoil of the first half of the twentieth century and its implications for the management of the British life offices kept valuation, surplus and investment strategy at the top of the thought-leadership agenda. New mortality tables continued to be developed, but the first half of the twentieth century did not witness any notable fundamental breakthroughs in thinking on the construction of the tables.

The historical developments in these three broad areas of actuarial thought—mortality modelling, valuation and surplus, and investment strategy for life assurance liabilities—over the nineteenth and first half of the twentieth century are discussed in turn below.

Developments in Mortality Modelling, 1808–1881

Chapter 2 noted that Richard Price's Northampton table has been described as a mortality benchmark that stood for the next 100 years and that according to Hacking 'the Institute of Actuaries did not do anything much better until

1869'. The truth is more nuanced. The next significant development in mortality modelling to follow Price was made by Joshua Milne in 1815. Milne was then actuary at the Sun Life Assurance Society, one of the many new life assurance offices established in the early 1800s following the success of the Equitable.

Milne's two-volume tome *A Treatise on the Valuation of Annuities and Assurances on Lives and Survivorships* largely consisted of mathematical bookwork on the valuation of various forms of annuities and assurances.[2] But the inclusion of his Carlisle mortality table made it a historic actuarial publication. The Carlisle table was, like Price's Northampton table, based on observations from local Bills of Mortality data (the Carlisle table was based on data from 1779 to 1787 inclusive). However, Milne's Carlisle data was superior to the data used in any prior mortality table in one key respect: it included population data categorised by year of age (it was also categorised by sex, and whilst Milne did analyse and discuss some of the differences in mortality behaviour observed for the two sexes, he did not publish separate male and female mortality rates in his Carlisle table). The 'Exposed to Risk' element of the mortality calculation therefore did not need to be estimated with the use of assumptions about population stability and immigration/emigration patterns. It could now be directly observed. This naturally supported a more refined and accurate estimation of experienced mortality rates.

The Carlisle data had another advantage over Northampton: it was somewhat less out of date. The Northampton data was based on mortality data from 1741 to 1770, whereas the Carlisle table used data from 1779 to 1787. Life expectancy had improved at a significant rate over the second half of the eighteenth century, particularly for infants. Milne noted that, even over his observation period, infant mortality rates were rapidly changing, largely due to the impact of social health policy initiatives such as smallpox inoculation. The Carlisle data included some data on cause of death, and Milne noted that the number of deaths from smallpox in the Carlisle Bills was 90 in 1779 and only 141 for the remaining eight years from 1780 to 1787. Dr Heysham, a local doctor who prepared some of the Carlisle data noted that in 1779 that 'several hundreds were inoculated in the neighbourhood of Carlisle, and it is a pleasing truth that not one of them died'.[3] Almost all of these smallpox deaths would have occurred at age five or under. In the Carlisle data, only 712 deaths of age five and under occurred over the full nine-year data period. If the smallpox death rate had continued at its 1779 rate, more than 700 would have died from smallpox alone. This reduction in smallpox mortality therefore had a substantial impact on the overall infant mortality rates of the period.

Figure 3.1 compares the Carlisle and Northampton tables.

[2] Milne (1815).

[3] Milne (1815), p. 735.

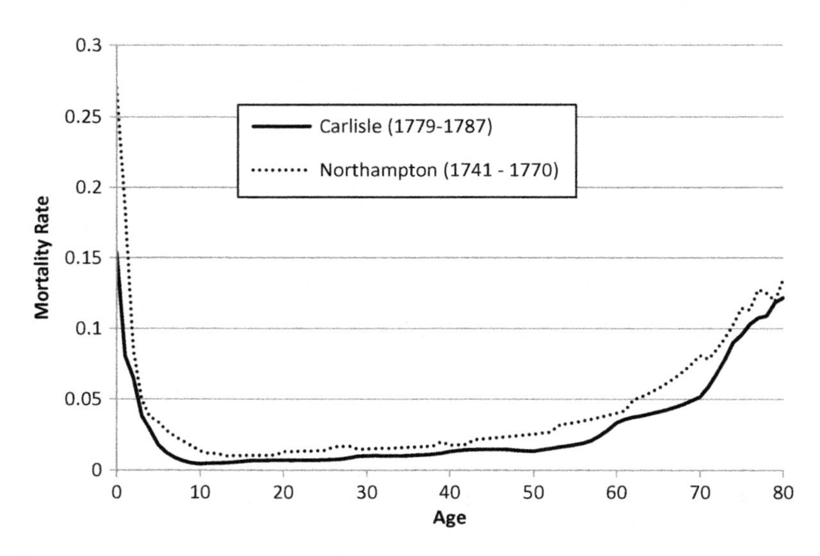

Fig. 3.1 The Carlisle and Northampton tables

The Carlisle table produced lower mortality rates than the Northampton table for all ages from birth to aged 80. This reduction in estimated mortality is attributable to two effects: the later data period of the Carlisle table and the improvements in life expectancy that occurred over the second half of the eighteenth century; and Price's assumption of a stable population from Northampton in estimating Exposed to Risk, when in fact its population was rising (as noted above, Milne did not require any such assumption for Carlisle due to the availability of population data).

The Carlisle table was undoubtedly an improvement on Price's Northampton table, and it was widely recognised as such in the actuarial and academic circles of the period. For example, Professor Augustus De Morgan, an academic mathematician and influential writer on life contingencies (and the great grandson of James Dodson!), wrote of the Carlisle table in 1838 'they are to be considered the best existing tables of healthy life which have been constructed in England'.[4] Yet life offices never applied it to the pricing of life contingencies to the degree that they used Price's table. This was at least partly because, as we shall see later, life offices increasingly focused on the use of mortality data directly observed on assured lives rather than the general population. Meanwhile, at the Equitable, an aging William Morgan, increasingly set in his ways, would never abandon his uncle's Northampton table. As we shall see

[4] De Morgan (1838), p. 167.

in the next section, Morgan's intransigence was set to become an increasing source of contention and controversy within actuarial circles and beyond.

John Finlaison and the British Government's Life Annuity Issuance (1808–1829)

The aftermath of the British Government's life annuity issuance of 1808 provides a colourful and historically important example of how the eighteenth-century foundations of mortality thinking were built upon and extended by some new and highly able actuarial thinkers in ways that prompted some disagreement with the earlier actuarial generation. This particular example of the changing of the guard of actuarial thought leadership involved some public controversy, prompted Cabinet-level political reaction and was accompanied by increasing personal animosity.

Some 120 years earlier, the British Government had experimented with the use of life annuities as a form of debt funding for its European wars. It found, somewhat peculiarly, that the scheme was both quite unpopular with investors and yet highly unprofitable for the government issuer, relative to simply issuing fixed long-term bonds. The government subsequently tried on occasion to use life annuities to raise debt funds over the course of the eighteenth century, but the issuances were very small in volume. For example, a life annuity was issued in 1779 with the aim of raising up to £7 million, but only 133 annuities were sold raising a sum of only around £150,000.[5] In 1808, in the midst of the Napoleonic wars and with Britain under serious threat of invasion, the government tried again with renewed vigour. The profitability to the issuer was similarly dismal to the 1692 issuance, but this time the market was better prepared to spot a bargain.

William Morgan, as the actuarial leader of the pre-eminent and dominant life assurance company of the age, advised the government on the mortality basis to be used for the pricing of these life annuities. His advice was to do what he did: use the Northampton table. The exact form of the advice that Morgan provided is not available, but with the benefit of hindsight his recommendations might appear surprising, or indeed rather odd. Morgan was well aware at this time that the Equitable's mortality experience was significantly lighter than that implied by the Northampton table. He would have known that if the annuity purchasers had experienced mortality rates similar to those experienced at the Equitable, life annuities priced with the Northampton table

[5] Leeson (1968), p. 9.

would prove highly profitable for the annuitant investor and unprofitable for the issuer. However, Morgan's view was steadfast and consistent throughout his entire career: he believed that the Northampton table was a fundamentally accurate description of the law of human mortality; and the lighter mortality experienced by the Equitable on its assured lives was due to the selection effects of its rigorous life assurance underwriting practices.

Even if Morgan's belief was in fact correct, it ignored the self-selection effect that is inevitably to be observed in annuity experience: only the healthiest of lives would be nominated for annuity contracts, and this selection effect was likely to be even stronger than the impact of the underwriting safeguards that Equitable applied to its assured lives. The potential for annuity selection was exacerbated by the ability of the annuity purchaser to nominate the life on which the annuity is written—annuities were not considered as insurance in the context of the 1775 act that created the concept of insurable interest. In Morgan's defence, however, it should be noted that the Equitable had been open for annuity business for many years and used the Northampton table as the premium basis throughout. This had hardly stimulated an overwhelming public demand for annuities (they consistently sold in tiny volumes compared to assurance business). Indeed, the Equitable had managed to sell such small volumes of annuities on the Northampton basis that there does not appear to be any record of an analysis of its profitability.

And so, in 1808, the government offered life annuities at prices based on the Northampton table, and this offer remained open for many years to follow. Ten years later, several million pounds had been raised by the annuity scheme, requiring annual annuity payments of £640,000.[6] Around this time, a civil servant at the Admiralty named John Finlaison wrote a letter to the Chancellor of the Exchequer expressing his concern at the losses that he believed were arising from the overly generous annuity pricing basis. Finlaison estimated that losses of £8,000 per month were being incurred as a result.

Finlaison would, in 1848, become the first President of the Institute of Actuaries. For an actuary of the period, he had an unconventional career and background. He never worked in a life office, and instead spent his entire career as a civil servant. He first made his mark in 1805 as the second clerk of the Commission of Revising and Digesting the Civil Affairs of the Navy. As noted in a recent history of the role of Britain's civil service in the Napoleonic Wars: 'Finlaison, born in Caithness, the son of a fisherman, had come to London after an education in Edinburgh; at this time only twenty-three, but already demonstrating a formidable logical brain and an immense capacity for

[6] Francis (1853), p. 203.

work.'[7] He worked at the Admiralty from 1809 to 1822, where he joined the newly established Admiralty Record Office and was responsible for a much-improved parliamentary reporting of the naval accounts and expenditure. He then moved to the Treasury and was appointed the Actuary of the National Debt, a position he held for the next 29 years.

No action was taken by the chancellor in 1819 when Finlaison first raised his concerns about the pricing of the government's life annuities. Empowered by his appointment as Actuary of the National Debt, he was eventually asked by the Treasury to provide a full analysis and he delivered a parliamentary report on 31 March 1829.[8] Finlaison's report contained a comprehensive analysis of the mortality experience of the annuities and tontines written by British (or English) governments since 1693. His work was ground-breaking in many ways. Most pertinently, it showed that the view he had expressed to the chancellor ten years earlier was right: the mortality experience of the government life annuities was much lighter than the Northampton table that had been used in pricing them for the previous 20 years. This is illustrated by Fig. 3.2, which also shows Milne's Carlisle table for comparison.

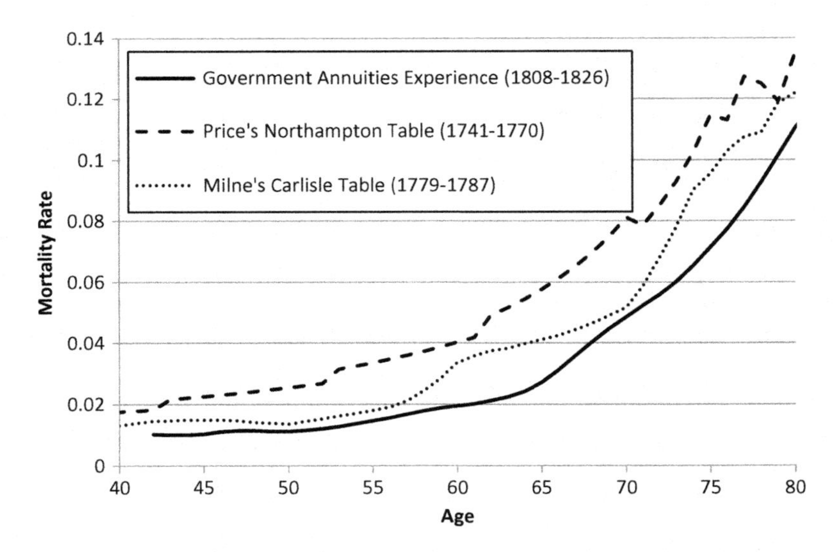

Fig. 3.2 The government life annuity mortality experience and Northampton and Carlisle tables

[7] Knight (2013), p. 325.
[8] Finlaison (1829).

Finlaison's analysis evidenced the profound differences between the government annuity mortality experience and the mortality rates assumed in the Northampton table: for example, at age 60, the Northampton table was implying a mortality rate that was fully double that experienced by the government annuitants! The Carlisle table provided a significantly closer guide, but still overestimated the mortality rate experience (which could be explained by annuity anti-selection effects and by life expectancy improvements in the decades between the two sets of observations). Finlaison's analysis implied that the government annuity prices should be increased by around one quarter. His report had an immediate impact: an Act of Parliament was passed in 1830 which upwardly revised the government life annuity prices in accordance with his recommendations.

Even under the revised annuity pricing basis, scope for anti-selection was still significant, especially at older ages: a 90-year-old life could obtain an annuity rate of 62 % (payable in quarterly instalments). Francis' 1853 book provides amusing tales of the creative anti-selection speculation that this generated:[9]

> The shrewd gentlemen of the Stock Exchange immediately saw and seized the advantage. Agents were employed to seek out in Scotland and elsewhere robust men of ninety years of age, to select none but those who were free from the hard labour which tells on advanced life ... Wherever a person was found at the age of ninety, touched gently by the hand of time, he was sure to be discovered by the agents of the money market ... The inhabitants of the rural districts of Scotland, of Westmoreland, and of Cumberland, were surprised by the sudden and extraordinary attention paid to many of their aged members. If they were sick, the surgeon attended them at the cost of some good genius; and if they were poor, the comforts of life were granted them. In one village the clergyman was empowered to supply the wants of three old hale fishermen during the winter season, to the envy of his sick and ailing parishioners. In another, all the cottagers were rendered jealous by the incessant watchful attention paid to a nonagenarian by the magnate of the place. It was whispered by the less favoured that he had been given a home near the great house; that the cook had orders to supply him with whatever was nice and nourishing; that the laird had been heard to say he took a great interest in his life, and that he even allowed the doctor twenty-five golden guineas a year, so long as he kept his ancient patient alive.

Clearly there was still much to learn about annuity market behaviour! Nonetheless, the new pricing basis undoubtedly brought the cost of servicing new government life annuities closer to that of long-dated gilts.

[9] Francis (1853), Chapter XII.

William Morgan resented and resisted the implication that his advice had unnecessarily cost the government huge sums of money, commenting:

> By the assistance of these tables (Finlaison's tables of the government annuity mortality experience), the notable discovery has been made of the loss sustained by the public of many thousands every week ... With the least knowledge of the subject, it might easily have been seen that the use of the Northampton table could not possibly be attended by such a loss, and that the extent of any loss (if it really existed), instead of six years, did not require as many hours to ascertain it.

Finlaison and Morgan also adopted opposing views on other actuarial debates of the time. Whilst Morgan argued that selection was the primary factor explaining the discrepancy between the Equitable's experienced mortality and the Northampton table, Finlaison's position was 'that there is very little if any advantage at all in favour of selection'.[10] Whereas Morgan argued for the use of statistical rates of sickness in friendly societies, Finlaison argued that 'life and death are subject to a known law of nature, but that sickness is not, so that the occurrence of the one event may be foreseen and ascertained, but not so the other'.[11] On these other points, history proved Morgan the more accurate. The following section will discuss how selection effects were measured by actuaries using techniques developed over the nineteenth century that showed them to be highly material. The estimation of rates of sickness was also developed into an increasingly reliable science by actuaries over the course of the nineteenth century, culminating in the Manchester Unity sickness tables that were published in 1903 and remained in use through much of the twentieth century.[12]

Beyond the impact it had on government policy, Finlaison's report was an important milestone for actuarial thought. It was noted in Chap. 1 that a government annuity mortality experience analysis was first performed by the Dutchman Johannes Hudde in the 1660s. Finlaison noted that Kersse (again in Holland) and Deparcieux (in France) had performed analysis of government annuity and tontine experience in the 1740s. But the scale and rigour of Finlaison's analysis was of a different order to these earlier works. His study also provided the era's most comprehensive analysis of how British mortality behaviour varied by sex. The chronological span of Finlaison's analysis provided insight into how mortality rates had decreased over the eighteenth century. It was commonly understood that life expectancy had improved

[10] Finlaison (1829), p. 19.

[11] Parliamentary Papers (1825).

[12] Watson (1903).

substantially over this period, but Finlaison was able to provide greater detail of how mortality rates had changed differently across ages. He also noted that this had implications for how historical mortality observations should be used in constructing forward-looking estimates:

> If the diminution (in mortality rates) had been doubtful, or very slight in degree, it would follow as a matter of course, that a combination of all the facts observed upon would have formed the proper basis for life annuities and insurances; but this being quite otherwise, there was no choice but a combination of two or three observations, the latest in point of time.

Finlaison also applied a smoothing or graduation technique to the fitting of the tables derived from mortality experience. Graduation was starting to emerge as an important topic in mortality table construction at this time and Finlaison was a pioneer of its real-life application. More generally, Finlaison's detailed work with experience data rather than population data anticipated and perhaps partly motivated the next important development in life office mortality modelling.

Life Office Experience Tables and Selection (1829–1881)

Despite John Finlaison's scepticism, by the 1820s there was a strengthening conviction amongst life office actuaries that the selection effects of life underwriting practices were likely to have a material impact on the mortality rates that their business experienced. But no quantification of the impact of selection had been made. How could selection effects be measured? If selection effects existed, did they have a permanent impact on the mortality behaviour of policyholders relative to the broader population, or a more temporary one? If temporary, for how long did the effect persist? These questions could only be answered through an analysis of the life offices' mortality experience data.

The first analysis of mortality experience at a British life office was conducted by William Morgan in 1776 at the behest of Richard Price. As noted earlier, this investigation led Morgan to conclude that 'the probabilities of life in the Society had been higher than those in Mr Dodson's Table, from which its premiums were computed, in the proportion of three to two'. Fifty years later, as the life assurer with by far the most data on the mortality rates experienced in its business, the Equitable was naturally positioned to take the lead on further exploring selection effects. But William Morgan resisted the development or application of mortality tables based on Equitable's mortality experience for as long as possible. Like most of Morgan's later resistance to change, it was at least partly motivated by reasoned argument as well as mere intransigence. Morgan held a firm belief that the Equitable's lighter mortality experience relative to the Northampton table was due to a transient selection effect which would be imprudent to pass on to policyholders via the premium basis. In 1828, Morgan also argued that moving the valuation of Equitable's business onto the much-lighter experience mortality basis would motivate the release of an inequitable amount of surplus to older policyholders relative to younger policyholders.[13]

For once, Morgan's resistance proved futile. Charles Babbage and Griffith Davies, both leading actuaries of the time, each published their own versions of Equitable experience mortality tables based on the limited statements Morgan had made about the Equitable experience relative to the Northampton table. An Equitable member formally proposed that the Equitable's liability valuations (and hence assessments of surplus) should be based on the experienced mortality rates. Whilst the Equitable's legal counsel found in favour of maintaining the status quo, the pressure from Equitable policyholders was sustained and Morgan recognised that it might be more counterproductive to continue to resist. He finally published an Equitable experience mortality table in 1829, though this was clearly done through gritted teeth and with heavy caveats: 'The following tables ... are still deduced from documents, particularly in the earlier parts of life, much too defective to be depended upon, and have been constructed merely for the purpose of effecting some calculations directed to be made by the last General Court of the Society.'[14]

Morgan's table confirmed the substantial extent to which the Equitable's experience was lighter than the Northampton table that it used in its pricing and valuations. However, the information provided by Morgan could not provide much new insight on the *source* of this differential in mortality. Was it simply that the segment of society that purchased Equitable life assurance tended to be richer and healthier than the population at large? Or had mortality significantly improved since the time of Price's data due to improvements in economic and social conditions? Or were the Equitable's rigorous underwriting processes and the rejection of lives with above average mortality rates the main reason for the lighter mortality experience? To answer these questions, a more comprehensive analysis of the behaviour of the mortality experience would be necessary (particularly if and how mortality experience varied as a function of the duration of policy). But the reluctant Morgan would not sanction any further exploratory work on the Equitable experience data.

---

[13] See Ogborn (1962), p. 200.

[14] Morgan (1829), Appendix.

William Morgan retired in 1830 and passed away in 1833, aged 83. His son, Arthur, who was a member of the Equitable's actuarial staff when Morgan senior retired, was elected actuary. Arthur Morgan's first major task was to construct mortality tables of the Equitable experience on a more thorough basis than his father's somewhat half-hearted effort of 1829. His analysis included all 21,398 lives who had been members of the Equitable during its first 67 year years since inception (1762–1829). The data included a total of 5,144 deaths. This work appears to have been regarded highly by his actuarial peers—he was elected as a Fellow of the Royal Society in 1835 primarily for his work on the experience tables, and his Royal Society sponsors included leading actuaries of the day such as Benjamin Gompertz and Griffith Davies.[15]

Arthur Morgan's Equitable experience analysis was the first statistical analysis of mortality rates that considered the behaviour of mortality rates as a function of the duration of membership as well as age of member. This demonstrated that the Equitable's experienced mortality rates had varied significantly with duration of policy. For example, a 40-year-old who had been admitted to the society within the previous five years experienced a mortality rate of around 5 %, whilst a member of the same age who had been a member for between 15 and 20 years experienced a mortality rate of around 7.5 %.[16] This strongly supported the hypothesis that the selection effect of underwriting was a significant driver of the gap between the mortality experience and the pricing basis, and that the selection effect, whilst substantial, was temporary.

Arthur Morgan made his experience analysis publicly available, providing the opportunity for other actuaries to contribute to the investigation of the data. T.R. Edmonds, a widely published economist and demographer who had been appointed actuary of Legal & General in 1832, published an important paper in *The Lancet* in 1837 that discussed the implications of Morgan's results.[17] Edmonds argued that the bulk of Equitable's mortality profit had likely arisen from this selection effect—the mortality rates of long duration policies were not significantly different to the premium basis. This had been William Morgan's consistent position. Edmonds also warned about expecting this effect to be a permanent feature of the future of life assurance:

[15] Ogborn (1962), p. 215.

[16] Edmonds (1837), p. 159.

[17] Edmonds (1837).

> At the present day, such is the competition among various life offices, that a large proportion of lives proposed for insurance are now accepted, who would formerly have been rejected.[18]

In 1841 Thomas Galloway, the actuary of the Amicable, followed Morgan's example and published the mortality experience analysis of his office. This data covered the period 1808–1841.[19] The Amicable was the second-largest British life office of the time and the volume of the data was approximately one third of that available to the Equitable (1,792 deaths). Galloway performed some innovative analysis that suggested that the selection effect was not unique to the Equitable. He divided his dataset into two parts: one for policyholders who were already members at the start of 1808; and another for those who joined in 1808 or thereafter. His reasoning for this was that the office's pricing and underwriting standards were fundamentally altered in 1808. Galloway wanted to measure the selection impact of this change in underwriting practice. His two mortality experience tables are shown below in Fig. 3.3.

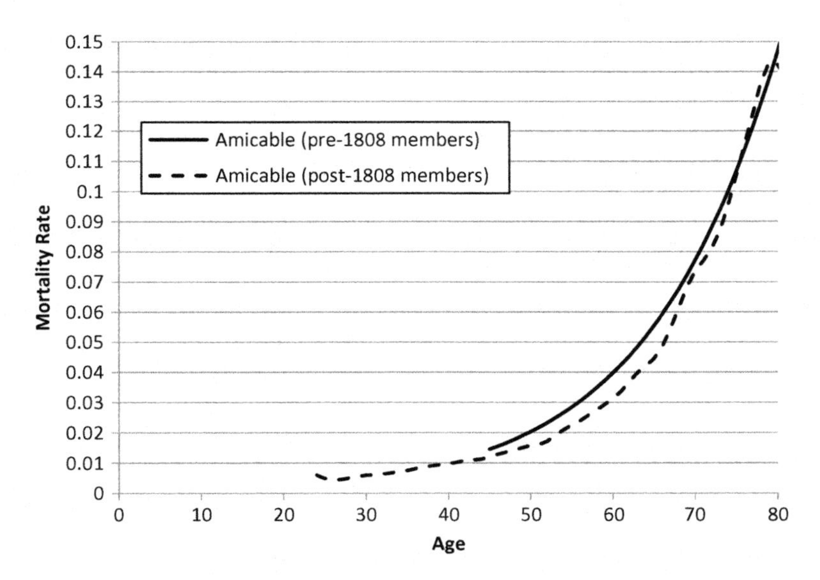

Fig. 3.3 Amicable mortality experience (1808–1841)

[18] Edmonds (1837), p. 162.
[19] Galloway (1841).

The post-1808 members experienced materially lower mortality rates than the pre-1808 members, particularly between the ages of 50 and 70. Galloway's experience analysis provided a measure of the impact of changes in underwriting practices on mortality experience. This measure of the effect of selection would, however, have been somewhat over-stated if life expectancies had improved over the 33-year period, as the average calendar year of experience of, say, a 60 year-old would have been earlier for the pre-1808 members than for the post-1808 members.[20] Galloway did not note this possible effect, which itself highlights that no systematic allowance for mortality improvements tended to be considered during this era.

The next important step in the development of experience mortality tables was initiated at a meeting held at the London Coffee House on Ludgate Hill on 19 March 1838. There, the actuaries of many of the leading British life offices met and agreed to jointly share their experience data for the purposes of producing mortality tables based on their pooled experience. A committee of senior actuaries including Benjamin Gompertz, Joshua Milne and others was appointed to oversee the effort. Seventeen life offices submitted data that totalled 83,905 policies and 3,928 deaths. The Equitable and Amicable were amongst the contributors to the combined experience, but they did not contribute the full data that they used in their own individual experience studies, partly because the Seventeen Offices dataset only used policies written on single lives (to avoid potential double-counting of lives). The combined analysis was eventually published in 1843 and the mortality tables became known as the Seventeen Offices tables.[21]

The average duration of the policies in the combined data set was only eight years. The committee decided that this data could not support an analysis of mortality rate as a function of duration of policy as well as age of policyholder. This had two important implications for the results: given the published results of the Equitable and Amicable experience, it was reasonable to assume that the relative immaturity of the business being analysed meant that long-term mortality rates experienced by life offices were likely to be higher than those estimated from this dataset; and, secondly, it restricted the ability to quantify selection effects. The committee had also set out to analyse cause of death but found that 'the returns ... were deficient in so many of the Lists, that is was not considered desirable to make any classifications of them'.[22]

[20] I am grateful to D.C.E. Wilson for highlighting this point to me.

[21] Ansell et al. (1843).

[22] Ansell et al. (1843), p. xi.

These limitations meant that no new fundamental insights were obtained from the combined data. The report's main conclusion was that the combined experience was very similar to that published by the Equitable a decade earlier. Nonetheless, the Seventeen Offices table is notable for representing the beginning of a long and important tradition of actuaries pooling experience data for the purposes of combined mortality studies. The table itself was also used widely in life office premium bases both in Britain and overseas.

The established life offices naturally accumulated more experience data as the years moved on, and by 1869 the available data could facilitate a much richer combined experience analysis. A committee of some of the eminent actuaries of their generation—Brown, Sprague, Bailey, Woolhouse and others—superintended the development of a new set of combined tables.[23] This time 20 life offices contributed their experience, providing data on 160,426 lives and 26,721 deaths. The 1869 tables are most notable for their treatment of selection. As we saw above, Galloway's Amicable analysis of 1843 had statistically demonstrated that rigorous underwriting, and the selection of healthy lives for assurance, could generate a significantly lighter mortality experience. But questions remained. In particular, would this selection effect be experienced for the entire duration of the policy, or would its effect be transient, only impacting on mortality for the first few years of the policy's duration? Arthur Morgan's Equitable experience analysis suggested the latter, but no other evidence had yet been produced to corroborate this. Samuel Brown, the President of the Institute at the time, even worried that the mortality rates of select lives may ultimately be poorer than other lives[24]:

> It is essential to the prosperity and good management of an office to know not merely whether the selection of lives affords unusual profits within a short period after admission of the assured, or how long that period continues, but also whether it may not require some reserves out of those profits to meet a permanent excess of mortality hereafter.

When the Seventeen Offices tables were constructed, actuaries were asking these questions, but the volume and spread of data could not support a rigorous answer. By 1869 the British life offices had been in existence long enough to generate the experience data required to build a two-dimensional mortality table that recorded mortality as a function of the age of the policyholder and the number of years the policy had been in-force. This data

[23] Institute of Actuaries (1869).

[24] Institute of Actuaries (1869), p. 19.

analysis yielded significant new insights and quantifications on the duration of selection effects: 'For all practical purposes, the benefit of selection may perhaps be said to be lost after the fifth year of assurance.'[25] It also highlighted how substantial the selection effect was in the early years of the policy. For example, the mortality rates of 40–44 year-old healthy males in the first year of assurance were found to be only 37 % of the mortality rate of the same category in their fifth year or later.[26]

By the mid-nineteenth century, significant life assurance businesses were developing outside Europe, particularly in the British colonies and the United States of America. In the USA, a small number of life assurers had existed since the second half of the eighteenth century, but it was legislation in New York State in 1840 that provided a catalyst for rapid growth in the sector. It was estimated that 50,000 life assurance policies were in-force in 1859, and that this number increased to 800,000 by 1872.[27] During this period, in the absence of any local data, American actuaries used British mortality tables, most typically Price's Northampton table, Milnes' Carlisle table or the Seventeen Offices table.

In 1859, the first US life office experience table was produced by Sheppard Homans, based on the experience of the Mutual Life Insurance Company of New York, of which he was the actuary. This table was updated in 1868 and was commonly referred to as the American Experience Mortality Table. This table was used widely by US life offices over the following fifty years.[28] In 1881, American actuaries set a new world record for the amount of experience data used in a mortality table when the Thirty American Offices' Table was published.[29] This table was based on a 30-year experience up to the end of 1874, and, as its name suggests, 30 life offices contributed their mortality experience. It included experience data on over one million policies and 46,543 deaths. It claimed to represent a full three quarters of all experience of life offices in the United States over the 30-year period considered.

Like the Institute of Actuaries table that came before it, the Thirty American Offices' Table provided strong evidence of the impact of underwriting and selection on mortality experience. For example, for males aged 40–45, the mortality rate experienced in the first year of the policy was found to be around two thirds of the rate experienced for policies' with durations of five years or

[25] Institute of Actuaries (1869), p. 25.
[26] Institute of Actuaries (1869), p. 22.
[27] Meech (1881), Preface, p. 4.
[28] Maclean (1948), p. 284.
[29] Meech (1881).

more.[30] In general, the results were consistent with the Institute's conclusions that the selection effect was evident for the first five years of policies' duration and was not significant thereafter.

By 1881, the practice of pooling experience mortality data was well-established in developed life assurance markets around the world. Actuarial thinking moved on from the quantification of selection effects to considering the implications that this should have for pricing and reserving.

Graduation of Mortality Tables (1825–1867)

The use of pooled mortality data increased the sample sizes available for use in the development of mortality tables. But statistical noise was inevitably still found in mortality rate estimates, and this was especially true when rates were estimated as a function of both age of policyholder and duration of policy. Visual inspection of the Carlisle table highlighted that it could also be true for the contemporary tables based on population data. Some back-of-the-envelope calculations could easily highlight why. A total of 1,840 deaths were observed over the nine-year span used in the Carlisle data set. 173 of those deaths were at ages between 60 and 70.[31] So there were less than 20 deaths observed at age 60: a single observation of one more death of a 60-year-old within the nine-year period would result in a 5 % proportional increase in the mortality rate estimate. In the Carlisle table, the mortality rate of 60–70-year-olds increases by an average proportion of 4 % per year of age. So the 'noise-to-signal' ratio was high. Data samples of this size would unavoidably result in material irregularities in the observed progression of mortality rates as a function of age.

Actuarial thinking on the smoothing, or graduation, of age-dependent mortality rate estimates started to develop in the 1820s. Broadly speaking, two types of approach emerged: a non-parametric smoothing approach that set the mortality rate of a given age as a weighted average of the 'raw' mortality rates estimated across a range of ages centred around the given age; and a parametric approach that specified a functional form for how mortality rates varied as a function of age (and potentially policy duration), and then fitted the parameters of that function to the observed mortality rates. These two approaches were focused on the same objective, but were philosophically different. The latter approach aimed to provide an explanatory 'law of mortality'

---

[30] Meech (1881), p. 31.

[31] Milne (1815), p. 405.

that was consistent with the data, whilst the former merely tried to remove the 'noise' in the statistical samples of mortality rates. In a wider context than actuarial thought, this was an era of exploration in the application of statistical approaches to social science, and the notion of scientific 'laws' that could explain social phenomena was in keeping with the zeitgeist of the first half of the nineteenth century. Perhaps partly for this reason, the explanatory approach caught the actuarial imagination and became the dominant method. As we shall see below, it also had more grounded actuarial advantages.

To gain an understanding of these approaches and how they were implemented, we need to recall the overall state of development of statistics during this era. As we know from Chap. 2, the method of least squares had been developed and placed in a statistically rigorous context by Legendre, Laplace and Gauss a decade or so earlier. But this method had not yet been used to develop a best fit for a function that describes how a dependent variable (in this case, the mortality rate) behaves alongside an independent variable (in this case, age). Perhaps surprisingly in retrospect, it took another half-century before a statistical theory of regression was developed and applied when Francis Galton considered the hereditary dependencies of plants.[32] And so, lacking a statistical framework to consider how information from neighbouring estimates could statistically improve the estimate of a given point, the methods proposed for fitting the 'smoothed' rates were inevitably somewhat ad hoc. There was also some debate about *what* should be smoothed. Efforts were focused on the smoothing of the mortality rates, but some thinkers of the time argued that it would be preferable to apply the smoothing process directly to the ultimate variable of interest: the value of a life contingency. Writing in 1838, the influential Augustus De Morgan argued:

> The events of single years are subject to considerable error, and generally present such varieties of fluctuation, that it has become usual to take some arbitrary and purely hypothetical mode of introducing regularity. This practice cannot be too strongly condemned, since the tables thereby lose some of their physical facts, without any advantage ultimately gained. For if by using the raw result of experiments, tables of annuities were rendered unequal and irregular, it would be as easy, and much more safe, to apply the arbitrary method of correction to the money results themselves, than to introduce it at a previous stage of the process.[33]

[32] See Stigler (1986) Chapter 8.

[33] De Morgan (1838), p. 162.

This perspective did not prevail: virtually all actuarial thinking of the time focused on smoothing the observed mortality rates. John Finlaison, in his parliamentary report of 1829 on government annuity pricing, employed a couple of non-parametric smoothing formulae in his experience mortality tables, such as:

$$P_x(smoothed) = \frac{5P_x + 4P_{x-1} + 4P_{x+1} + 3P_{x-2} + 3P_{x+2} + 2P_{x-3} + 2P_{x+3} + P_{x-4} + P_{x+4}}{25}$$

where P_x is the observed probability of a life aged *x* surviving one year.

In Finlaison's approach, each smoothed mortality rate is set as a weighted-average of the 'raw' mortality rate observations found for ages up to four years older and younger than the age of the rate being estimated. The weights used vary inversely with the distance of the age of the observation from the age being estimated. The above formula and its logic are quite similar to a modern-day local or non-parametric regression method.

In 1839, W.S.B. Woolhouse, in the early years of a long and distinguished career as an actuary and mathematician, suggested another non-parametric smoothing approach in his paper on the observed mortality rates of the Indian army.[34] Woolhouse's approach made adjustments to the observed mortality rates such that the time progression of the lives remaining in a closed population would have a regular pattern. Statistically, he aimed to ensure that the pattern of the fourth order of differences in the progression of lives remained as smooth as possible. He devised an iterative arithmetic process that could be implemented to obtain this objective.[35] This approach entailed more arithmetic manipulation than Finlaison's smoothing approach, but had a more explicit statistical objective.

Finlaison and Woolhouse appeared satisfied with the performance of these methods. The Woolhouse method had a greater impact on actuarial practice and it was applied in the development of the Seventeen Offices tables of 1843. But others had less success with these approaches. In his Amicable experience analysis of 1841, Galloway attempted to use a smoothing formula of the kind proposed by Finlaison but found that 'considerable anomalies remained'.[36] He therefore reverted to the use of a parametric function, and this increasingly became the standard actuarial practice.

---

[34] Woolhouse (1839).

[35] Woolhouse (1839), p. 7.

[36] Galloway (1841), p. ix.

The parametric function approach had another advantage over the non-parametric smoothing methods. Since the time of de Moivre, it was recognised that specifying a particular mathematical form for the behaviour of mortality could provide an annuity pricing formula that involved significantly less arithmetic operation than the explicit calculation of expected cash-flows directly from a given set of mortality rates. De Moivre's assumption of arithmetic decrements in deaths could be viewed as the first parametric form of mortality graduation. For de Moivre, the object of this assumption was entirely motivated by the improvement in the efficiency of the annuity pricing calculation. One hundred years later, enough progress had been made in arithmetic computation to make full calculation of single-life annuity prices accessible. But the calculation of the prices of annuities that were written on two or three lives was still very challenging. So in the early nineteenth century actuaries such as Gompertz hoped that parametric functions for mortality rates could kill two birds with one stone: providing an appropriate way of smoothing out the noise in sampled mortality rates, and providing efficient formulae for the pricing of complex annuities.

Gompertz started a revolution in graduation thinking in 1825 when he developed a parametric function that was intended to be consistent with the fundamental characteristics of how mortality should behave as a function of age (a 'law of human mortality')[37]:

> It is possible that death may be the consequence of two generally co-existing causes; the one, chance, without previous disposition to death or deterioration; the other, a deterioration, or an increased inability to withstand destruction.[38]

This suggested there were two forms of exposure to mortality: one that was constant across all ages; and the other that increased with age. To model the behaviour of the age-varying component of the mortality rate, he assumed 'the average exhaustions of a man's power to avoid death were such that at the end of equal infinitely small intervals of time, he lost equal portions of his remaining power to oppose destruction which he had at the commencement of those intervals'.[39]

The above statement implied that the age-dependent component of the mortality rate increased exponentially with age. Gompertz then considered how well this assumed mortality behaviour could fit to standard mortality

[37] Gompertz (1825).
[38] Gompertz (1825), p. 517.
[39] Gompertz (1825), p. 518.

tables such as Price's Northampton table and Milne's Carlisle table. Curiously, when Gompertz came to the application of his formula, he chose to omit the *age-independent* mortality component that he had earlier described. Hence a 'Gompertz function' only includes an age-dependent exposure, even though he expressly identifies a constant age-independent source of mortality ('chance without ... deterioration').

Gompertz originally expressed his function in terms of the number living at age x, l_x, rather than the force of mortality. T.R. Edmonds, writing a few years later in 1832,[40] defined the force of mortality, μ_x, as the continuously compounded rate of mortality and expressed the formula in those terms. The modern form of the Gompertz function would typically be written:

$$\mu_x = \alpha e^{\beta x}$$

It was implemented by Edmonds without recourse to the exponential function as:

$$\mu_x = \alpha \beta^x$$

Today, such a function would typically be fitted by finding the parameters that minimise the squared errors between the observed and fitted mortality rates. As discussed above, this regression-style approach to function fitting had not yet been applied to statistical problems, and Gompertz used a more heuristic approach. He considered the change in mortality rates that applied over the ten-year gap between ages 15 and 25, and over the ten-year gap between ages 45 and 55. These observations could be used to solve two simultaneous equations that uniquely determined the two parameters of the function.

Whilst the parameters would fit exactly to those equations, they would not produce an exact fit to all the mortality rates of the table. However, he was pleased with the quality of fit he obtained when he compared his function to the observed rates of the Northampton table between ages fifteen and 60, noting: 'This equation between number living and the age is deserving of attention, because it appears corroborated during a long portion of life by experience'.[41] Figure 3.4 shows a Gompertz function fitted to the Northampton data for ages 15–60.

[40] Edmonds (1832).

[41] Gompertz (1825), p. 519.

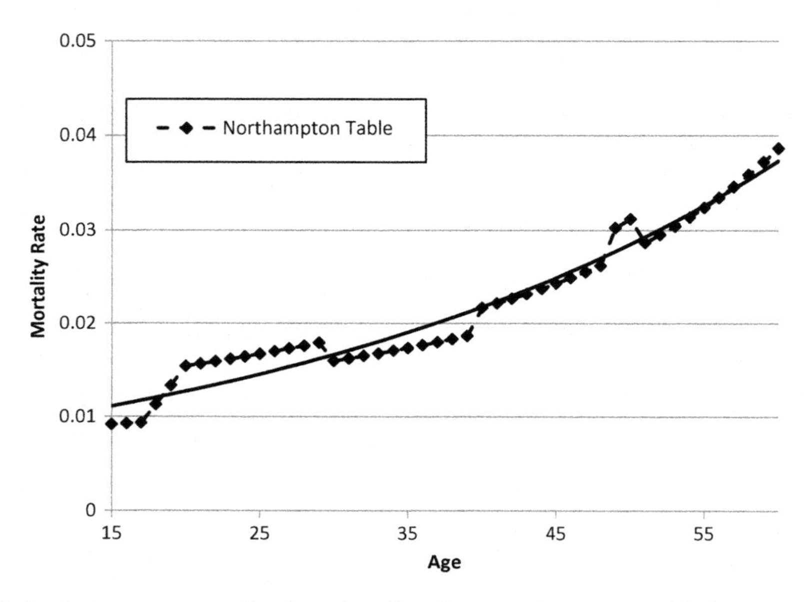

Fig. 3.4 Gompertz mortality function fitted to Northampton table (ages 15–60)

The above chart illustrates how the Gompertz function can provide a very useful graduation of the Northampton table over this range of ages. However, the limitations of the function become apparent when a broader range of ages is considered. Figure 3.5 fits to the 15–80 age range and again shows the fitted function values and observed mortality rates.

Clearly, the two-parameter function is unable to provide an adequate fit over this wider age range. The same conclusion was reached when Gompertz applied the function to other standard mortality tables. Nonetheless, the Gompertz law of human mortality was widely celebrated by actuaries and medical thinkers of the time. De Morgan wrote of Gompertz: 'As this ingenious paper contains a deduction from a principle of high probability, and terminates in a conclusion which accords in a great degree with observed facts, it must always be considered a very remarkable page in the history of the enquiry before us.'[42]

It was clear, however, that as a practical actuarial tool, improvements were required. William Makeham, another senior actuary of his generation (and also a noted mathematician), proposed a natural generalisation of the Gompertz law in papers published in 1859[43] and 1867:[44] he suggested including the age-independent parameter that Gompertz had described but chosen to ignore in his formula. The force of mortality could then be written as:

[42] De Morgan (1839).

[43] Makeham (1859).

[44] Makeham (1867).

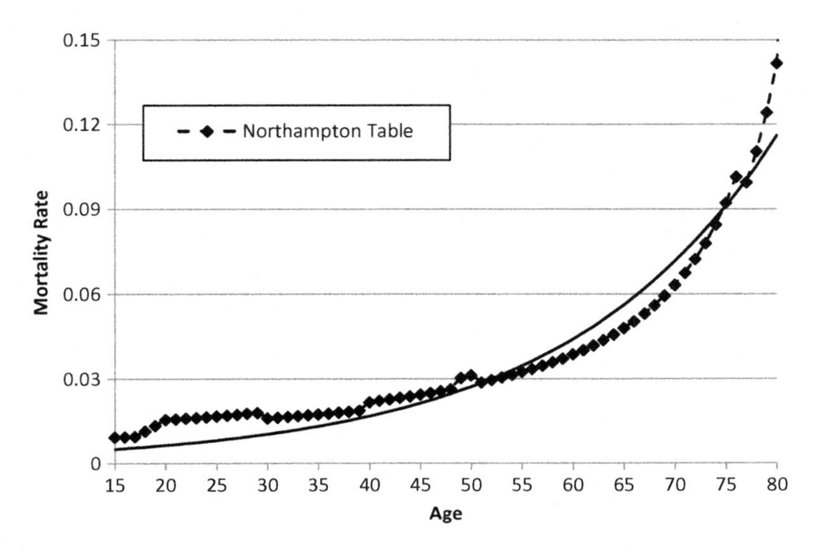

Fig. 3.5 Gompertz mortality function fitted to Northampton table (ages 15–80)

$$\mu_x = \alpha\beta^x + \gamma$$

This was, of course, exactly in keeping with Gompertz's original description of the effects of mortality on age. We can only speculate as to why Gompertz did not proceed directly to this functional form in his paper of 40 years earlier. It clearly would have resulted in a more complicated fitting process, but his heuristic fitting approach could be extended to the three-parameter case in a straightforward way. Makeham found that the additional degree of freedom permitted significantly more accurate fits to mortality tables, as illustrated below in Fig. 3.6.

Whilst the additional degree of freedom naturally resulted in better fits to mortality data, Makeham was at pains to emphasise that his extension was not an arbitrary additional degree of freedom but rather something that captured the characteristics of a fundamental law of mortality. His extension, he wrote, 'in no way interferes with the philosophical principle upon which Mr Gompertz has shown his theory to be based: a feature which distinguishes his formula from all others which have hitherto been proposed, and which doubtless accounts for the favourable reception it has met with from the highest scientific authorities'.[45]

The three-parameter Makeham function could provide a good fit to mortality data within the range of ages of primary importance to life assurers,

[45] Makeham (1867), p. 333.

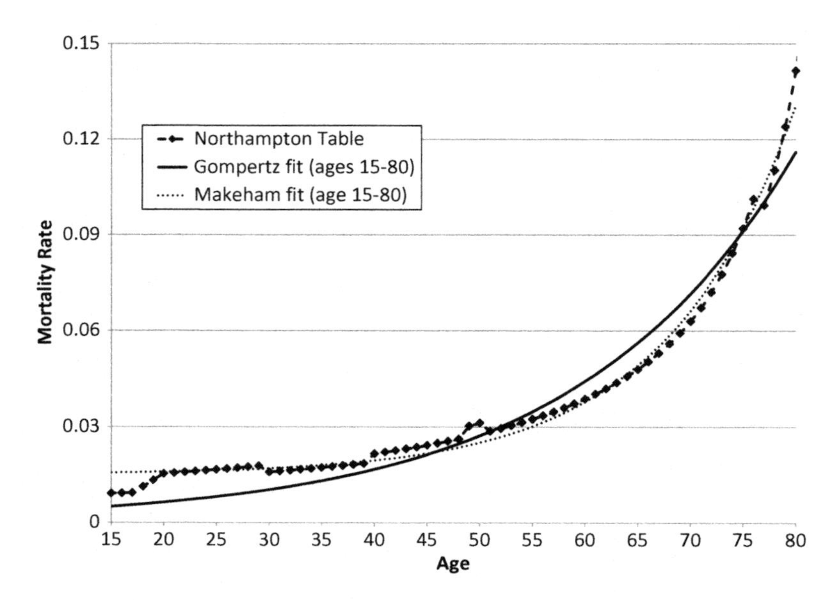

Fig. 3.6 Makeham and Gompertz mortality functions fitted to Northampton table (ages 15–80)

but Makeham and Gompertz both recognised that such functions could not provide a satisfactory fit across the entire spectrum of ages in mortality tables, particularly at very young and very old ages. They each proposed various extensions. Gompertz proposed using piece-wise functions with different parameterisations of Gompertz functions applied to specified age bands.[46] Makeham proposed extending the function further by adding polynomial terms.[47] Elsewhere in Europe, other mathematically inclined actuaries developed similar functions for graduation purposes. For example, the leading nineteenth-century Danish actuary T.N. Thiele proposed a seven-parameter exponential function in 1871.[48]

The significant developments in statistics that occurred in the early twentieth century provided a further fillip to actuarial research in mortality graduation. Elderton[49] and Ogborn[50] each considered the application of Pearson frequency curves to mortality table graduation, but neither was able to demonstrate any significant advance in performance relative to the Gompertz-Makeham framework. It remains today an important and well-used part of the toolkit of mortality actuaries and demographers.

[46] Gompertz (1871).

[47] Makeham (1889).

[48] Thiele (1871).

[49] Elderton (1934).

[50] Ogborn (1953).

Actuarial Thought on Investment Strategy, 1858–1952

British actuarial thought on investment strategy for life offices began to emerge, somewhat tentatively, in the third quarter of the nineteenth century. Barely a handful of investment papers were published in the *Journal of the Institute of Actuaries* and *Transactions of the Faculty of Actuaries* in the second half of the nineteenth century. These papers, however, tended to be written by senior and influential actuaries. They were leading the profession into a new field that would become ever more significant in twentieth century actuarial thought and practice.

There were several drivers for the increasing interest in this topic in the mid-nineteenth century. The experience of the previous decades had imbued actuaries with a growing sense of confidence in their modelling of mortality rates. A long track record of success in predicting mortality rate behaviour and profiting from it had been established. The plagues and epidemics of the seventeenth and eighteenth centuries were, by then, largely distant memories. Mortality behaviour appeared to be governed by laws that were well understood by the actuarial profession. With this mastery came the realisation that the increasingly competitive life assurance landscape could not yield the level of mortality profits that had been earned on assurance business over the previous 100 years. Mortality protection was becoming commoditised.

This was accompanied by an increasing sense that canny investment decision-making could be an important source of profit and differentiation for life assurers. In 1858, actuaries would read in the *Journal*:

> The two principal sources of profit to an Assurance company are the selection of lives and the accumulation of the excess of the premiums at a higher rate of interest than is assumed as the basis of the Company's operations ... With regard to the [selection of lives], it is impossible not to feel that the margin of profit has very much diminished of late years ... This makes the question of the rate of interest obtained from investments so much the more important: it is, in fact, the principal source of profit.[51]

And in 1862:

> Of the two main elements on which all life assurance transactions depend – the rate of mortality and the rate of interest – the latter, I think, affords more scope for the exercise of judgement and skill than the former ... there can be no doubt

[51] Brown (1858), pp. 241–42.

> that the amount of interest realised on the assets can be materially influenced by the degree of judgement and knowledge brought to bear upon the subject.[52]

These statements could be viewed as the very first steps on a long slippery slope that led to twentieth-century British actuaries taking actions which suggested that they regarded geared exposure to financial market risks as an equivalent and equally sustainable form of risk taking to the underwriting of largely diversifiable mortality risk. That was not, however, the vision of the nineteenth-century British actuarial profession. Rather, these early considerations of the asset side of the balance sheet can be viewed as a natural broadening of the actuarial horizon as the skills, experience, confidence and influence of the profession grew. It was also perhaps inevitable in the context of the long-term trend in British life assurance whereby the products' investment element became ever more significant relative to the protection element.

In the early nineteenth century, British life offices invested in two main asset classes: long-term government bonds (usually consols, which were perpetuities) and mortgages on freehold property (including agricultural lands). Long-term government bonds served two purposes for life offices. First, whilst the emergence of concepts such as duration and interest rate immunisation were many years away, there was an intuitive understanding that the long-term nature of the liabilities demanded that some funds should be invested in assets that paid similarly long-term cashflows. Second, in this era, the possibility of an epidemic or plague, whilst viewed as increasingly unlikely, was still a credible contingency. Funds needed to be readily convertible to cash in such circumstances, and so the liquidity offered by government bonds was valued.[53] The main attraction of mortgages was that they offered higher yields than government bonds—high-quality mortgages typically yielded between 50 and 100 basis points more than long-term government bonds.[54] The mortgages, however, offered neither liquidity nor long-term fixed cashflows, and these characteristics constrained the amounts that were invested in the asset class.

By the middle of the nineteenth century, the concerns about the possibility of epidemics had receded and with it so did the requirement for asset liquidity. The amounts invested in mortgages correspondingly rose. This investment strategy stands up well to modern scrutiny: it was a strategy that paid respect to the liabilities' interest rate duration and liquidity characteristics and which aimed to obtain additional yield through reducing excess asset liquidity rather than by materially increasing market or credit risk.

---

[52] Bailey (1862), p. 143.

[53] Deuchar (1890), p. 451.

[54] Brown (1858), pp. 245, 253–54.

The first actuarial paper on life assurance investments was published by Samuel Brown in 1858.[55] Brown was a senior figure in the actuarial profession of the time. He was actuary of the Guardian Assurance Company, a well-established British life office, and he would go on to be President of the Institute of Actuaries between 1867 and 1870. Brown did not propose a doctrine for investment strategy. He restricted himself to providing a commentary on the main asset classes that life assurance funds invested in, and undertook a small survey of the asset allocations of four major British life assurance offices of the time (Eagle Life, London Life, Rock Life and Metropolitan Life). The survey showed that, across the four offices, the amounts invested in mortgages varied between 46 % and 79 % of their total assets.

By the 1850s, low interest rates had rendered long-term government bonds increasingly unattractive—the consol yield had fallen from almost 5 % in 1810 to barely above 3 % by 1855.[56] The 3 % yield level was important to actuaries because virtually all with-profit business had been written using a 3 % interest rate in the premium basis. If funds could not generate a 3 % return, the life offices' profitability, and ultimately their solvency, was threatened. This fall in yield triggered significant reductions in allocations to consols—by 1858 the allocations had been reduced from perhaps as much as 50 % earlier in the century to between 5 % and 15 % across the four offices. The funds were mainly switched into high-quality bonds (such as colonial sovereign bonds) and debentures (mainly on railways and docks). These asset classes had increased in size and liquidity over the first half of the nineteenth century with the continuing progress of the industrial revolution and the British Empire. Debentures typically offered between 50 and 150 basis points more than consols.[57] They also provided more liquidity and more duration than mortgages, but were accompanied by a more material default risk. This is arguably the first example of a phenomenon that reoccurs in actuarial history: long periods of falling long-term interest rates (for example, in Britain in 1930–1945) tended to compel institutions to bet their way out of their potential unprofitability by increasing investment risk rather than locking in a loss.

This brief sketch of the nineteenth-century asset allocation practices of British life offices highlights a number of asset-liability management considerations that were the natural domain of the actuary: how much asset liquidity was required by life assurance liabilities? How much investment and credit risk should be sought by the asset strategy and how did this choice interact

[55] Brown (1858).

[56] Homer and Sylla (1996), p. 182.

[57] Brown (1858), pp. 253–54.

with pricing, reserving and bonus policy? Did the very long-term nature of the liabilities create particular forms of risk that would help determine a desirable asset profile? The history of actuarial consideration of questions such as these is explored below.

Actuaries and Investment Liquidity (1862–1933)

A second paper on life office investments followed Samuel Brown's survey a few years later in 1862. Arthur Hutcheson Bailey's 'On the Principles on which the Funds of Life Assurance Societies should be Invested' was the first actuarial paper that analysed investment strategy in the context of life assurance liabilities.[58] At five pages in length, it also has the distinction of being one of the shortest ever actuarial papers. It became a classic paper and was core reading in British actuarial examinations for at least 50 years following its publication. Bailey was the actuary of London Assurance, a significant life office of the period (and one of the very oldest, having been established by Royal Charter back in 1720). Like Samuel Brown, Bailey too went on to become President of the Institute of Actuaries, in Bailey's case from 1878 to 1882.

Bailey considered investment strategy in the context of the characteristics of life assurers' liabilities. The nature of these liabilities would inevitably place some constraints on the features of an appropriate strategy. Bailey, however, also recognised that these liabilities could create some unique comparative advantages for life assurers relative to other financial institutions:

> They (life assurers) engage to pay fixed sums of money at periods generally long distance from the time when the contracts are entered into ... the probable amount of demands on their resources can be calculated from time to time within not very wide limits. Life assurance Societies, unlike banks and commercial enterprises generally, are not exposed to sudden or unusual demands on their resources in times of panic and financial difficulty.[59]

Put simply, life assurance liabilities are long-term, predictable and illiquid, especially when considered relative to those of banks. Bailey took it as a given that illiquid assets offered additional compensation to investors for sacrificing liquidity. He hence argued that life assurers should take full advantage of their unusual capacity to bear asset illiquidity:

[58] Bailey (1862).
[59] Bailey (1862), p. 143.

> The much larger proportion [of life office assets] may safely be invested in securities that are *not readily convertible*; and it is desirable...that it should be so invested, because such securities, being unsuited for private individuals and trustees, *command a higher rate of interest in consequence* [emphasis added].[60]

Mortgages, which Bailey defined as 'every species of loan secure on tangible property', were the main asset class he identified for investment in 'securities that are not readily convertible'. He recognised that the size of these asset classes may be exhausted by the growing funds of life assurers and hence that alternative illiquid asset classes would need to be found. He identified loans for land development under recent Land Drainage Acts as a natural candidate—essentially, nineteenth century infrastructure debt.

Bailey viewed the illiquidity premium as *the* way in which life assurers should look to enhance their investment yield. He did not believe that life assurers should attempt to increase investment returns by taking equity risk or substantial credit risk. His starting investment principle was 'That the first consideration should invariably be the security of the capital',[61] and the notion of life assurers investing in equity investment was anathema to Bailey: 'It is, I think, generally admitted that the ordinary stock and shares of trading and other companies are not eligible for Assurance Societies' investments, as being too speculative.'[62]

Bailey's thinking stands the test of time strikingly well. To the modern actuarial reader there are clear similarities with Bailey's arguments and contemporary ideas on investment policy for long-term non-profit life business such as fixed annuities. The concepts of competitive advantage through unique long-term illiquid funding and the availability of asset illiquidity premia are increasingly central to modern actuarial thinking on investment strategy for illiquid liabilities such as these. Asset classes such as mortgages and loans remain the natural asset classes for life assurers intending to access illiquidity premia.

Bailey's investment doctrine was largely consistent with the prevailing mid-nineteenth-century asset allocation practices described above. By 1890, life office allocations to mortgages were around 52 % and approximately 22 % was invested in various forms of illiquid loans. Another 6 % was invested directly in property. So by 1890, fully 80 % of funds were invested in illiquid asset classes. The remaining assets were invested in a range of asset classes such as British, foreign and colonial government securities (8 %), debentures (5 %) and cash and deposits (2 %).[63]

[60] Bailey (1862), p. 144.
[61] Bailey (1862), p. 143.
[62] Bailey (1862), p. 145.
[63] Mackenzie (1891), p. 195.

The final quarter of the nineteenth century was the high watermark for illiquid asset investment by British life assurers. In subsequent years, British life offices started to tilt their asset allocation towards more liquid assets. The final decade of the nineteenth century marked a decisive shift away from mortgages. Allocations to more liquid stock exchange securities such as long-term debentures and government bonds correspondingly increased. Between 1890 and 1900, the percentage of life assurance assets invested in stock exchange securities increased from around 18 % to 39 %. By 1910, this percentage had reached 49 %; by 1920 it was 60 %.[64] A similar trend was observed amongst the US life offices, and indeed started a decade earlier following the US banking crisis of the early 1870s. The US offices, however, retained a larger mortgage allocation than the British offices after the switch. Between 1874 and 1920, US life offices' investments in mortgages had been reduced from 54 % of assets to 34 %.[65]

This great rotation from illiquid to liquid assets that occurred between 1890 and 1920 was not solely due to a desire for greater asset liquidity, though that was a factor. Other drivers were also important. There was a substantial fall in the value of British agricultural land during this era (perhaps as much as 30 %). This may have awakened life offices to the potential risks inherent in the mortgage asset class. It also reduced the value of the land upon which new mortgages could be offered. Increasing competition amongst mortgage lenders emerged, with legislative changes making the provision of mortgages more accessible to other investors such as trust funds. Mortgage yields fell.[66] In summary, there was a reduction in the illiquidity premium on offer, a reduction in the size of the market, and perhaps a re-appraisal of the risks associated with the asset class. All whilst the size of funds available for investment by life offices was increasing substantially.

The First World War also had a direct and significant impact on asset allocations. In particular, it resulted in a very substantial increase in life offices' allocations to British government securities. Life offices were not legally compelled to buy government debt, but they reached a form of patriotic gentleman's agreement with the government where they agreed to buy gilt issues during the war.[67] As a result, investment in British government bonds increased from 1 % to 35 % of life office assets between 1915 and 1920. However, this was not the main driver of the move from illiquid to liquid assets: much of the increase in gilt allocations was financed by reductions in

[64] Murray (1937), p. 259.

[65] C.M. Gulland, in Discussion, Murray (1937), p. 274.

[66] Mackenzie (1891), p. 198.

[67] Whyte (1947), p. 223.

liquid debenture securities (the asset allocation of which was reduced from 30 % to 12 %). Nonetheless, the scale of the gilt allocation necessarily further reduced illiquid asset holdings. Mortgage allocations fell from 21 % to 15 % between 1915 and 1920.[68]

Changes in the liquidity profile of life assurance liabilities played a significant part in motivating this fundamental shift in asset allocation and liquidity profile. The liquidity requirements of the liabilities were growing by the end of the nineteenth century. Unlike 100 years earlier, these liquidity requirements were not driven by the risk of mortality spikes arising from epidemics (whilst the influenza epidemic of 1918/19 did generate an unexpected concentration of claims, its impact was still relatively small in comparison to premium incomes).[69] Instead, the liabilities' liquidity profile was being changed by surrender practices. Surrender rates had increased. In the fast-growing industrial assurance sector of the 1920s, double-digit surrender rates were the norm.[70] Historical anecdotes from the USA also highlighted how life liabilities may unexpectedly experience sudden liquidity demands. In 1873, a US banking liquidity crisis had created a surge of demand for the surrender of life assurance policies as these policies became the policyholders' most direct route to cash. In British life assurance, there was an increasing practice, driven by competitive forces, of offering guaranteed surrender values (legislation passed in 1929 may also have encouraged guarantees in surrender values on some types of business).[71] A 1933 paper by William Penman noted that 'if very large sums can be withdrawn, virtually without penalty and at very short notice, it becomes necessary to revise materially our views of Life Assurance finance ... it seems to me that the investment position, to the extent that guaranteed surrender values are made excessively liberal, approximates to that of a bank.'[72]

These changes in surrender practices materially impacted on the desired liquidity profile of the assets. With-profit life policies were never again thought of as illiquid. Today, actuaries' illiquid investment strategies are largely focused on liability types such as non-profit fixed annuities that do not offer surrender values.

Actuaries and Interest Rate Risk (1858–1952)

Actuarial awareness of interest rate risk was first heightened by the economic experience of the early nineteenth century. Interest rates increased significantly

[68] Dodds (1979), Table A.3.

[69] Clayton and Osborn (1965), p. 65.

[70] Clayton and Osborn (1965), p. 64.

[71] Laing in Discussion, Penman (1933), p. 421.

[72] Penman (1933), p. 391.

during the Napoleonic Wars, and then experienced a long and substantial decline for the following next forty years (Fig. 3.7). In 1838, Augustus De Morgan wrote:

> I consider the fluctuations of mortality as very little to be feared, compared with those of the rate of interest ... the rate of interest has been halved within the memory of man, and a heavy war might double it again.[73]

Whilst Bailey's thinking on illiquid investment strategies strikes a contemporary chord, nineteenth-century actuarial thought on interest rate risk was several steps removed from the sophisticated and quantitative techniques that abound in twenty-first-century yield curve risk management practices. But from the time of Richard Price and William Morgan, there was a general appreciation that funds backing long-term fixed liabilities should include assets that deliver long-term fixed cashflows. The perpetual format of the standard long-term British government securities of the time (consols) confounded any prototypical matching method. It also created a strong distinction in the mind of the actuary between income and capital: with a risk-free perpetuity, income is guaranteed, but the capital is only ever returned through the sale of the bond at the then-prevailing market price. In his 1858 paper, Samuel Brown's discussion of consols noted that 'the certainty of the income is dearly purchased at

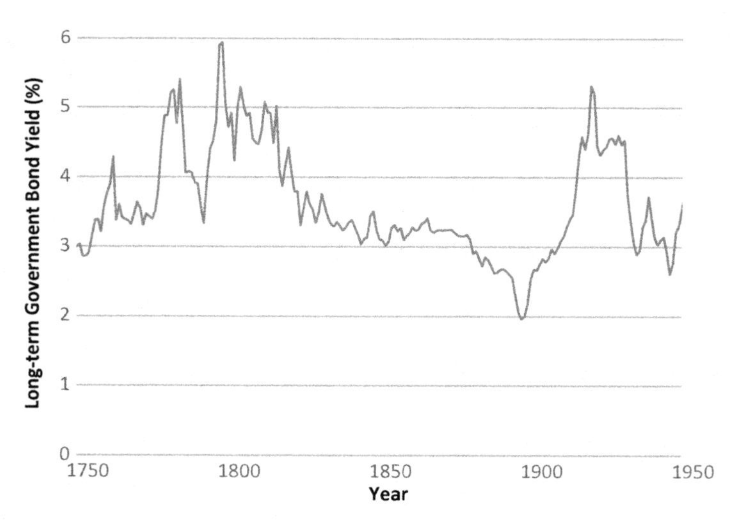

Fig. 3.7 UK long-term government bond yield (1750–1950) (Mitchell (1988))

[73] De Morgan (1838), p. 260.

the price of the uncertainty of the return of their capital' and 'so many events cause fluctuations in the Funds [consols], that he never knows what amount of his capital he may be able to recover when he needs it'.[74]

Bailey's analysis of long-term government bonds as an asset class for life assurance liabilities continued in the vein started by Brown, and was even more emphatic:

> I believe it to be true that the English Funds [consols] are altogether unsuited for life assurance investments. For income they offer probably the best security the world has yet seen; but with us that is a secondary consideration; the capital, the security of which is our first object, is subject to very inconvenient fluctuations in value.[75]

There was no substantive development in technical actuarial thinking on interest rate risk beyond Brown and Bailey's observations until the 1930s, when several papers emerged following the economic turmoil of the First World War and the Great Depression. However, it has been argued that an increasing awareness of the reinvestment risk created by the potential for further falls in interest rates may have been another motivating factor in the switch from mortgages to bonds during the 1890–1920s period.[76]

British long-term interest rates had fallen from 5 % at the end of the Napoleonic Wars to 2.5 % by the start of the twentieth century. These falls were almost fully reversed over the traumatic first quarter of the twentieth century, peaking at 4.5 % in 1925. Rates then started to fall sharply again as government policy attempted to stimulate economic growth during the depression era. This volatility motivated the first wave of actuarial ideas on what would now be called asset-liability management. In 1933, Charles Coutts, then President-Elect of the Institute, opined that 'an attempt should be made to "marry" the liabilities and the assets as far as possible'.[77]

There was a growing recognition amongst the profession that the investment strategy for assets backing long-term life liabilities was more than merely incidental to actuarial work. But no actuarial framework for investment strategy existed beyond Bailey's 70-year-old investment principles. The profession rose to the challenge and a number of important papers were presented on interest rate risk and asset-liability management over the following 20 years. In the 1930s, there was a lack of precision both in the objectives and the desirable

[74] Brown (1858), p. 244.
[75] Bailey (1862), p. 144.
[76] See Murray (1937), p. 249.
[77] Coutts in Discussion, Penman (1933), p. 425.

features of strategies that would successful 'marry' assets and liabilities. But clarity gradually emerged. In the Institute's sessional meeting that discussed an investment paper by A.C. Murray in 1937, J.D. Binns submitted:

> Regarding the office as a closed fund, owing to the incidence of future income it will be found that assets should be made to mature at *rather later dates* than the corresponding liabilities. Then should interest rates *fall* the future income will be accumulated at a lower rate of interest, but this would be compensated by the appreciation shown by the assets at the time that the corresponding liabilities fall due (and *vice versa* if interest rates go the other way).[78]

This is a non-mathematical anticipation of Redington's immunisation theory (which would be presented to the Institute fifteen years later). Redington would provide a revolutionary degree of clarity, but the crucial essence of his immunisation idea was bubbling under within the actuarial profession for decades prior to his paper.

Actuaries were not the only ones thinking about interest rate risk at this time. Economists of the period were also making important strides in their measurement of interest rate behaviour. The most notable contribution was from Frederick Macaulay, the Canadian-born economist. He spent much of his professional career in the USA as a researcher at the National Bureau of Economic Research. He was the son of T.B. Macaulay, the influential actuary of Sun Life of Canada (and son of a Scots immigrant).

Frederick Macaulay undertook ground-breaking empirical analysis of the long-term behaviour of yield curves in the USA since the time of the US Civil War. In his statistical analysis of the relative behaviour of bonds of different maturities, he recognised the inadequacy of a bond's term to maturity as a measure of its 'longness'. He saw that an accurate measure must take account of the size and timing of its coupon payments as well as its maturity. Macaulay introduced the term 'duration' to measure the 'longness' of the bond and defined it as the weighted average of the time of the cashflows where the weights are the present values of the cashflows.[79] He did not deduce this definition from a set of first principles or objectives, but arrived at it in an ad hoc way, noting it was an intuitive and well-behaved measure:

> It would seem almost natural to assume that the "duration" of any loan involving more than one future payment should be some sort of weighted average of the maturities of the individual loans that correspond to each future payment.

[78] J.D. Binns in Discussion, Murray (1937), p. 278.

[79] Macaulay (1938), pp. 45–46.

> Two sets of weights immediately present themselves – the *present* and the *future* values of the various individual loans ... Future value weighting seems clearly inadmissible. It gives absurdly long "durations" ... The argument for present value seems strong.[80]

Macaulay did not recognise that his 'duration' measure was also the first derivative of the bond price with respect to changes in the interest rate. He did, however, find that it was extremely useful in describing the empirical relative behaviour of different bonds: 'The concept of "duration" throws a flood of light on the fluctuations of bond yields in the actual market.'[81]

Macaulay's duration measure had no discernible impact on economic research or financial market practice for over 30 years. Then, in the early 1970s, the confluence of the rapid growth in fixed income securities with important breakthroughs in quantitative financial economics research suddenly placed it firmly at the centre of fixed income risk management. The actuarial profession appears to have been wholly unaware of Macaulay's work in the years following its publication. Macaulay himself had no appreciation of its implication as a risk and asset-liability management tool. His discussion of duration covered a handful of pages in a 600-page tome on nineteenth-century US interest rate behaviour. In these circumstances, it is perhaps unsurprising that the latent usefulness of Macaulay's duration measure in asset-liability management passed by the actuarial profession (and everyone else).

Following the hiatus of the Second World War, the British actuarial profession embarked on its most fertile period of thought leadership activity since the third quarter of the nineteenth century. Two important papers on interest rate management were published in 1952—Redington's famous paper on immunisation theory[82] and a paper by Haynes and Kirton[83]—that tackled similar subject matter. Whilst these papers would have been somewhat controversial at the time of publication or at the least outside the norms of contemporary thinking, with historical hindsight they can now be viewed as the natural culmination of the developing interest rate and asset-liability management thinking of the previous 20 years (in particular, the Penman and Murray papers of the 1930s and their respective Staple Inn Discussions).[84]

Frank Redington spent his entire working career at Prudential Assurance, a leading UK life assurer, joining in 1928 after graduating in mathematics at

[80] Macaulay (1938), p. 47.

[81] Macaulay (1938), p. 50.

[82] Redington (1952).

[83] Haynes and Kirton (1952).

[84] Penman (1933), Murray (1937).

Cambridge. He became chief actuary there in 1951 and remained in that position until his retirement in 1968. He was President of the Institute of Actuaries from 1958 to 1960. He is most famed, both in actuarial circles and in the broader financial community, for his work on duration and immunisation. His contribution to actuarial thought-leadership and practice, however, was much broader than immunisation theory. He published widely throughout his career on a range of actuarial topics, most notably on with-profit reserving, and valuation and bonus policy. Indeed, the paper in which he introduces his immunisation theory was entitled 'Review of the Principles of Life Office Valuations', and the section on immunisation was a last-minute addition to it.[85]

Redington's immunisation theory took the amorphous actuarial intuition on interest rate matching that had gradually developed over the previous quarter of a century and crystallised it into some clear mathematical statements. In this work, Redington produced the same measure of duration that Frederick Macaulay published some 25 years earlier. Redington was not, however, aware of Macaulay's work, and Redington's approach was the polar opposite to that taken by Macaulay: where Macaulay had been ad hoc and empirical, Redington was mathematical and deductive. Macaulay had been interested in explaining the historical price changes of different bonds. Redington was interested in matching assets and liabilities, where 'matching' meant 'the distribution of the term of the assets in relation to the term of the liabilities in such a way as to reduce the possibility of loss arising from a change in interest rates'.[86]

It is often the hallmark of a great idea that it appears disarmingly simple after someone has thought of it. Redington's immunisation theory belongs in that category. He was aware that the results of the mathematical treatment he was about to introduce were intuitive or even inevitable: 'in the broadest sense, it is apparent without mathematical proof that if the liability-outgo and asset-proceeds are to be equally sensitive to changes in the rate of interest they must have roughly the same mean terms'.[87] Writing 30 years after the publication of the paper, Redington explained how his fundamental development of the theory arose:

> 'As is so often the case once you look at a problem straight between the eyes and with undivided attention you find yourself staring the answer in the face. The actual assets and liabilities were known facts … if their present values were equal at one rate of interest and remained equal on a shift in the rate of interest this

[85] Redington (1982), pp. 84–85

[86] Redington (1952), p. 289.

[87] Redington (1952), pp. 289–90.

was only the layman's way of saying that the differentials of the present values with regard to the rate of interest must be equal. The basic equation ... followed immediately as an elementary piece of differential calculus.'[88]

Redington set out the Taylor series expansion of the change in assets and liability values for a given change ε in the interest rate i (which, importantly, is assumed to be constant as a function of the term of bond). From this expansion he was able to argue that a satisfactory immunisation policy is one where two conditions hold:

$$\frac{dV_A}{di} = \frac{dV_L}{di}$$

and

$$\frac{d^2V_A}{di^2} > \frac{d^2V_L}{di^2}$$

where V_A is the value of the asset portfolio and V_L is the present value of the liability cashflows.

Assuming asset and liability cashflows are fixed, we can write the asset and liability values as:

$$V_A = \sum v^t A_t$$
$$V_L = \sum v^t L_t$$

where $v = 1/(1 + i)$ and A_t and L_t are the asset and liability cashflow respectively due at time t. Taking the derivatives of the above asset and liability values with respect to i, the first immunisation condition implies:

$$\sum tv^t A_t = \sum tv^t L_t$$

Redington noted that this condition is simply that 'the mean term of the value of the asset-proceeds must equal the mean term of the value of the liability out-go'.[89] The weights used in the calculation of the mean term are the ratios

[88] Redington (1982), pp. 84–85.
[89] Redington (1952), p. 290.

of the present value of the cashflow due at that time to the total present value. He did not know it at the time, but he had shown that Macaulay's duration measure was the first derivative of the bond price with respect to the interest rate, and that immunisation therefore required equating the duration of assets with liabilities (the term 'duration' never appears in Redington's paper).

Redington's second immunisation condition implies that a successful immunisation strategy should satisfy the secondary condition:

$$\sum t^2 v^t A_t > \sum t^2 v^t L_t$$

He intuitively describes this condition as 'the spread of the value of the asset-proceeds about the mean term should be greater than the spread of the value of the liability-outgo'.[90]

The major limitation of Redington's immunisation analysis was its assumption of a uniform interest rate applicable over all maturities (that is, a flat yield curve). This was a natural modelling assumption to make and Redington noted the 'practical complication' that 'Yields are not uniform for all terms of assets, and the differentials are not stable in time'.[91] But he did not ponder much on the limitations that this assumption placed on the implications of his analysis. This is doubtless easy to see in hindsight after several intervening decades of quantitative yield curve modelling development, but it is nonetheless true that the direct implications of Redington's equations could lead to dangerously misleading insights. Redington's two immunisation equations imply that the optimal asset strategy would entail equating asset and liability duration whilst maximising the spread of asset cashflows around the liability duration. This implies that the best asset strategy would be the mix of cash and perpetuity (or, even better, very long-dated zero-coupon bond) that has the same duration as the liability cashflows. Such a strategy, however, is merely maximally exploiting the limitations of the uniform yield assumption. If we consider the possibility that movements in yields may be imperfectly correlated over the term structure, or that long-term yields will be less volatile than short-term yields, we can see that the cash/perpetuity strategy would gear up on risks that are not recognised by Redington's analysis. This is not a point that he directly addressed.

Haynes and Kirton independently published a paper on cashflow matching only a month prior to Redington's immunisation paper.[92] Their paper did not

[90] Redington (1952), p. 291.
[91] Redington (1952), p. 294.
[92] Haynes and Kirton (1952).

develop the duration or immunisation concepts, but they did find that asset portfolios that used combinations of cash and perpetuity ('dead-short and dead-long assets' in their terminology) were optimal across a range of parallel yield curve stresses amongst the bond portfolio choices they considered. This result was entirely consistent with Redington's pair of immunisation equations. Haynes and Kirton attached the explicit caveat that 'at this stage we would emphasise that any such combination of dead-short and dead-long investments offers no protection against a change in interest rates which is not uniform for all terms'.[93]

Redington's immunisation theory delivered a thunderbolt of much-needed clarity to the management of interest rate risk inherent in long-term liability business. Thirty years later, he wrote 'It is really extraordinary that the elementary matching principle to which I have given the rather impetuous title of immunization should have escaped the profession for so long'.[94]

For life offices, however, it did not resolve the question of *which* liabilities it ought to be applied to. In particular, in the context of with-profit business, should an immunisation strategy be pursued that immunises liabilities inclusive of anticipated future bonuses, or only the guaranteed liabilities that have accrued to date? In his paper, Redington explores, without explicitly advocating, the strategy that immunises both accrued liabilities and future bonuses (and future regular premiums could be treated as negative liability cashflows). This strategy of attempting to 'lock in' all future bonuses at a with-profit policy's inception met with the disapproval of many of his peers. It did not appear consistent with policyholders' expectations of participation in future increases in interest rates, if such rate increases should arise over the lifetime of their policy.

In 1953, the year following Redington's paper, two senior life actuaries, G.V. Bayley and Wilfred Perks, presented a paper[95] to the institute that advocated applying immunisation only to the 'paid-up' policy values (i.e. the sum assured that would be provided in the event that the policyholder chose not to pay any further regular premiums). This would mean that future regular premiums, by being invested at the prevailing interest rate when they are received, could participate in any intervening increase in the interest rate. However, this failed to address the fundamental interest rate solvency risk: under this strategy, falls in interest rates below that assumed in the premium basis could result in future premiums being inadequate to fund the contractual liabilities.

[93] Haynes and Kirton (1952), p. 155.
[94] Redington (1952), p. 90.
[95] Bayley and Perks (1953).

Haynes and Kirton's paper advocated what is perhaps the most natural strategy: match the accrued guarantees with bonds, and invest the remainder of the funds in risky assets. Whilst sounding straightforward, even this strategy was not without complications in the context of regular premium with-profit business. For example, in the early years of a with-profit policy, a realistic valuation of future regular premiums could exceed the present value of the guaranteed sum assured, implying that all of the accrued assets could be invested in risky asset classes such as equities. This was generally viewed amongst life actuaries as being uncomfortably imprudent. However, if a with-profit fund was considered in aggregate, as was arguably reasonable for a pooled investment vehicle, then this strategy could result in with-profit equity/bond allocations that were sensible and sustainable.

Finally, it is interesting to consider from a historical perspective if Redington's concept of matching asset and liability value sensitivities can be viewed as an antecedent to the dynamic replication concept that underlies derivative pricing and hedging. Dynamic hedging assumes the continuous matching of the instantaneous sensitivities of asset values to changes in underlying fundamental variables. Redington noted that immunisation was not a static solution but rather 'the equations define the position at a moment in time. *Their solutions change continuously* [emphasis added]'[96]

Redington's paper presented some charts that showed how example asset and liability values changed as a function of interest rates, showing that there are levels of interest rate at which they are solvent (i.e. assets exceed liabilities) and levels of interest rate at which they are not. The liability valuation curves included heuristic allowances for the effects of cash maturity guarantees and annuity options. But he did not explicitly pursue the possibility that the asset value curve could be altered by dynamically updating the asset duration to match the liabilities' new duration following an interest rate shift. In the discussion that followed the Institute's meeting at which the paper was presented, A.T. Haynes (of Haynes and Kirton) commented:

> Where the asset and liability curves might cross owing to a change in the rate of interest ... the position could be converted immediately by altering the assets in such a way as to immunise the liabilities. The asset curve would then follow the same form as the liability curve ...[97]

[96] Redington (1952), p. 292.

[97] Haynes in Discussion, Redington (1952), p. 331.

This is an explicit description of the concept of the dynamic hedging of a non-linear liability, over 20 years before Black and Scholes used such a concept to price derivatives! However, the embryonic dynamic replication thinking of Redington and Haynes here is entirely in the context of asset-liability management: it did not occur to them that such a replication concept could be used to derive arbitrage-free prices for the options that were increasingly being embedded in life assurance contracts. As we shall discuss later, this topic would confound the profession for many decades to come.

Actuaries and Equities (1912–1948)

We saw above how both British and American life offices shifted substantially away from illiquid mortgages and loans and moved further into liquid government and corporate bonds between 1890 and 1920 (the US shift starting some fifteen years earlier). Over the following 30 years, another radical change in actuarial investment thinking occurred: equities emerged as a major asset class for British life offices. In aggregate, British life offices increased their equity allocations from under 2 % in 1920 to 21 % by 1952.[98] This time, British and American life office investment practices diverged. The British offices had more surplus capital and a more flexible with-profit bonus distribution system than their US counterparts. These features allowed the British offices more freedom to tolerate and manage the greater volatility associated with equity investment. The American offices were also under a more prescriptive and restrictive regulatory system that varied somewhat from state to state but which generally curtailed equity investment (Fig. 3.8).

Nineteenth-century actuarial orthodoxy had been unambiguous: equities were an unsuitable asset class for life assurance business. Bailey thought it self-evident that equities were too risky for such liabilities. Other nineteenth-century actuarial papers followed this doctrine. In 1891, Mackenzie wrote that 'ordinary shares ... should be rigidly excluded'.[99]

Actuaries' first intellectual embrace of equity investment occurred in the 1920s. This was in part a reaction to the poor performance of the long-term government bonds that life assurers had invested so heavily in during the First World War. The high-inflation post-war environment had resulted in substantial falls in long-term bond prices and consequent losses for life offices. As noted by *The Economist*: 'From the beginning of 1916 to the end of 1920,

[98] Dodds (1979), Table A.3.

[99] Mackenzie (1891), p. 205.

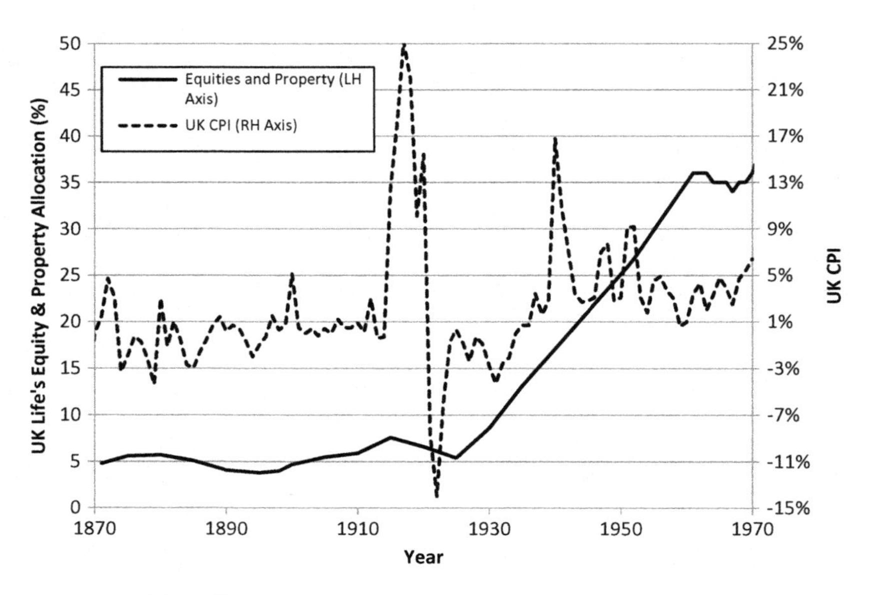

Fig. 3.8 UK life offices' equity and property allocations and UK inflation (1870–1970) (Dodds (1979))

[bond] depreciation ... has been enormous and unprecedented ... and has wiped out nearly whole of the surpluses which would otherwise have been available for bonuses to policyholders.'

The extraordinary burst of inflation that occurred during and after the First World War was an important factor in motivating actuaries' reappraisal of equities as a potentially useful asset class for with-profit business. The six years from the start of 1915 to the end of 1920 produced annual CPI increases of 13 %, 18 %, 25 %, 22 %, 10 % and 15 % respectively.[100] Whilst with-profit guarantees were written in nominal money terms, there was a conviction that bonuses should be used to ensure the policy delivered a satisfactory real return (especially for endowment business where the pay-out would be used to secure a pension income). The impact of unexpected inflation shocks on the performance of long-term nominal bonds in the early 1920s helped to support the argument amongst actuaries that equities, as a 'real' asset class, could offer some form of inflation risk protection. As the chairman of the Pearl put it in 1928:[101]

[100] Office of National Statistics Composite Consumer Price Index.

[101] Scott (2002), p. 80.

> The effects of the War in matters of finance have taught us ... that it may be safer to have a proportion of our investments ... in first-class ordinary shares, rather than entirely on a fixed monetary payment such as is given by ... gilt-edged investments.

The increase in long-term bond yields that occurred in the early 1920s generated disturbing falls in the market values of life office asset portfolios. It probably did not have an adverse impact on life offices' overall economic balance sheets, however, as the assets were unlikely to be substantially, if at all, longer than life office liabilities. In this era, actuaries were reluctant to increase the discount rate used in liability valuation unless absolutely necessary, and so the stated surplus positions in the early 1920s were likely to have been a prudent interpretation of the economic reality. Later in the 1920s, bond yields started to fall again. In the midst of economic depression and a policy of cheap money, consol yields fell from over 4.5 % in 1925 to 3.5 % by the mid-1930s. Whilst this resulted in a recovery in bond prices, it resulted in a new problem: such low rates made it increasingly difficult to for bond investments to sustain historical bonus rates, although the competitive pressure to do so was intense. John Maynard Keynes's speech at the 1927 annual general meeting of National Mutual captured this sentiment:

> Policyholders have become accustomed to bonuses which presume, in most cases ... a net rate of interest of not much less than 4.5% after payment of income tax. If an attempt is made to put British Government securities on a basis appreciably below 4 per cent net, it is certain that ... *a powerful stimulus will be brought into play to find more profitable outlets for invested funds* [emphasis added].[102]

The British with-profit policy was by this time unambiguously positioned as a long-term investment contract. As such it increasingly competed with other investment vehicles such as investment trusts. Investment yields and bonus rates had become increasingly important marketing statistics for with-profit funds both in competing with each other and with other investment offerings. A post-tax consol yield below 3 % put in serious peril not only the ability to make bonus distributions commensurate with a 4.5 % net return, but also the ability to meet the guaranteed liabilities (which had typically been set using a 3 % premium rate). In this context, the 4–5 % dividend yield offered by equities was very alluring.

[102] Scott (2002), quoting from Keynes (1983).

The First World War had created an unintended concentration of British life offices' assets in long-term government bonds. The post-war economic environment and its consequent impact on both realised and prospective life assurance investment performance created a context where life assurers and actuaries were ready for another significant change in asset allocation profile. A stream of research, from both within the actuarial profession and outside it, emerged in the 1920s that provided an intellectual basis for the new openness to equity investing. But the seeds of this thinking can be found earlier.

George May, an actuary at the Prudential in the UK, presented a notable paper[103] to the Institute of Actuaries in 1912. May had a highly distinguished career. He was secretary of the Prudential from 1915 to 1931; he also held significant advisory roles with the British government between 1915 and 1941. He was made a Baron in 1935. In his 1912 paper, he did not explicitly advocate investing in equities, but his was the first actuarial paper to present a technical justification for life offices taking on more investment risk. He paved the way for the change in actuarial attitudes to investment that would emerge more explicitly over the following ten to 20 years.

His rationale for increasing investment risk was twofold. First, he argued historical analysis suggested that 'by careful selection, relatively high yielding investments can be obtained which ... will show a margin considerably greater than is required to cover the risk run'.[104] He didn't, however, provide any statistical analysis or historical statistics, claiming that 'I do not feel it desirable to introduce examples into this paper; further, I am sure that it is only by personal investigation that one is in a position to form a reliable judgment'.[105]

The second pillar of his argument for increasing investment in riskier assets was that risk should be measured at the total portfolio level rather than at the individual security level. He pointed out that individually risky assets may diversify so that, at the margin, they do not materially increase the overall portfolio risk. He suggested that Bailey's already historical investment principles should be extended with a further law: 'That in order to minimise the result of temporary fluctuations and to secure the safety of the capital in the best way, it is desirable to spread these investments over as large an area as possible'.[106] Actuaries were naturally familiar with the diversification concept in mortality risk. Indeed, it was the fundamental principle of insurance business. But this was the first time the diversification concept had been used to argue that life

[103] May (1912).
[104] May (1912), p. 135.
[105] May (1912), p. 143.
[106] May (1912), p. 153.

offices should be more willing to invest in high-yielding securities that are significantly riskier than those that had been typically held by life assurers.

May's advocacy of the adoption of greater investment risk by life offices anticipated the post-First World War zeitgeist. This experience highlighted the benefit of diversification (or the risk of a lack of it). Life offices were ready to invest in something else. During the decade of the 1920s, three men in particular had an especially great influence on British actuaries' thinking on life office equity investment: John Maynard Keynes, the famous British economist; Edgar Lawrence Smith, an American economist and investment manager; and H.E. Raynes, the actuary of the Legal & General Assurance Society.

John Maynard Keynes joined the board of National Mutual, a medium-sized British life office, in 1919 and became its chairman in 1921, holding that position until 1938. He also had a less substantial role on the board of another life office, Provincial Insurance, during this period. In the early days of his involvement with National Mutual, Keynes was not yet the global statesman he would become in his elder years. His revolutionary *General Theory of Employment, Interest and Money* was still some fifteen years away from publication. He had, however, played a prominent role as an economic advisor at the Treasury during the First World War, and his 1920 book, *The Economic Consequences of the Peace*, was a bestseller that raised his profile and marked him out as an influential commentator on economics and politics. Although he only ever played a part-time role there, he had a major influence at National Mutual, especially on investment policy, where the office pursued a distinctive strategy. In the actuarial journals of the period, National Mutual was often referred to as 'Keynes' office'.

Throughout his adult life, Keynes was an active investor. As well as his role at life offices, he managed money for himself and his friends, was involved in various investment trusts, and he managed the endowment fund of King's College, Cambridge, where he was a Fellow. Keynes was a man of strong opinions, often flavoured by a strain of iconoclasm, and he always had the courage to back his convictions. This is reflected in his approach to investment. As an investor, Keynes has been described as a 'scientific gambler'.[107] His investment philosophy was active and aggressive. As might be expected of a man of Keynes's breadth, his investment views transcended any specific type of asset and he regularly took short-term positions in commodities, currencies, equities and bonds as dictated by his macroeconomic and geopolitical outlook. He had a noticeable preference for investing in liquid securities: illiquid assets could not support the short-term active speculation that was at the heart of

[107] Skidelsky (2003), p. 520.

Keynes's investment style. When applied to life assurance, this was the polar opposite of Bailey's conception of life offices reaping illiquidity premia from assets that would never be sold.

National Mutual was not one of the largest offices, but this appears to have suited Keynes, providing him with an institution which had a certain critical mass but which could be made more malleable to his vision than the larger offices. His vision was of mutual life assurance as the leading savings vehicle for the middle classes. He strongly advocated the promotion of endowment assurance as a more savings-orientated alternative to whole-of-life assurance. He argued that life offices should pursue an 'active investment policy' and he piloted such a policy at National Mutual. He viewed investment as the critical function of a life office. In January 1922, in the first of his chairman's speeches at the annual meetings of National Mutual, he said:

> I venture to say that at the present time life assurance societies must stand or fall mainly with the success or failure of their investment policy. The labours of the great actuaries of the 19th century have carried actuarial science to a point where great improvements or striking innovations are no longer likely ... But investment, on the other hand, provides both more pitfalls and also more opportunities than formerly.[108]

Most significantly of all, Keynes held a strong view that the investment policy of a life office should feature material allocations to equities. In his speech at National Mutual's annual meeting of 1928, he explained:

> The arguments in favour of holding a certain proportion of ordinary shares are ... to be found in the advantage of spreading the fund between assets such as bonds, expressed in money values, and assets representing real values ... the second reason is the fact they are undoubtedly under-valued relatively to bonds after making all due allowance for risk.[109]

National Mutual was a pioneer of life office equity investment. Prior to Keynes's involvement, National Mutual's equity allocation stood at 3 %. By 1927, it had increased to 18 %. The other British life offices had not yet followed suit. In 1926, the average equity allocation across the industry stood at only 4 %. But they noticed, and in time, they followed. His heavy emphasis on investment strategy and appetite for investment risk never sat completely comfortably with this generation of the actuarial profession. As late as 1942, F.C. Scott, the managing director of Provincial Insurance wrote to Keynes:

[108] Keynes (1983), p. 118.
[109] Keynes (1983), pp. 156, 158.

> My conception of our responsibility is that our first primary duty is to so invest our funds, which are the security we offer our policyholders for the fulfilment of our contracts, that the public confidence will never be impaired.... I am sometimes tempted to think that you would pervert my morals and encourage me in my secret vice to regard ... the investment of our funds, not as incidental to our main business – insurance – but as an equally important contributor to the profits of the Company.[110]

Keynes would indeed pervert his morals, and most of the rest of the actuarial profession. But he was not the only important actor in this scene. Edgar Lawrence Smith was an economist at Columbia University and a Wall Street investment manager. His pioneering short book, *Common Stocks as Long Term Investments*, was published at the end of 1924. It, perhaps inevitably, resonated with Keynes, who wrote a positive review of the book in the *Nation and Athenaeum*. The book also registered with the British actuarial profession, and was referred to in the papers and discussions of the profession in the second half of the 1920s and beyond.

Smith's book was the first significant quantitative comparative analysis of the long-term empirical returns of equities and bonds. Smith considered the returns that had been produced from diversified portfolios of equities and high-quality bonds over various 20-year holding periods between 1866 and 1922. He demonstrated that the long-term returns of equities had consistently exceeded those of high-quality bonds. He also showed that the *income* received from equities had also tended to exceed that generated by bonds over long-term investment horizons. Neither of these observations would necessarily have been especially surprising to financial practitioners. But a more provocative conclusion of his analysis was that equities would even tend to outperform bonds in periods of low inflation or deflation as well as in inflationary periods. It was generally accepted wisdom that equities were a 'real' asset class whilst bonds were 'money' assets, and as such equities would outperform in times of unexpectedly high inflation and vice versa. Smith argued that the evidence suggested that over a horizon of 20 years equities would outperform bonds under any economic conditions with a high degree of likelihood.

Smith produced a novel argument for why equities performed so robustly. He argued that the corporate practice of only paying out a small proportion of annual earnings as dividends and reinvesting the remainder facilitated a compound interest effect in future earnings which allowed dividends to be smoothed over time, and hence protected equity values from economic

[110] Keynes (1983), p. 80.

downswings and deflation. Such an argument had much resonance with Keynes and the British actuarial profession of the time, but the idea that dividend policy could play such a pivotal and positive role in equity value creation would be rejected by the financial economists that would emerge a few decades later.

Smith's data analysis did not include any statistical calculations such as confidence intervals or significance tests. His paper did not overly focus on equity tail risk or the possibility that the 20-year outperformance could only be obtained by enduring an uncomfortably volatile path. However, he did include a short section that considered how long a period of equity underperformance an investor may have to suffer. He considered investing in equities at the start of 86 different years from 1837 to 1922 and found that only four of the 86 investment projections resulted in equities producing a negative return over the first four years. He concluded that there was a 78 % probability of equities delivering a positive return over one year, an 87 % chance over two years, and a 94 % chance over four years. To modern eyes, this may appear to be a stretching use of overlapping data periods and a significant understatement of equity risk (the assumption of a normally distributed log annual return with a mean of 7.5 %[111] and volatility of 18 % implies a probability of around 20 % that equities would produce a negative return over a four-year period, assuming annual returns are independent over time).

Finally, it is important to note that Smith's advocacy of equities as a long-term investment was primarily targeted at the savings strategies of individual investors. It was expressly not intended for insurance companies where 'their investments are all made to protect dollar obligations. Their sole concern is so to conduct their affairs that they will always have available more dollars than they will be called upon, under their commitments, to pay out, at the same time deriving a profit above operating expenses. It is natural and right that (they) should concentrate their attention upon promissory obligations payable in a fixed number of dollars.'[112]

But by the 1920s, for better or worse, British with-profits business had moved substantially beyond the provision of fixed monetary amounts in the event of life contingencies. Endowment assurance and large 'bonus loadings' in with-profit premium rates had furthered the evolution of life assurance into a long-term investment vehicle where mortality protection was a secondary feature. Perhaps perversely, the high degree of actuarial caution in the setting of with-profits premium rates had created a vacuum that risk-taking filled.

[111] Assuming a cash rate of 5 %, an equity risk premium of 4 % implies an (arithmetic) expected return of 9 % and hence a geometric expected return of c. 7.5 %.

[112] Smith (1924), p. 12.

Keynes and Edgar Lawrence Smith were both outsiders to the British actuarial profession. H.E. Raynes, however, as the actuary of the Legal & General Assurance Society, was in its inner sanctum. He presented two influential papers to the Institute of Actuaries in 1928 and 1937 that advocated a greater role for ordinary shares in life office asset allocations.[113] Raynes was doubtless influenced by the practices of Keynes and the writings of Smith (both are referred to either directly in his papers or in the formal discussion of the papers that followed their presentation to the profession). But he was not an immediate convert. As late as mid-1927, Raynes 'contended that considerations of security excluded ordinary shares as investments for life offices'.[114]

He underwent his Damascene conversion when he attended an International Congress of Actuaries in Canada in 1927. He had discussions with T.B. Macaulay, the actuary of Sun Life of Canada (and father of Frederick Macaulay of bond duration fame), in which Macaulay 'strongly advocated investment of a proportion of funds in common stocks as a measure to combat the trouble of a depreciation of currency [i.e. inflation]'.[115] This inspired Raynes to develop a piece of empirical analysis of UK equity data that was very similar in spirit to Edgar Lawrence Smith's US equity work, if less comprehensive. The paper, 'The Place of Ordinary Stocks and Shares (as distinct from Fixed Interest bearing Securities) in the Investment of Life Assurance Funds', was presented to the Institute of Actuaries in 1928. A follow-up paper, 'Equities and Fixed Interest Stocks during Twenty-Five Years', provided further statistical analysis to the first paper, but no new conclusions, and was published in 1937.

Raynes analysed the investment performances of a diversified portfolio of UK equities and a portfolio of debentures over the fifteen-year period from 1912 to 1927. Like Smith, Raynes found that equities strongly outperformed the bond portfolio, both in terms of capital appreciation and income. He found this was true during periods of rising interest rates and falling interest rates. Raynes concluded that equities had a significant role to play in life office asset allocation for the two distinct reasons that Keynes forwarded in his chairman's speech of 1928: inflation protection as per T.B. Macaulay's argument; and for the more fundamental reason that they appeared good value to the long-term investor who could ride out the higher short-term volatility that comes with equity investment.

Given Raynes's role as a leading life actuary, his paper is surprisingly light on the risk implications of equity investment, and how equity asset allocation

[113] Raynes (1928), Raynes (1937).

[114] G.H. Recknell in Discussion, Raynes (1928), p. 35.

[115] Raynes (1928), p. 22.

might be considered in the context of the fixed liabilities of life assurance contracts. Interestingly, it was a visitor to the actuarial meeting, the eminent economist Ralph Hawtrey, who was prepared to look beyond the quantitative results of a relatively small data sample, suggesting that 'if [insurance companies] were going to take the risk of investment in ordinary shares, then they ought to have a correspondingly larger capital'.[116]

Perhaps with the benefit of hindsight, in reviewing Raynes's papers and their accompanying actuarial meeting discussions, a creeping sense of hubris and complacency can be detected emerging in actuarial thinking and attitudes. Early in the 1928 paper Raynes writes:

> Whether greater security to the capital of a fund is given by the investment of a proportion in ordinary stocks and shares might be considered by deductive reasoning from the principles of economics, but while no doubt the result of such a process would lead to the same result, a statistical investigation will, I think, prove more convincing to actuaries.[117]

This sentence creates the sense that actuaries thought themselves above the theories and deductions of others such as economists, and instead would use their uniquely endowed capacities of observation to judge what was right. This sense of hubris and complacency can be increasingly identified when considering some of the doctrines of the actuarial profession in the second half of the twentieth century. Arguably, it started when actuaries were seduced by the cult of equities in the 1920s.

Thus, the intellectual groundwork for greater life office equity allocations was laid in the decade of the 1920s by the triumvirate of Keynes, Smith and Raynes. Lewis G. Whyte, a senior Scots actuary, wrote in 1947 that 'by the time Mr Raynes' paper ... was submitted to the Institute of Actuaries in 1927, the movement [to invest life office funds in equities] had received a large measure of professional approval'.[118] But equity asset allocations in aggregate across the British life offices remained fairly modest through the 1920s. A few life offices such as National Mutual, Scottish Widows and Equity & Law led the way with larger equity allocations than the rest of the industry. Equity allocations gradually climbed during the 1930s in an environment of enduring low government bond yields. The post-Second World War economic environment then provided equity allocations with a significant further fillip. This was driven by near necessity. The 'cheap money' era of Hugh Dalton, the Labour chancellor of the exchequer, drove consol yields down to 3.1 %

---

[116] R.G. Hawtrey, in Discussion, Raynes (1928), p. 42.
[117] Raynes (1928), p. 22.
[118] Whyte (1947), p. 221.

in 1945, presenting a real threat to life offices' ability to meet the 3 % (after-tax) return assumption almost universally used in with-profit premium rates at the time. As in the 1850s, low interest rates had prompted life offices to take more investment risk. J.B.H. Pegler, the investment secretary of Clerical Medical presented a paper[119] to the Institute of Actuaries in 1948 that argued that this environment demanded a less risk-averse attitude to investment: 'if the assets are invested, no matter how safely, to provide a yield less than that assumed in the calculation of premium rates, the 'soundness' of the investments is somewhat illusory'.[120]

Pegler proposed investment principles that turned Bailey's risk-minimising ethos on its head. He suggested: 'It should be the aim of life office investment policy to invest its funds to earn the maximum expected yield thereon' whilst, in a nod to George May, 'Investments should be spread over the widest possible range in order to secure the advantages of favourable, and minimize the disadvantages, of unfavourable, political and economic trends'. These principles were perfectly reasonable long-term investment objectives of an investment strategy, but the disregard for how such a strategy should ensure assets could meet guaranteed liabilities is striking. With the benefit of hindsight, it is hard to resist the thought that actuaries were starting to surrender rigour and prudence under the significant competitive pressures of their marketing departments.

Valuation and Surplus, 1800–1952

The need for a valuation of life assurance liabilities arose with the introduction of the regular premium whole-of-life assurance policy. Level premiums for an increasing risk implied that a portion of the premiums received in the policy's early years must be reserved to fund the larger frequency of claims that was expected to occur in later years. A liability valuation determined how much needed to be reserved. This was well-understood by James Dodson in his original designs of whole-of-life assurance in the mid-1750s. What he did not anticipate was that by the end of the century a test of whether adequate reserves were being held would be necessary for two distinct purposes: to determine if the fund was solvent (i.e. if it had sufficient assets to meet its contractual liabilities); and to determine how much surplus could be distributed to the with-profit policyholders (and how the distribution should be equitably allocated across different policyholders).

[119] Pegler (1948).

[120] Pegler (1948), p. 180.

Through most of the near-150-year period between the Napoleonic Wars and end of the Second World War, the prudent with-profit premium rates of the established British life offices had led to an embarrassment of riches. The demonstration of solvency with respect to life office's contractual liabilities could be done with relative ease and was not a particular focus of actuarial minds. The other side of this coin was that significant surpluses were being generated that had to be distributed to with-profit policyholders, at least in part.

With-profit funds were a form of pooled investment vehicle. There were no set rules on how the earned surplus should be distributed between current policyholders and the 'estate' of the office; or, once the policyholders' allocated surplus was determined, how it should be distributed across the different generations of current policyholders. Questions such as these featured prominently in the thinking of the British actuarial profession from the 1850s onwards, and triggered some of its most impassioned debates. Fundamentally, two distinct tasks arose in determining with-profit bonuses: determining how much surplus had arisen over a period by measuring the changes in the values of assets and liabilities; given the determination of the surplus, decide how much to distribute to existing policyholders and how to distribute it amongst them. We will return to methods for distributing the surplus at the end of this section. First we consider how actuarial thinking developed on how the size of surplus was determined through the valuations of assets and liabilities.

Liability Valuation (1800–1952)

The actuarial valuation of liabilities was, and arguably still is, the quintessential actuarial task of life actuaries. It has been a topic of much debate throughout most of the history of actuarial thought in both life assurance and defined benefit pensions. The second half of the nineteenth century was arguably the period of its greatest technical development in life assurance. Throughout this period, the liability valuation was calculated prospectively as the present value of a set of projected cashflows. Technical debates centred around a handful of methodology choices for the projection and discounting of these cashflows. The major (and interrelated) valuation methodology topics were (and, to some extent, still are!):

- The treatment of future premiums in the liability valuation. All liability valuations could be described as the difference between the present value of the assurance claim plus related expenses; and the present value of the future premiums due over the remaining term of the policy. But there was significant debate over the treatment of future premiums, and, in particular, whether it was appropriate to recognise the profit margins in future premiums in today's liability valuation.

- Related to the above treatment of profit margins was the treatment of expenses. In particular, should future expenses be explicitly modelled in the valuation, or be implicitly allowed for in the treatment of future premiums.
- Should the mortality and interest rate assumptions used in the valuation be best estimates of future experience and market conditions; or should they include margins for prudence; or be tied to the basis that was currently in use for pricing the product; or that was used in pricing the product at the time of the valuation.
- Should the with-profit liability valuation include an explicit allowance for future bonuses or only the accrued contractual obligation.

As discussed in Chap. 2, the first life office liability valuations were performed by William Morgan in the 1780s. The solvency of the Equitable was never doubted during this period, and the object of his valuations was primarily to limit the pressure for faster distribution of surplus. At this time, Morgan's liability valuation basis was the same as the Equitable's premium basis (a 3 % interest rate and the Northampton mortality table). At the time of the liability valuation, the premium basis was arguably already recognised as being significantly more prudent than a best estimate of future experience. The liability valuations included the bonuses already accrued, but made no provision for future bonuses. The value of the future (gross) premiums was deducted from the reserve. He did not explicitly reserve for future expenses. He likely believed that the margins in the liability valuation (premium) basis adequately allowed for them.

Over the course of the first half of the nineteenth century, a number of other prototypical actuaries, both within the emerging group of life offices (Joshua Milne, T.R. Edmonds and others such as David Jones, Charles Babbage and Griffith Davies), and outside it (such as John Finlaison and Augustus De Morgan) were actively considering all aspects of nascent life office actuarial practice. The concept of a *net premium valuation* emerged as an alternative conceptual approach to Morgan's liability valuation approach by the middle of the century (though as we discuss below, Morgan's approach could be cast as a form of net premium valuation). It is difficult to precisely date the first conception of the net premium valuation, especially as there were no professional actuarial journals in the first half of the nineteenth century. However, we know it was established in actuarial thinking by 1853 as it was discussed at length in the testimonies of leading actuaries such as Finlaison and Edmonds to the 1853 Parliamentary Select Committee on Assurance Associations.[121]

To understand the workings of the net premium valuation and its rationale, it is perhaps easiest to start from a consideration of the more intuitive, 'gross premium' valuation. We can loosely define the gross premium valuation as a liability present value that is assessed using best estimate assumptions for all future experience, including an explicit projection of future expenses, and which includes the contractual premiums in the valuation. Consider a new without-profit regular premium whole-of-life policy. The office sets the regular premium for the contract such that it expects to generate a profit. If the actuary performed a gross premium valuation of the liability at the point of inception of the policy, it would therefore have a negative value: the present value of the future regular premiums would exceed the present value of the sum assured. Negative valuations created difficulties. Most obviously, the contract could be surrendered and the surrender value must be non-negative. More generally, it was argued that it was inappropriate to immediately recognise the profit that was expected to be generated over the lifetime of the policy in the current valuation of the liability.

The net premium valuation method aimed to only recognise the profit loading in the premiums at the point at which each premium is received. The liability valuation would be made on the assumption that the future regular premiums would be 'net premiums'—defined by Bailey in 1878 as 'the premium which, according to the best judgement that can be formed, will suffice for the risk but leave no margin beyond'.[122] The net premium would therefore be calculated on best estimate assumptions for the future behaviour of the risks underlying the contract, but would make no allowance for expenses—the assumption being that expenses are adequately covered by the profit margin in the gross premiums that is being discarded in the net premium valuation. By construction, a net premium valuation would produce a liability value of zero at the inception of the policy.

There was much heated debate in the mid-nineteenth century over the relative merits of using Morgan's method of valuing liabilities using gross premiums and the premium basis or valuing using net premiums and a realistic valuation basis. This was further confused by the observation that Morgan's method could be cast as a net premium method, in the sense that the gross premiums were the net premiums implied by the premium basis (but not by a 'best estimate' basis). Like any net premium valuation, Morgan's approach

[121] Parliamentary Report (1853).

[122] Bailey (1878), p. 118.

would generate a liability value of zero at inception. But it did so by putting the 'wrong' parameters into the net premium valuation formula, and this made the progression of the reserves through the life of the policy less natural or transparent than a 'pure' net premium valuation (where the reserve progresses by reserving the net premium portion of the gross premiums as they are received and the remainder is used for expenses and released to surplus).

The founding of the Institute of Actuaries in 1848 and Faculty of Actuaries in 1856 provided an important professional infrastructure for the development of British actuarial thought and practice. It is no coincidence that the third quarter of the nineteenth century is one of the most fertile periods of development in actuarial thought. In particular, their professional journals, the *Journal of the Institute of Actuaries* and the *Transactions of the Faculty of Actuaries*, and the practice of presenting these papers at formal professional sessional meetings were powerful vehicles in disseminating and debating actuarial theory and practice. A series of Institute papers by Charles Jellicoe in the early 1850s,[123] followed by a paper by Thomas Sprague in 1863, helped to establish the net premium valuation as the fundamental piece of actuarial machinery for liability valuation.

The net premium valuation quickly became an article of faith for the actuarial profession, but, to the profession's credit, its limitations were highlighted and debated as early as 1870. The discussions of these limitations produced unusually ill-tempered exchanges amongst actuaries. Sprague, by then a pre-eminent actuary and vice-president of the Institute, wrote in 1870 of 'the intolerant manner in which some of the advocates of the net premium method of valuation insist upon its being the one method that should be employed in all cases'.[124] Clearly, there has always been something about liability reserving methods that ignites actuarial passions like few other things can!

The net premium valuation, in its original form, was parsimonious and elegant. When its assumptions for interest rates and mortality rates held true, surplus emerged from the contract in a most natural way: when a regular premium was received, the liability value would increase by the net premium, whilst the asset value (before the deduction of expenses) would increase by the actual premium. Thus the profit loading in the premium would be recognised with the receipt of each premium. If the net premium portion of the actual premium was reserved when it was received, and mortality and interest rate experience was as assumed in the net premium valuation, then the reserve would be exactly sufficient to meet claims as they fell due. But the net

[123] Jellicoe (1851), Jellicoe (1852), Sprague (1863).

[124] Sprague (1870), p. 411.

premium valuation had some counterintuitive features. As we have seen, it did not explicitly allow for future expenses—the difference between the actual and net premium was expected to take account of those. But there was nothing fundamental in the method that could capture the impact of unexpected increases in future expectations for expenses (which would be especially problematic if the new expectation for expenses exceeded the margin of the actual premium over the net premium).

Indeed, it might be argued that the elegance of the net premium valuation could quickly unravel if *anything* unexpected happened. Suppose interest rates fell after the inception of a contract. If the net premium valuation used the old, higher interest rate, this would be imprudent, as assets could not be expected to earn that rate. But if a new lower interest rate was used in the net premium valuation, this resulted in a higher net premium being used in the valuation, even though the terms of the contract and the actual size of the premium payable had not changed (a feature succinctly described by Redington as 'a technical idiosyncrasy which has no counterpart in the facts'[125]). Again, there was a risk that this could result in using net premiums in the valuation that exceeded the actual premiums net of expected future expenses.

A number of papers were presented to the Institute around 1870 that raised objections to the net premium valuation concept. Some of these merely highlighted that there were some actuarial tasks where the net premium valuation was clearly inappropriate. Sprague highlighted that in a valuation of liabilities for the purposes of determining a transfer value from business to another, it was 'manifestly unjust'[126] to ignore the actual premiums that were a contractual reality of the business. He also highlighted that there were US regulations that required net premium valuations to be made on a prescribed basis that produced higher net premiums than the actual premiums that were being charged in the marketplace, which he described as a 'manifest absurdity'. Makeham presented a paper that went further than Sprague, arguing that the net premium valuation approach was not even appropriate for its core application of determining surplus distribution.[127] Sprague also noted that the superintendent of the New York Insurance Department had conceded that 'valuations for Government purposes will … ultimately have to be made upon actual premiums received by the different companies'.[128] A firmer regulatory context was similarly emerging in the UK following the failure of the

[125] Redington (1952), p. 307.
[126] Sprague (1870), p. 414.
[127] Makeham (1870).
[128] Sprague (1870), p. 417.

Albert Insurance Company in 1869—the 1870 Life Assurance Companies Act created, for the first time, a regulatory requirement for British life offices to assess prospective reserves on life policies. The methodology, however, was not prescribed to the same degree as in the USA. British actuaries retained more freedom to choose how to determine their reserves, but were compelled to disclose their methods. Thus the British life industry's 'freedom with publicity' regulatory regime was born.

The net premium valuation retained a loyal following, despite the above criticisms. In 1878, the influential Arthur Bailey (the pioneer of actuarial thinking on investments) presented a staunch defence of the net premium valuation. As might be expected of Bailey, his position was sophisticated and nuanced. He argued that the net premium valuation should be made only with best estimates of future experience, and that the premium basis should play no direct role. But he also argued that the implementation of the net premium valuation method should include a comparison of the actual premiums and net premiums of the contract 'in order to ascertain what portion [of the actual premium] represents the real prime cost of the risk [i.e. the net premium]'.[129] His advocacy of the net premium valuation was explicitly 'when the object is to determine the amount of divisible surplus'.[130]

The debate rumbled on, and over the following decades a number of practical complexities emerged that further challenged the implementation of an elegant, parsimonious net premium valuation. The net premium valuation implicitly assumed that the expenses associated with the policy are incurred uniformly over the product's duration (or, more accurately, over the premium-paying term). But practices in the acquisition of new business were changing such that much of the expense associated with a policy was incurred at its inception in sales commission payments. This meant that a standard net premium valuation could result in a loss being recognised at inception, even though the policy was profitable. A fundamental attraction of the net premium method was that it deferred the recognition of profit, but it was arguably now overstating this deferral by initially stating a loss.

The practice of paying significant up-front commissions for new business sales had first emerged in Germany, and it was the President of the German Life Institute, Dr Zillmer, who first proposed an actuarial adjustment to the net premium valuation that would allow for the change in the expense pattern.[131] His proposal was first made in 1863 and it essentially allowed for an

[129] Bailey (1878), p. 119.
[130] Bailey (1878), pp. 125–26.
[131] Zillmer (1863).

asset to be created that recognised that an initial expense had been incurred that would otherwise be due later in the contract. Zillmer's work did not have an immediate impact in Britain as the tendency at the time was for sales commissions to be paid with each regular premium received rather than at the policy inception. But by 1870 British commercial practices had increasingly moved towards the up-front commission model, and net premium valuations started to generate losses on new business. Sprague's 1870 paper introduced Zillmer's approach to the Institute and commended it as 'a very able work'.[132] It was broadly well-received as sound actuarial practice in Britain and the USA, though there were reservations that it could be imprudently misused or abused. US regulators were particularly sceptical and 'commuted commisions' were not admissible as assets in the US regulatory returns.

Zillmer's logic could be shoe-horned into the net premium valuation method. It detracted from one of its fundamental advantages—parsimony—but it did not deliver a fatal blow. A more serious and fundamental challenge to the logic of the net premium valuation arose in its application to with-profit business and its bonus system. The logic of the net premium valuation as the natural method for recognising the emergence of surplus demanded the use of best estimates of future experience for mortality and interest rates in the valuation. This ensured that the surplus that emerged would simply be the difference between the actual and net premiums (providing the experience was consistent with the best estimate used in the valuation basis). By this time, with-profit premium rates typically included 'bonus loadings' that set them at significantly more expensive levels than without-profit premium rates. The premium rates were set by using interest rate and mortality assumptions that included substantial margins relative to the experience that was expected. This, in itself, was not a fundamental issue for the use of net premium valuations: if the net premium valuation was performed using best estimate interest rate and mortality assumptions, the net premium would be significantly lower than the actual premium and a surplus would gradually emerge with the receipt of each regular premium. However, assuming the experience assumptions were borne out in practice, such an approach would result in the same cash amount of surplus emerging with each premium. This was not consistent with the way that surplus was distributed to with-profit policies—the approach of adding a reversionary bonus to the sum assured meant that an increasing cash value of surplus would be added as policy duration increased. The cost of bonus was not perfectly aligned with the premium pattern. So, for growing with-profit businesses writing high volumes of new

[132] Sprague (1870), p. 420.

business, this implementation of the net premium valuation could result in a larger surplus being recognised than was desirable. In order to defer the recognition of this surplus, it became standard practice in the final quarter of the nineteenth century to add loadings to the assumptions used in the net premium valuation—the assumed interest rate would be artificially lowered by as much as 1.5 % below the interest rate that was expected to be earned.[133]

Valuation practice had come full circle! This was very similar to Morgan's approach of 100 years earlier of using the premium basis to produce a bastardised form of net premium valuation. The approach 'worked' in the sense that it produced a slower emergence of surplus that was commensurate with the reversionary bonus system. But it worked by putting the wrong (interest rate) parameters into the wrong (net premium) formula. It was now very hard to explain what the net premium valuation was intended to mean. There was an increasing sense amongst many in the actuarial profession that something more logical and 'scientific' ought to be done. This movement arguably began with Sprague's 1870 paper, but it was not until 1907 that an alternative reserving system was clearly proposed when C.R.V. Coutts, the assistant actuary of the National Mutual, presented his paper 'Bonus Reserve Valuations'.

Coutts proposed replacing the net premium valuation machinery with a more direct and transparent valuation method. He argued that the actual gross premiums charged by the contract should be used in its valuation, together with explicit estimates of future expenses and best estimates of interest rates and mortality. Of course, for with-profit business the presence of bonus loadings in the premium basis would result in negative liability values, especially at early policy durations. He proposed to address this by a controversial means: he argued that the future reversionary bonuses that the office intended to pay should be explicitly reserved for as a part of the current liability: 'in the valuation balance sheet, the present value of the bonus at the rate to be declared at a given valuation should be treated as a liability for the future existence of the policies'.[134] To the layperson, this may seem a very natural approach. But it created a great deal of tension and controversy within the actuarial profession. There were two main concerns—one technical, and one relating to policyholder communication.

The technical concern was that the bonus reserve valuation required several accurate estimates of the various dimensions of future experience, and the valuation result would be highly sensitive to those assumptions. The margins in with-profit premium rates and net premium valuations were large

[133] Coutts (1908), p. 162.

[134] Coutts (1908), p. 161.

and approximate, and the notion of, for example, having to estimate interest rates at a granularity of 1/16 of 1 % in valuations created some consternation amongst actuaries.

The second concern related to the management of policyholders' expectations: if actuaries started to explicitly reserve for future bonus as though it was a promise, did that turn it into a promise? In the discussion at Staple Inn following the presentation of Coutts's paper, S.J.H.W. Allin commented, 'they should be very careful about doing anything which tended to give policyholders the idea that the whole of the funds of the company belonged to them to do as they liked with ... It was even possible that policyholders, if their bonuses were reduced ... might insist that the contract between them and the company had been broken'.[135] Of course, the application of a net premium valuation with bonus loadings in the assumptions also reserved for future bonus, but this was an implicit approach buried in the actuarial machinery—it was believed that this opaqueness reduced the risk of encouraging inappropriately demanding expectations from policyholder.

The net premium and gross premium methods went on to live in a somewhat uncomfortable coexistence. The stability of the net premium valuation was attractive to actuaries in measuring the emergence of distributable surplus from one valuation period to the next; gross premium valuations (without future bonuses) would be the preferred approach for testing the solvency or absolute adequacy of the reserves to meet liabilities. Redington's seminal 1952 paper that introduced his immunisation theory was actually a paper primarily about reviewing these with-profit liability valuation methods. In this review, he emphasised that the distinct questions that actuaries attempt to answer could not be answered by the same valuation:

> A valuation has two main purposes, and the fundamental difficulty is that these two purposes are in conflict. The first and primary purpose is to ensure that the office is solvent. The second is to allow the surplus to emerge in an equitable way suited to the bonus system. The solvency criterion leads to a changing valuation basis ... On the other hand, the pursuit of equity of emergence of surplus tends to lead to stable valuation bases, influenced mainly by retrospective considerations.[136]

Later in the paper, he elaborates further:

> Surplus can be estimated only by a passive valuation policy which leaves the valuation basis both of assets and liabilities unchanged ... Yet when we turn to

[135] Coutts (1908), p. 170.

[136] Redington (1952), p. 298.

the question of solvency there is no word to be said in favour of a passive policy; it is only an active policy, paying full regard to existing (and estimated future) experience, which has any significance.[137]

This provided an intellectual basis for performing distinct actuarial valuations in distinct ways for distinct purposes. It was broadly representative of the British actuarial practices that had emerged over the first half of the twentieth century, although there was not universal agreement within the profession on the respective roles of the different valuation methods. The Staple Inn discussion of Redington's paper makes clear that there remained many life actuaries who had decided that the net premium valuation should be regarded as a piece of actuarial antiquity. But time has proven it to be a resilient idea. At the start of the twenty-first century, it remained enshrined in the regulatory solvency regimes of many countries around the world.

Asset Valuation (1829–1952)

Nineteenth-century actuarial thinking on valuation was resolutely focused on the liability side of the balance sheet. Actuaries generally took asset values as they were given to them by the life office's investment department. This started to change in the mid-twentieth century, most notably in the late 1940s and early 1950s, when a number of sophisticated actuarial papers were published discussing how assets should be valued for the purposes of measuring solvency and surplus. The increased holdings in risky asset classes such as equities and property, together with the substantial movements in interest rates that were experienced in the 1930s and 1940s, were both significant in motivating this new actuarial focus on the asset side of the balance sheet.

Early nineteenth-century actuaries were aware that asset valuation could play an important role in the assessment of surplus. William Morgan, in his address to the Equitable general court in 1829, stated:

'That the Society may probably have the right to estimate its Stock in the Public Funds at the price it bears at the time of determining the surplus, I do not pretend to dispute ... The great fluctuation in the value of the funded property [long-term government bonds] will always render the surplus very precarious. It might perhaps be desirable that the Stock in the Funds should be invariably fixed at some mean price, rather than that the amount of the surplus should be suffered to depend on property perpetually changing its real value.'[138]

---

[137] Redington (1952), p. 304.

[138] Ogborn (1962), p. 202.

This is an interesting early example of actuaries' disdain for market prices. But it is important to understand the context of this statement. Morgan was specifically considering the determination of surplus that had emerged between two valuation dates for the purpose of setting the bonuses that would be distributed to with-profit policyholders. His comment is essentially a reflection on the amount of 'bonus smoothing' that he believed was desirable to have in with-profit distributions over time and between generations. This need not have any direct implication for his view of whether market values were a sound indicator of economic worth or whether or not they should be used when considering the absolute solvency of life offices.

The actuarial preference for the asset side of the balance sheet to use something more stable than market values was well-established by the early twentieth century. The standard practice was to hold the lower of an asset's book value and market value, a practice summarised as 'always write down, never write up'. A rationale for this approach can be found in R.K. Lochhead's 1930s textbook for actuarial students:

> The established practice of British Offices is to write down security values when prudence dictates such a course and only in very exceptional circumstances to take credit for capital appreciation apart from realized profits on maturity or sale. The justification for this procedure lies in the necessity for creating a reserve against adverse movements in the value of assets, so endeavouring to secure that the life assurance fund shall be represented by assets of equal value.[139]

Lochhead's justification is notably focused on solvency rather than the equitable distribution of surplus. It is also interesting to observe that his justification can be construed as a prudential solvency approach where assets are valued at market value less a reserve for risk. More generally, a number of possible justifications were forwarded in support of the 'always write down, never write up' approach, relating to both measuring surplus and ensuring future solvency: it produced a prudential valuation (any excess of market value over book value was a 'hidden reserve'); it could achieve more consistency with the net premium valuation of liabilities (which tended to use 'sticky' interest rate assumptions that would only be revised when absolutely necessary); it smoothed the release of surplus to policyholders over time and produced a less volatile public balance sheet.

As mentioned in the liability valuation discussion above, a gross premium valuation became the standard actuarial practice for the internal assessment

[139] Lochhead (1932), p. 66.

of the solvency position of with-profit funds. In that context, and in the context of surplus assessment using a bonus reserve valuation, the use of market values for assets was more typical, and was increasingly viewed as the natural valuation approach to take in such cases, as was argued by T.R. Suttie in his important Institute paper of 1944:[140]

> There does not seem to be any object in making a bonus reserve valuation unless it is based upon the true facts, i.e. the actual premiums to be received, the actual renewal expenses, the experience rate of interest, and the market value of assets. Hidden reserves seem entirely contrary to the principles of a bonus reserve valuation ... The position is quite different under a net premium valuation. The valuation basis being more or less fixed, hidden reserves are necessary, not only to avoid violent fluctuations in the rate of bonus if appreciation or depreciation occurs, but also to enable the results of the published net premium valuation to correspond with those of the unpublished bonus reserve valuation.[141]

The net premium valuation was still used widely, both in measurement of surplus for bonus determination, and in the published regulatory returns of life offices. The 'always write down, never write up' asset valuation method was invariably used with net premium valuations in the first half of the twentieth century. As the asset and investment thinking of the profession developed in the 1940s, the justifications for its use came under inevitable scrutiny. It made the relative treatment of assets and liabilities somewhat opaque, particularly in the context of volatile long-term interest rates. In the event of a general fall in the level of interest rates, assets' book values could substantially understate the market value of long bonds, creating a substantial hidden reserve. Meanwhile, the actuary would be under pressure to reduce the valuation interest rate used for the liabilities to reflect the reality of the new market prices. But the assumed valuation interest rate was at the discretion of the actuary and need not be directly related to the market yield. The net impact of the rate fall would therefore be confused or even arbitrary. Similarly, in the event of an interest rate rise, the assets would be written down to their new market values, and the actuary would have some choice over whether this was offset by a reduction in the size of the estate or by increasing the liability valuation interest rate. It was the role of actuaries to use their judgement in these choices so as to meet their twin objectives of ensuring the solvency of the fund and delivering an equitable distribution of surplus to different generations of policyholders (and holders of different types of with-profit policy). But it is easy

---

[140] Suttie (1944).

[141] Suttie, in Discussion, Suttie (1944), p. 226.

to see how a combination of a liability valuation that assumes premiums that have no direct relationship to the actual premiums that will be received and that uses an interest rate that is not directly related to that which is expected to be earned, together with an asset valuation that bears no direct relationship to current asset market values, could make non-actuaries somewhat suspicious of these actuarial dark arts.

There was a more specific issue with the logic of the 'always write down, never write up' asset valuation approach: if a reserve against adverse asset movements was desirable, as per Lochhead's textbook, why was the difference between market values and book values the right size for this reserve? If such a reserve was important, should a more risk-based and scientific approach not be used to determine its necessary size? This line of argument was developed in a couple of papers presented in 1947–1948—Pegler's Institute paper that arose in the discussion of equity investment above; and a Faculty paper written by the investment actuary Lewis Whyte.[142]

Pegler's 1948 paper that advocated the maximising of expected investment returns also advocated what was, for the time, a novel approach to dealing with equity valuation and the distribution of surplus arising from equity investment. He argued that equities should always be valued at market value, dismissing book values as 'illogical' and based on 'fallacies'.[143] The recognition of the surplus created by an increase in market values could be partly deferred by creating an explicit reserve for depreciation. This reserve would be set based on an assumption about how far equity market values could fall. Pegler suggested an assumed fall of 50 % in equity market values might be a reasonable basis, though he conceded the choice would be subjective. He argued for a similar approach to be applied to fixed interest securities, and suggested a long-term interest rate rise of 0.5 % would be a reasonable assumption for determining the depreciation reserve. The sizes of these two 'stresses' are interesting: both were 'plucked out the air' rather than being the result of any exhaustive statistical study, but both are reasonably consistent with the range of stress sizes applied to equities and long-term interest rates in twenty-first-century insurance solvency assessments.

In his paper, Whyte argued for an asset valuation approach that had a similar effect to Pegler's market value plus depreciation reserve approach. Whyte proposed that assets should be valued at a 'notional value', which was 'defined to reflect its fair value … less an appropriate margin for adverse price fluctuations'.[144] In the context of setting a notional price for equity holdings he

[142] Whyte (1947).
[143] Pegler (1948), pp. 192–93.
[144] Whyte (1947), p. 235.

wrote 'it would be helpful to have in mind a price below which the security in question would be considered as unlikely to fall',[145] though he did not venture any quantitative suggestions. For long-term government bond yields, he, like Pegler, suggested a margin of 0.5 % (Pegler's paper came after Whyte's and Pegler's assumption was influenced by Whyte's).

Both these approaches are notable for advocating a prospective risk-based asset reserve rather than an arbitrary 'hidden reserve' based on historical transaction prices. This concept was fairly revolutionary, and inevitably was resisted by the more traditional thinkers. The depreciation reserve approach also required difficult assumptions to be made about how much depreciation would be reasonable to reserve for. A parallel to the net premium valuation versus bonus reserve valuation debate can be observed here—actuaries were reticent to adopt new methods that required them to make new, difficult and explicit assumptions relative to the more opaque requirements of their established methods.

The application of the depreciation reserve concept to bonds was especially confusing. Actuaries understood that from an asset-liability perspective they were often more exposed to falls rather than rises in interest rates. So it did not make sense to hold a reserve based on how much bond values would fall if rates went up. These were teething issues that would be ironed out in the coming decades. Pegler and Whyte's contributions should be recognised as important groundwork for the later development of solvency capital systems that were based on the assessed riskiness of assets' market values.

The Equitable Distribution of Surplus (1800–1944)

Above, we identified that, historically, a central purpose of the actuarial valuations of assets and liabilities has been to determine the amount of surplus that emerged between two valuation dates, and hence what was available to be distributed to with-profit policyholders as bonus. The measure of the emergence of surplus was not necessarily intended to most accurately capture the economic profit that had emerged over the period. There was a (sometimes implicit) objective to distribute more or less than the amount of profit that had emerged—to attempt to 'smooth' volatile returns over time, or simply because there was a view that some surplus should be retained as a contribution to the estate. At the Equitable, William Morgan had established in 1800 that a maximum limit of two thirds of the surplus that had been assessed to have emerged between valuations could be distributed to policyholders.

[145] Whyte (1947), p. 236.

But this was never a universally accepted standard and a number of actuaries felt that the Equitable had inequitably under-distributed bonus. Charles Jellicoe, the first editor of the *Journal of the Institute of Actuaries*, argued in 1851 that the whole of the earned surplus should be distributed to the policyholders, reasoning 'The practice of reserving a part of it seems unnecessary, for there is not only the excess included in the future premiums, but very generally a subscribed capital to fall back upon, beside the fund reserved for the liability.'[146]

How much of the surplus should be distributed to the current generation of policyholders, and how it should be distributed amongst those policyholders, were decisions primarily driven by a consideration of what was *equitable*. The concept of policyholder equity has long been amorphous and ambiguous, especially for British life offices. What exactly was meant by equity in the context of a pooled system of investment such as with-profits? When unexpected events happened, how much cross-subsidy between the lucky and unlucky policyholders was 'equitable'? As R.H. Storr-Best put it in 1962: 'The problems [with equitable surplus distribution] that have arisen result from the conflict of these two ideas—on the one hand, the averaging of experience and on the other the return to the policyholder of what is his due. The interpretation of equity depends on the emphasis placed on each idea'.[147] Thus policyholder equity was a matter of interpretation, precedent and, ultimately, actuarial judgement.

As well as considering *how much* bonus should be distributed to each policyholder, the *form* the bonus took was another important element in surplus distribution. Chapter 2 discussed how William Morgan, in accordance with Richard Price's advice, pioneered the distribution of surplus through an irreversible addition to the sum assured (instead of by cash dividend). This practice was adopted by the significant majority of British with-profit offices during the nineteenth century, but not universally. Of 59 British life offices writing with-profit business in 1870, 47 were found to be distributing surplus by reversionary bonus, with the remaining twelve doing so by cash dividend or by abatement of future premiums. Around half of the offices distributing by reversionary bonus did so by a *uniform* rate applied to all with-profit policyholders' sums assured. This was clearly the simplest mode of bonus, but it constrained the amount of variation in the relative size of bonus that could be distributed to different policyholders. The remaining half varied the reversionary bonus according to age and/or the duration of the contract.[148]

[146] Jellicoe (1851), p. 26.

[147] R.H. Storr-Best in Discussion, Cox and Storr-Best (1962a), p. 40.

[148] Cox and Storr-Best (1962a), p. 47.

By the third quarter of the nineteenth century, the equitable distribution of surplus had become one of the most active areas of actuarial research. It featured heavily in the early volumes of the *Journal of the Institute of Actuaries*, and was a favoured topic of Charles Jellicoe, its first editor (Jellicoe went on to become President of the Institute between 1860 and 1867). Jellicoe held the view that bonuses should be distributed in close accordance with the profit that had been contributed by the specific policy: 'The surplus then, being satisfactorily ascertained, it appears only consistent with justice and proprietary that it or its value, as exactly as possible, should be returned to the contributors.'[149]

This view was concurrently becoming an established principle in the USA. The US with-profits system was still nascent at this time. Considerable freedom therefore existed to devise a surplus distribution system without the encumbrance of precedence or pre-existing policyholder expectations. Sheppard Homans, the chief actuary of Mutual Life Insurance Company of New York, one of the largest US life offices, played a leading role in the development of the theory and practice of equitable surplus distribution in the USA. Homans would go on to have a distinguished actuarial career, becoming the first President of the Actuarial Society of America in 1889, and, as we saw earlier in this chapter, he published the first mortality table based on US life office experience in 1868. He was invited to present a paper to the Institute of Actuaries on his thinking on surplus distribution. His paper, 'On the Equitable Distribution of Surplus', was published in the *Journal* in 1863.

Homans's paper starts by noting that the experience of the previous five years at the Mutual Life of New York had been very profitable—mortality experience had been around 70 % of the Carlisle table (and the table approximated to the Mutual Life's premium mortality basis), whilst investment returns had been around 6.5 % compared to a premium basis of 4 %. He noted that this size of surplus was unexpected and could not be expected to regularly recur. Homans's approach to determining how surplus should be allocated amongst policyholders started with the principle 'that each participant should be benefited in proportion to the excess of his payments over and above the actual cost of insurance'. Homans did not dwell on this objective, writing that it was 'seemingly recognised by all actuaries'.[150] Whilst that was something of a simplification, his principle was consistent with the views published by Charles Jellicoe over the preceding decade.

---

[149] Jellicoe (1851), p. 26.

[150] Homans (1863), p. 122.

Homans's principle still left plenty of latitude. Any insurance relies on pooling and cross-subsidy, so the calculation of an 'actual cost of insurance' experienced by an individual depends on a definition of what risks are pooled by whom over what period. The bonus allocation system advocated by Homans involved splitting the life fund into separate 'sub-funds' with each sub-fund including a cohort of policies of lives of the same age of policyholder and same policy duration. The cost of mortality insurance was assessed within each of these sub-funds based on the mortality experience of the cohort. A surplus calculation was then made for each based on this assessed cost of insurance, together with the fund's earned rate of interest and the change in the cohort's liability reserve over the assessment period. He did not suggest any smoothing of the mortality insurance cost over time or any cross-subsidy between cohorts: if the experienced mortality rate in a given year of the cohort of 50 year-old policyholders with a policy duration of two years was half of that anticipated in the premium basis, and the experienced mortality rate of 61 year-old mortality rate with policy duration of one year was the same as that anticipated in the premium basis, the first cohort would receive a contribution to dividend from this particular source whereas the second cohort would not.

Homans's contribution method was first implemented by himself at Mutual Life in the 1860s, and was then widely adopted by the US life offices. It did not have a direct impact on bonus practices at the British offices at the time. The reversionary bonus system had never aimed to deliver the granularity of cohort-specific surplus distribution that Homans's system achieved. Over the subsequent decades, the British method of bonus distribution moved further towards the uniform reversionary bonus system, and hence further from a method that could deliver the level of granularity in surplus distribution that was implied by the contribution method (and that was present in the USA). The British actuarial profession, however, continued to wrestle with its own interpretation of equitable distribution of surplus over the following decades. The contribution concept constantly resonated in the profession's collective conscience as an important reference point during these deliberations.

Surplus distribution was an ongoing topic of actuarial thought through the final quarter of the nineteenth century. H.W. Andras, the actuary of the University Life Assurance Society, presented a notable paper to the Institute in 1896 that shone further light on the performance of British bonus methods in the context of the contribution concept.[151] He analysed the uniform reversionary bonus system and its practice in Britain at the time, considering how well it could achieve equitable outcomes for policyholders. Andras concluded that:

[151] Andras (1896).

> This system of bonus distribution ... although having the advantage of simplicity, is of too rigid a nature to be suitable to the varying circumstances of life offices, *if we regard as important the equitable distribution of profit amongst those who have contributed to it* [emphasis added].[152]

Andras's conclusion was partly based on the difficulty of using a uniform bonus rate to distribute profit equitably when the in-force policies had been written at different premium rates. This was one of the reasons why British with-profit rates were subsequently so 'sticky' through the first half of the twentieth century, despite the substantial changes in interest rate environment over that period. However, Andras balanced his conclusion by noting that 'in offices experiencing average results ... it [the uniform compound bonus rate] would give bonuses approximately the same as the contribution method, so that, in such cases, the merit of simplicity may be combined with the principles of equity'.[153]

As ever, the record of the Staple Inn discussion of the paper provides an insight into the state of thinking amongst the leaders of the profession. G.F. Hardy argued:

> [The American contribution method] was a highly scientific plan, but on the whole was not practicably workable – at all events in this country – the main objection being that the whole of the mortality fluctuations at individual ages reappeared in the shape of fluctuations in the bonus ... In the present state of our knowledge it was not known what the true rates of mortality were, and that was a great argument in favour of adopting a simple method like the compound reversionary bonus method, instead of the much more complex contribution plan.[154]

The above comments highlighted the fundamental tensions: the large amount of calculation and complexity the administration of the contribution method entailed was unappealing to British actuaries, but they were naturally anxious that their uniform bonus rate approach would not be so blunt a tool as to be self-evidently inadequate for delivering equitable treatment amongst policyholders. Furthermore, the British system embraced cross-subsidy between current policyholders and across policyholder generations in a way that US with-profits had never done—stability in bonus rates was a fundamental feature of the British with-profit system.

[152] Andras (1896), p. 359.

[153] Andras (1896), p. 359.

[154] In Discussion, Andras (1896), p. 368.

The challenges of ensuring an equitable distribution of surplus further increased for the British system of with-profits in the twentieth century, as with-profits business was increasingly run on steroids. By the second quarter of the twentieth century, 'bonus loadings' in British with-profit premium rates tended to be significantly larger than elsewhere in the world. The bonus loadings implied by the difference between with-profit and without-profit premium rates tripled between 1870 and 1940.[155] And as was discussed in this chapter, with-profit funds' allocations to risky assets such as equities and real estate increased significantly over the second quarter of the twentieth century. Obtaining equity amongst policyholders was further complicated by changing product mixes within with-profit funds (with endowment assurance becoming increasingly popular relative to whole-of-life assurance). These features ensured that the equitable distribution of surplus remained one of the major topics of British actuarial thought during the first half of the twentieth century.

Alongside these changes in with-profit funds' bonus loadings, investment strategy and product mix, with-profit business was also exposed to exceptional economic variability in the first half of the twentieth century, particularly in the aftermath of the First World War. With-profit funds tried to maintain very stable with-profit premium rates through this period, even in the presence of unprecedented gyrations in long-term interest rates. This meant that the 'bonus-earning' ability of policies written at different times could be substantially different. The historical priority given to stable bonus rates over contribution equity became an increasing concern, as was tactfully noted in Lochhead's actuarial student textbook:

> The question has been raised whether the subordination of other considerations to the stabilizing of bonuses is always perfectly fair as between policyholders of different generations.[156]

In his presidential address to the Institute of Actuaries in October 1943, H.E. Melville questioned whether current methods were generating equitable distributions of surplus. This prompted a number of papers on the topic to be presented to the Institute and Faculty in the subsequent few years, the most important of which was T.R. Suttie's 'The Treatment of Appreciation or Depreciation in The Assets of a Life Assurance Fund'.[157]

---

[155] Cox and Storr-Best (1962a), p. 66.

[156] Lochhead (1932), p. 75.

[157] Suttie (1944). See also Anderson (1944).

Suttie's objective was 'to discuss how appreciation or depreciation [in fixed interest securities] should be treated in order to obtain the maximum practical degree of equity between the different generations of policyholders in an office which distributes by means of a uniform reversionary bonus'.[158] His analysis included a historical review of the twentieth-century performance of the 'always write down, never write up' asset valuation practice from the perspective of policyholder equity. He focused on two periods of volatility. First, in the inflationary aftermath of the First World War:

> During the war of 1914–18 and the years immediately following, there was a very substantial rise in the level of interest rates and consequent depreciation of investments. It would appear that in many cases life offices wrote down their assets to their new market values and continued to make their valuations on the same net premium basis as had been used in the past ... this procedure benefited the more recent policies at the expense of the older. Further, if the level of premiums charged for new policies remained unaltered, it benefited policies entering after the depreciation occurred, while if the premiums were reduced on account of the increase in the rates of interest obtainable, new policies were even more favourably treated.[159]

And second, the impact of the Great Depression:

> In 1932 ... there was a reduction in the level of interest rates and consequent appreciation in investments, but it would appear the majority of offices did not write up the assets to their new market values so that the difference between book values and the market values constituted a hidden reserve which was not made available for distribution in bonuses. This again benefited the more recent policies at the expense of the older policies, and greatly benefited the policies entering after the appreciation unless the premiums for new policies were substantially increased.[160]

Suttie particularly focused on the high interest rate scenario. He considered how current bonuses and assumed future bonus rates (in the context of a bonus reserve valuation) should be impacted by a substantial rise in interest rates, given the objective of ensuring equity between different generations of policyholders. He used a model office to examine how different allocations to current and assumed future bonus rates impacted on the different policyholders in his office. He concluded:

---

[158] Suttie (1944), p. 203.
[159] Suttie (1944), p. 214.
[160] Suttie (1944), p. 214.

The most equitable method of dealing with [bond asset] depreciation under a uniform reversionary bonus system is to write down the assets to their new market values and to increase the valuation rate of interest to that obtainable under the new conditions, valuing as a liability a future rate of bonus at the rate which the existing premium scale will support under the new conditions ... If these arguments are accepted, depreciation will involve a decrease in the rate of bonus declared in respect of the valuation period during which the change in value occurs ... while, if conditions again become stable, future bonuses will be at a higher rate than in the past.[161]

But he went on to note that 'Such abrupt changes in the level of bonus rates are contrary to the usual practice of offices and it would undoubtedly be difficult for one office to follow such a policy.'[162]

Suttie's analysis was essentially showing how a contribution approach to bonus distribution would behave through time in an interest rate stress scenario. Perhaps predictably, the discussion following his presentation of the paper cleaved to the more conventional route of setting the current and future bonus rates at equal levels that would result in a reduction in the bonus reserve valuation that absorbed the asset depreciation. In essence, his peers advocated spreading the recognition of the market value impact of the interest rate increase over the full run-off of the existing business. Concerns were voiced over the variability of bonus that would be implied by Suttie's proposed approach:

He did not know what would be said by the holder of a policy nearing maturity if, just before maturity, there were a fall in prices and, as a consequence, a reduction of the sum payable at maturity. He felt that such a type of distribution would be bound to cause dissatisfaction.[163]

Stability in bonus rates still trumped all else in British actuarial thinking on surplus distribution. Unlike in the United States, cross-generational smoothing had become a fundamental feature of with-profits. This cross-subsidy could apply not only between different generations of existing policyholders, but also between policyholders and the estate, as was expressed at the Staple Inn meeting:

---

[161] Suttie (1944), p. 213.

[162] Suttie (1944), p. 213.

[163] M.E. Ogborn, in Discussion, Suttie (1944), p. 220.

> With most offices, part of the profits of the past had been used to create contingency reserves which should be drawn upon to deal with the depreciation of the assets ... the aim should be to *smooth out the effect of the depreciation* by declaring a current bonus at a reasonable level and by rebuilding the contingency reserves gradually in the future [emphasis added].[164]

The British actuarial profession's interpretation of equitable bonus policy remained much broader, and its implementation more heuristic, than the contribution method that dominated bonus methods in the USA. This was partly a function of legacy: the British life offices, arguably and perhaps ironically due to a lack of policyholder equity in the past, had accumulated significant free reserves in their funds. These estates permitted the actuary a considerable latitude in bonus policy. The inevitable arbitrariness entailed in deciding what to do with the estate provided another reason not to be overly meticulous about how to distribute the surplus generated by existing policyholders. The British approach relied heavily on the judgement and expertise of the actuarial profession, and the public's trust in it. In that sense it was perhaps an approach that reflected the time and place of Britain in the first half of the twentieth century—a class-based society that placed unquestioning trust and deference in its institutions and professions. Times would change.

[164] L. Brown, in Discussion, Suttie (1944), p. 218.

4

A Brief History of Financial Economics for Actuaries

This chapter gives an historical overview of the emergence of some of the key ideas in financial economics (the branch of economics concerned with fields such as securities pricing, portfolio theory, corporate financial and investment theory and the behaviour of financial markets). Financial economics was not developed by actuaries. You may well ask why it is included in a history of actuarial thought. The answer is that, rather like probability and statistics, once its foundations had been developed, its ideas had important practical application in the world that actuaries occupied. As these ideas emerged, the actuarial profession had to determine how to incorporate the new insights provided by financial economics into its thinking and practices (and, indeed, where theory did not translate into practice). This was not an easy process, as these insights were often incongruent with the traditional actuarial perspective. It triggered great actuarial debate: was this incongruence merely a result of the theoretical and unworldly nature of the economic studies, or did it signal that some key areas of actuarial thought needed fundamental revision? This process is arguably still underway. It is one of the more interesting and fundamental aspects of the historical development of actuarial thought in the second half of the twentieth century. To really appreciate it, we need to understand something of the ideas of financial economics and how they originally developed, and this is the object of this part of the book.

Our historical survey of financial economics focuses mainly on the remarkable quarter of a century of activity bookended by the seminal papers of Harry Markowitz in 1952 and Oldrich Vasicek in 1978. This period saw the development of a handful of interrelated 'big ideas' in financial economics that

C. Turnbull, *A History of British Actuarial Thought*,
DOI 10.1007/978-3-319-33183-6_4

radically altered theoretical understanding and business practices in areas such as investment portfolio management, corporate finance, asset pricing and risk management generally and derivatives pricing and hedging in particular. We also note the important empirical research of the 1980s that challenged the 'classical' financial economic theories of the earlier decades and pointed to a richer, more complex economic reality.

The historical summary below is not intended to be an economics textbook. Rather it is meant to provide an overview of how and when the essence of the key ideas in financial economics emerged (particularly the ones that were potentially of practical relevance to actuarial thought). The discussion of how actuaries did and did not integrate these ideas into their thinking arises more fully in the later parts of the book, which focus on post-war British actuarial thought in life, pensions and general insurance.

Modigliani and Miller's Irrelevancy Propositions

Much of early financial economics considered what possible causes could and could not have an effect on the rational behaviours of individuals and firms in well-functioning markets. A number of 'irrelevance' and 'separation' theories followed where things that were intuitively believed to be important in matters such as corporate financial structure, corporate investment policy or individuals' portfolio composition were shown to be theoretically irrelevant to all or specific parts of a rational decision-making process. An early and important example of such a theory was provided by the Yale economics professor Irving Fisher in his seminal 1930 book *The Theory of Interest*.[1] Fisher substantially pre-dates the recognised classic period of financial economics that started in the 1950s—*The Theory of Interest* arguably provides a bridge from macroeconomics to the fledgling discipline of financial economics (that is, to the formal and rigorous economic treatment of the fields of investment and finance).

For financial economists, Fisher's book is best remembered for what has become known as Fisher's Separation Theorem. This theory states that a firm's value is maximised when it chooses to invest in projects that generate cash-flow streams with positive expected present values when discounted using market discount rates. This may appear as an unremarkable statement—but the key point was that the choice of discount rate should be determined only by the *market's* required return and should be independent of the particular risk and timing preferences of the firm's specific set of shareholders. These preferences are irrelevant to the shareholder's own wealth because, in

[1] Fisher (1930).

well-functioning markets, shareholders would be able to take actions (buying stock, selling stock, borrowing, lending) that would deliver to them whatever risk and timing preferences they wanted, *whilst preserving the value created by the firm's investment decisions.* Hence, all shareholders' wealth would be maximised when the firm chose to invest to maximise the present value at market discount rates, even if it resulted in a risk or income timing profile that was not desired by some, or even all, of the firm's actual shareholders. The firm's optimal choice of investments was therefore *separate* from the risk and income-timing preferences of its shareholders.

Fisher was not the only US economist of the 1930s who was focused on corporate investment theory and the valuation of the firm. In 1938, John Burr Williams published his Harvard doctoral thesis, *The Theory of Investment Value.*[2] According to Mark Rubinstein, the eminent late twentieth century financial economist and occasional historian, Williams's book 'contains what is probably the first exposition of the Modigliani–Miller (1958) proposition on the irrelevancy of capital structure'.[3] This principle is set out by Williams in his Law of Conservation of Investment Value:

> If the investment value of an enterprise as a whole is by definition the present worth of all its future distributions to security holders, whether on interest or dividend account, then this value in no [way] depends on what the company's capitalization [financial structure] is.[4]

Thus Williams's Law of Conservation of Investment Value said that the total value of the firm was determined by the cashflows its assets generate, and this was not affected by how those cashflows were split up and packaged into various forms of securities such as equities and debt. Williams never set out formally to prove his law: his justification was based on rhetorical persuasion rather than a rigorous demonstration deduced from a set of stated axioms. He reasoned that, 'Clearly if a single individual or single institutional investor owned all the bonds, stocks and warrants issued by a corporation, it would not matter to this investor what the company's capitalization was'.[5] Some 20 years later, Modigliani–Miller's famous paper was published in which Williams's argument was developed using a more formal economics.[6] Indeed, the Modigliani–Miller paper was arguably as important to financial

---

[2] Williams (1938).
[3] Rubinstein (2006), p. 79.
[4] Williams (1938), p. 72.
[5] Williams (1938), p. 72.
[6] Modigliani–Miller (1958).

economics for how it influenced the way in which its theories would be developed as for its specific content. In particular, Modigliani–Miller pioneered the formal use of no-arbitrage as a way of proving a financial theory. That is, the proof was based on the demonstration that if the theory did not hold, an arbitrage would exist such that some investors would be able to make limitless riskless profits at the expense of others: if it was an assumption of well-functioning markets that such opportunities do not exist, then the theory must hold in such a market. Virtually every notable theory of financial economics published ever since has used a no-arbitrage argument in its formulation. Modigliani–Miller's version of Williams's Law was given in their famous Proposition I:

> The market value of any firm is independent of its capital structure ... the average cost of capital to any firm is completely independent of its capital structure and is equal to the capitalization rate of a pure [unleveraged] equity stream of its class.[7]

Modigliani–Miller's no-arbitrage argument was built on the assumption that a group of firms in a given 'class' had perfectly correlated underlying assets. From this assumption, it was possible to show that different segments of the capital structure of the different firms in the class could be bought or sold to create an arbitrage if Proposition I did not hold. This is, of course, a somewhat artificial and theoretical arbitrage argument—in reality there are not many groups of firms with perfectly correlated assets—but it provided their argument with a new economic rigour. Like Williams's Law of Conservation of Investment Value, Proposition I said that the total size of the pie (the value of the firm) does not depend, in a well-functioning capital market, on how you split up the pie (between shareholders and bondholders and any other form of security holders). In essence, if we assume that the firm's asset cashflow is independent of its capital structure choice, then this is really only saying that present values are additive. For example, in a simple single-timestep model where the firm's (uncertain) asset cashflow X, is paid at the end of the period, and a debt of amount D is due to bond holders at this time, the respective pay-outs are:

$$\text{Shareholder pay-out} = \text{Max}\left(0, X - D\right)$$

$$\text{Bondholder pay-out} = \text{Min}\left(X, D\right)$$

$$\text{Total pay-out} = \text{Max}\left(0, X - D\right) + \text{Min}\left(X, D\right) = X$$

[7] Modigliani–Miller (1958), pp. 268–69.

The sum of the pay-outs is always X and the present value of the sum of the pay-outs is not a function of the choice of value for D. Proposition I also said that the choice of capital structure had no impact on the firm's total cost of capital (and hence on corporate investment choices): the cost of capital was solely a function of the riskiness of the investments that the firm made (the asset side of the corporate balance sheet), and this was not altered by the firm's capital structure (the liability side of the corporate balance sheet). In a similar vein to Fisher's argument, Modigliani–Miller said firms create value through the investments they make; these investments should be discounted at market discount rates commensurate with the risk of the investments; the firm's capital structure and the individual preferences of their capital owners were both irrelevant to these choices and values. The behaviour of the firm's cost of capital as a function of the level of debt in the capital structure according to Proposition I can be summarised by Fig. 4.1 below.

The cost of capital is constant across the choice of capital structure, and is therefore always equal to the cost of capital of a pure equity (unleveraged, zero debt) firm. As debt is introduced into the capital structure, three things happen that have a collectively offsetting impact on the firm's weighted-average cost of capital: the value of debt as a proportion of total assets increases; the expected

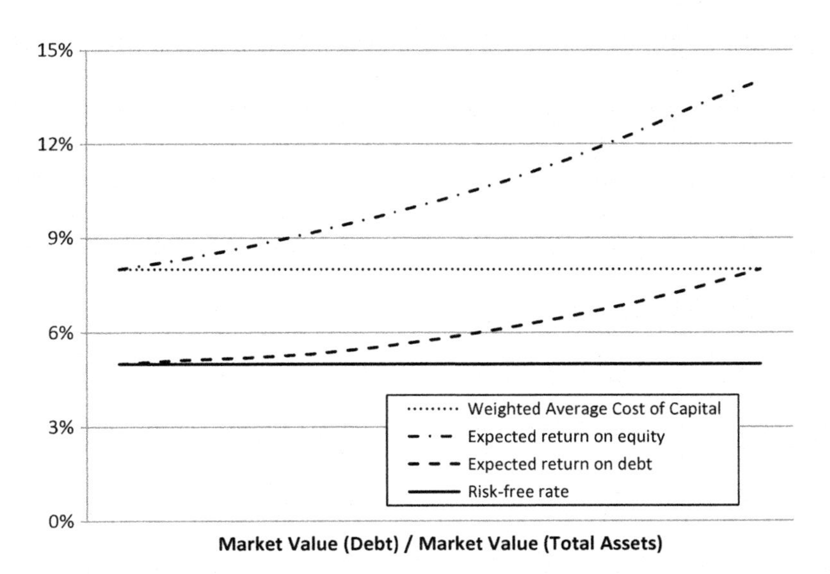

Fig. 4.1 Cost of capital and capital structure example (Modigliani–Miller proposition I)

return on debt increases (starting from the risk-free rate); and the expected return on equities increases (starting from the cost of capital of an unleveraged firm). In the limit, very high debt levels make default increasingly more likely and debt accordingly behaves more and more like (unleveraged) equity.

Modigliani–Miller did not say anything about *how* to value the respective shareholder and bondholder claims, or how the expected return on equities and debt specifically behaved. Their valuation statements were only relative: the sum of their pay-outs was not a function of the choice of D and the firm's cost of capital was not a function of D. Absolute statements about the valuation of equities and debt would require a specification for the stochastic process driving the underlying assets of the firm (and a theory of valuation of contingent claims that did not yet exist). Modigliani–Miller showed that these relative properties must hold true for any stochastic process that drives the asset values of the firm, and this had major ramifications for how management can and cannot create value for the firm's capital providers (in perfect market conditions).

Whilst their theory could be derived from a relatively simple arbitrage argument and could be shown to ultimately be a simple expression of the additivity of present values, it still ran counter to perceived wisdom—which was that firms should prefer to use debt rather than equity until required bond returns became prohibitive. Modigliani–Miller recognised that their assumption of perfect markets was a 'drastic simplification', and that real-life complexities may have an impact on optimal financial structure. Their paper prompted decades of further academic research into how market imperfections such as taxation,[8] bankruptcy costs,[9] management agency costs[10] and information-signalling[11] could make capital structure relevant to firm value. But their theory showed what did *not* matter: whilst expected equity returns would be enhanced through debt financing, this only systematically increased the riskiness of the equity return. Raising debt merely geared up equity returns in a way that individual shareholders could choose to do (or undo) with their own borrowing and lending (rather like in Fisher's Separation Theorem).

Modigliani and Miller published a second jointly-authored paper[12] in 1961 that provided the equivalent irrelevancy theorem for corporate dividend policy: in perfect capital markets, changes in dividend policy could not

---

[8] Miller (1977).

[9] Stiglitz (1972).

[10] Jensen and Meckling (1976).

[11] Ross (1977).

[12] Miller and Modigliani (1961).

have an effect on the overall return earned on a firm's equity, all other things being equal. Again their argument was that all value was determined on the asset side of the corporate balance sheet—whether firms chose to finance that investment through retained earnings or new equity issuance or other forms of capital raising was irrelevant to the economics of the value. Again they recognised that the theory was an abstraction—for example, in real life, dividend decisions were likely to have important informational content for shareholders —but they once again highlighted that where these decisions did matter, it was because of market imperfections or limitations rather than the fundamental economics. Miller-Modigliani were aware that what they were saying was beguilingly simple:

> Like many other propositions in economics, the irrelevance of dividend policy is "obvious once you think about it". It is, after all, merely one more instance of the general principle that there are no "financial illusions" in a rational and perfect economic environment. Values there are determined solely by "real" considerations – in this case the earning power of the firm's assets and its investment policy – and not by how the fruits of the earning power are "packaged" for distribution.[13]

And yet, it is not difficult to find examples of twentieth-century economists whose thinking on this topic was woolly at best. Recall Edgar Lawrence Smith's argument that US corporations' practice of retaining and reinvesting a significant portion of their earnings was a vital factor in the ability of equities to so strongly out-perform corporate bonds. Smith was not a substantial economist, but his views were endorsed by John Maynard Keynes, arguably the most influential economist of the century. In reviewing Smith's book, Keynes wrote of Smith's corporate-dividend-policy-as-explanation-of-equity-outperformance argument: 'I have kept until last what is perhaps Mr Smith's most important, and certainly his most novel, point. Well-managed industrial companies do not, as a rule, distribute to the shareholders the whole of their earned profits.'[14]

Miller and Modigliani brought a clarity to what matters—or rather to what should not matter—for capital structure, dividend policy and corporate investment policy. They did so by introducing a new formal rigour to the embryonic discipline of financial economics, profoundly influencing its subsequent practice. Franco Modigliani and Merton Miller received the Nobel Prize in Economic Sciences in 1985 and 1990 respectively, both in part for their two jointly written papers.

---

[13] Miller and Modigliani (1961), p. 414.

[14] Keynes (1983), p. 250.

Modern Portfolio Theory and the Capital Asset Pricing Model

1952 was a vintage year for quantitative finance. Chapter 3 has already identified one seminal moment: the publication of Frank Redington's paper on immunisation theory. A seminal concept in financial economics was introduced in this year by Harry Markowitz in his paper, *Portfolio Selection.*[15] Like Merton Miller, Markowitz spent much of his academic career at the University of Chicago. This paper is the starting point for the branch of financial economics known as modern portfolio theory. It also provided the foundation for major developments in asset pricing theory. It was the first important example of a quantitative, explicitly probabilistic analysis of the impact of financial market risk on rational investor behaviour.

Markowitz considered how rational investors would construct portfolios of risky investments. His analysis was built on two key assumptions: investors were risk-averse; and they cared only about the first two moments of the probability distribution of future investment returns (or that all probability distributions of security returns could be described fully by their first two moments). He also made the more basic assumption that investors were non-satiated (they would always prefer more wealth to less). These assumptions implied that for a given level of expected return on their asset portfolio, an investor would prefer a portfolio return distribution that minimised the standard deviation (or variance) of the return; or equivalently, for a given variance of return, they would prefer the portfolio return distribution that maximised the expected return. Portfolios that produced the minimum variance for a given level of expected return were termed *efficient portfolios* by Markowitz.

Markowitz used some elementary mathematical statistics to show that, whilst the expected portfolio return was simply a value-weighted average of the underlying expected returns of securities in the portfolio, the variance of the portfolio return was a more complicated expression that was dominated by the *covariances* between the securities. Markowitz showed that, given these expressions for portfolio expected return and portfolio variance of return, portfolio weights in the universe of available securities could be found that would generate *efficient portfolios.* Moreover, an *efficient frontier* of portfolios could be determined where each point on the frontier represented the portfolio with the lowest variance for a given expected return.

[15] Markowitz (1952).

The identification of these portfolio weights was a complex computational problem when the investment universe consisted of a large number of securities. But Markowitz was able to generate illustrative explicit solutions for the three-security case, and to make more general observations on the behaviour of the possible portfolio combinations. In particular, under the assumptions that there would be no short-selling of securities or gearing in the portfolio, and that every security's return was risky (i.e. had non-zero variance), he was able to develop the diagram in Fig. 4.2 below.

Figure 4.2 highlighted the power of the diversification of risk—far better risk/return combinations could be obtained by constructing efficient portfolios than holding individual securities. But Markowitz also showed that diversification was about more than merely holding many different securities—it was driven by the covariances amongst securities, not merely by the number of securities in the portfolio. And whilst imperfect correlation between securities was the mathematical key to diversification, the positive correlations that tended to be empirically observed meant that there was a limit to how much diversification could be generated:

> This presumption that the law of large numbers applies to a portfolio of securities cannot be accepted. The returns from securities are too intercorrelated. Diversification cannot eliminate all variance.[16]

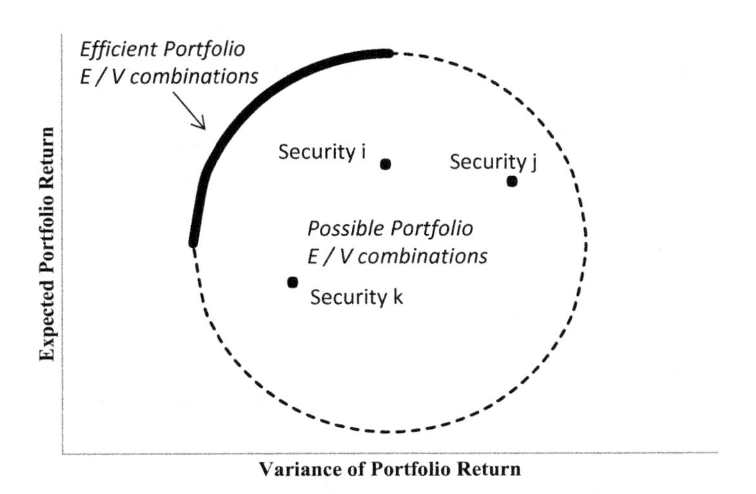

Fig. 4.2 Portfolio risk and return (Based on Markowitz (1952) Fig. 1))

[16] Markowitz (1952), p. 79.

Markowitz's approach echoed the arguments made by the British actuary George May 40 years earlier. In 1912, May urged actuaries and investment managers to give greater consideration to the benefits of diversification and hence to 'spread these investments over as large an area as possible'.[17] He emphasised that the riskiness of a security should not be measured by 'the safety of the capital in each investment by itself'.[18] He never used the words, but his argument that the 'stand-alone' risk of the security was not the relevant risk measure suggests the security's risk should be considered by its marginal contribution to the portfolio's riskiness, which is exactly what Markowitz's analysis implied.

Of course, May did not develop the mathematical analytical framework of Markowitz that would prove so durable as an investor tool, and so fundamental as a building block of asset pricing theory. For this contribution Markowitz received the 1990 Nobel Prize in Economic Sciences.

The next step in the development of modern portfolio theory was made by James Tobin, a leading US macroeconomist (this time from Yale University). In a 1958 paper he introduced what became known as the Tobin Separation Theorem.[19] Tobin extended Markowitz's analysis by considering how the presence of a risk-free (zero volatility) security impacted on the efficient frontier of portfolio choices. The risk-free security has zero variance and therefore naturally has zero covariance with all other securities. This seemingly trivial observation has profound implications for the efficient frontier: it results in a new, improved, linear efficient frontier and it implies that all efficient portfolios are a combination of the risk-free security and the risky efficient portfolio that is found at the tangent between the risky efficient frontier and the risk-free return. This linear efficient frontier can be extrapolated to the right of the risky asset portfolio by assuming negative positions in the risk-free security can be obtained (that is, the investor can borrow at the risk-free rate). Figure 4.3 below updates the previous diagram to reflect the impact of the presence of the risk-free asset.

The striking implication of Tobin's analysis was that all investors should hold the same sub-portfolio of risky assets, *irrespective of their risk preferences and wealth*. An investor would obtain his preferred level of portfolio risk by choosing the mix of this sub-portfolio of risky assets (labelled P in Fig. 4.3) and the risk-free security that produced their preferred level of portfolio volatility. Thus the choice of risky asset sub-portfolio is *separated* from the investor's risk preferences. This is the Tobin Separation Theorem. In 1981, Tobin also collected the Nobel Prize in Economics.

[17] May (1912), p. 153.
[18] May (1912), p. 136.
[19] Tobin (1958).

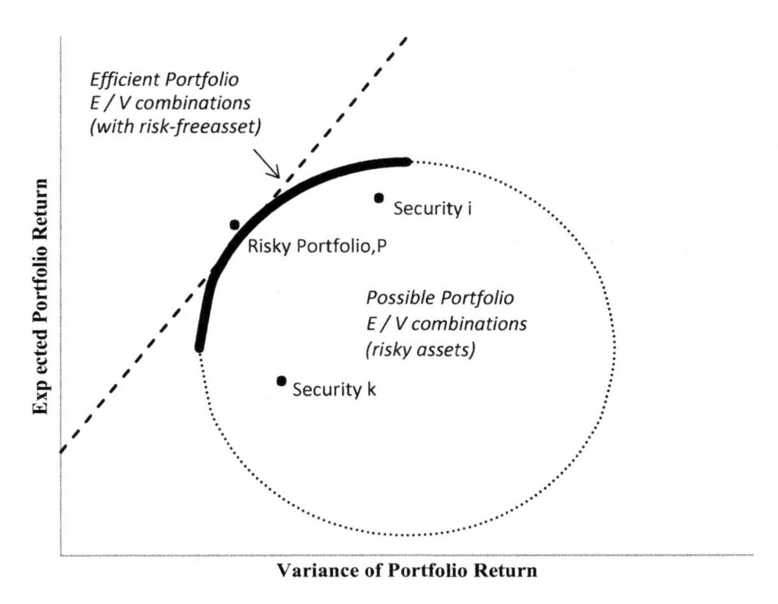

Fig. 4.3 Portfolio risk and return (with risk-free security)

So far, the portfolio theory story has been focused on how rational investors should behave in constructing investment portfolios. The story then moved onto a distinct new phase: financial economists next considered what this behaviour would imply for the *pricing* of risky assets. If investors followed the behaviour described by Markowitz and Tobin, what type of risk would impact on the price of an asset? In what way? At this time, there was no economic theory of how risk impacted on equilibrium asset prices. A great deal of economic theory explained how the risk-free interest rate was determined. And there was a general recognition that risky assets should offer a risk premium. But no one at this time (circa 1960) had an economic theory for how the risk premium on a particular security should be determined.

Several financial economists in the USA (in particular, Sharpe, Lintner, Mossin and Treynor) took the portfolio risk work of Markowitz and Tobin as a platform and built, independently and simultaneously, theories of risky asset pricing. Somewhat remarkably, they each produced more or less the same theory at more or less the same time, and this theory became known as the Capital Asset Pricing Model (CAPM). Rubinstein's *History* summarises this state of affairs:

> All four economists adopted nearly the same set of assumptions (mean-variance preferences, perfect and competitive markets, existence of a riskless security, and homogeneous expectations) and reached nearly the same two key conclusions:

(1) all investors, irrespective of differences in preferences and wealth, divide their wealth between the same two portfolios: cash and the market portfolio, and (2) equivalent versions of the CAPM pricing equation.[20]

A first glimpse of what was to come was offered by William Sharpe in his 1963 paper, 'A Simplified Model for Portfolio Analysis'.[21] This paper was still focused solely on portfolio theory rather than asset pricing theory and it addressed the practical application of Markowitz's portfolio construction optimisation—as Markowitz himself had noted, this would be a computationally intensive process when the universe of risky securities was large in number. Sharpe proposed restricting the covariance relationships between risky securities such that they could be summarised by a single common statistical factor plus a single independent source of risk per security. This 'Diagonal Model' was much easier to work with in terms of computing the composition of efficient portfolios. Sharpe also undertook some empirical analysis of 20 securities traded in the New York Stock Exchange between 1940 and 1951 and concluded that 'the diagonal model may be able to represent the relationship among [the realised returns of] securities rather well'.[22]

Sharpe's next paper[23] moved decisively into the territory of asset pricing theory: rather than taking asset prices as exogenous inputs as per portfolio theory, he considered how asset prices would behave if all investors rationally pursued the Markowitz/Tobin portfolio construction methods. Lintner's paper[24] was published the following year, and had already been written when Sharpe's paper was published, but as the first to publication, Sharpe generally receives the greatest credit for the CAPM's derivation (Sharpe received the Nobel Prize in Economics in 1990 for CAPM). At a high level, the fundamental insight that drove their asset pricing work was that if all investors held the risky asset portfolio P as implied by Markowitz and Tobin, then P must be the market value-weighted portfolio of all risky assets (generally referred to as the market portfolio). Thus risky asset prices must be determined by a risk pricing process that resulted in this occurring. This observation, together with the observation that the market portfolio is mean-variance efficient and is on the tangent with the linear efficient frontier, facilitated the geometric derivation of the famous CAPM pricing equation:

[20] Rubinstein (2006), p. 172.

[21] Sharpe (1963).

[22] Sharpe (1963), p. 292.

[23] Sharpe (1964).

[24] Lintner (1965).

$$r_i = r_f + \beta_i \left(r_m - r_f \right)$$

where:

r_i is the expected return on security *i*;
r_f is the risk-free rate of return;
r_m is the expected return on the market value-weighted portfolio of risky assets;
and $\beta_i = \left(\sigma_i / \sigma_m \right) \rho_{im}$

and where σ_I is the volatility of security i; σ_m is the volatility of the market portfolio return; and ρ_{im} is the correlation between the returns of security *i* and the market portfolio.

CAPM delivered a rigorous economic theory of how financial market prices would be impacted by risk. It was also a highly intuitive result. The CAPM stated that the expected return on risky assets would only be impacted by the systematic, non-diversifiable component of its volatility (as measured by its beta). Diversifiable risk would not be rewarded. This was an anticipated result: it was well-understood by this time that diversifiable risk was, by definition, risk whose impact could be easily mitigated by investors, and hence should not command a risk premium. Further, CAPM stated that the expected return would increase linearly in proportion to the security's risk (as measured by its beta). The linear risk premium was also tentatively anticipated by economists, but prior to CAPM they were unable to show why, or indeed in what component of risk the risk premium would be linearly dependent.

The CAPM implies that the realised return on a given security can be decomposed into two statistically independent components: an expected component of the return (which is conditional on the realised market portfolio return) and an unexpected component. Sharpe's Diagonal Model of joint realised security returns is consistent with this structure. Recall that in this model all stocks have an exposure to a common factor and an independent idiosyncratic risk (a correlation structure that was used to facilitate computationally efficient portfolio optimisations). However, the Diagonal Model is not the only correlation structure for realised returns that is consistent with CAPM. Empirically, significant correlations between the unexpected components of the realised returns of different securities tend to be observed in returns data (for example, across securities in the same industrial sector). This is not inconsistent with CAPM. CAPM does not restrict the realised return correlation structure such that the unexpected return components (relative to the return component explained by the market portfolio) are uncorrelated across all securities. It simply says that such correlations do not matter to the pricing of the securities. In CAPM, the only correlation that matters to the asset price is the correlation of the security with the market portfolio (which is reflected in the security's beta). That does not imply it need be the only significant correlation that can be found between realised security returns.

More general asset pricing models were developed following CAPM, such as Stephen Ross's Arbitrage Pricing Theory (APT).[25] This model allows for the possibility of a broader, indeed arbitrary set of statistical factors to drive realised security returns and to be potentially relevant to the pricing of assets. However, it does not provide a specification of how to determine what risk premiums should be commanded by such factors in equilibrium asset prices—it only states what the *relative* behaviour of different asset prices must be to avoid arbitrage for a given set of risk premium assumptions for the factors. CAPM can be considered as a special one-factor case of the APT, and one in which the risk premium structure is fully specified.

These key ideas of portfolio theory and asset pricing theory are now long-established cornerstones of investment education and practice: they are found in the syllabi of business schools and finance professions (including actuarial ones) around the world; an asset's beta is a universally recognised measure of risk (and required return) amongst investment professionals.

Option Pricing Theory

The publication of the option pricing work of Black, Scholes and Merton in 1973 ranks amongst the most famous and influential developments of financial economics. Their work threaded together three concepts that had individually emerged earlier in the twentieth century: Brownian motion as a description of the behaviour of security prices; valuation by the law of no-arbitrage; and the use of risk-neutral probabilities in valuation.

The earliest known example of the modelling of financial assets as a type of Brownian motion was produced by the French mathematician Louis Bachelier in his doctoral thesis, which was accepted at the Sorbonne in 1900.[26] In fact, the stochastic modelling of Bachelier's thesis is now generally recognised as the first mathematical modelling of Brownian motion *in any context*, beating

[25] Ross (1976).

[26] Bachelier (1900). For a usefully annotated English translation and accompanying historical discussion, see Davis and Etheridge (2006).

Albert Einstein's famous particle motion paper by a handful of years (Einstein was unaware of Bachelier's work).

Bachelier's motivation for developing a stochastic model of security prices was not merely to obtain probability distributions of future prices of fundamental assets like equities or bonds or their path behaviour. Such developments would in themselves be revolutionary for the time, but to Bachelier the stochastic description of assets was merely a means to an end: his thesis was not on how to model equities or bonds, but on how to *value options* (specifically options on French government perpetuities). He had the conceptual insight that the valuation of such options would require a continuous-time stochastic process to be specified for the path of the underlying asset. This is an astute insight given that no such form of mathematical model existed at the time. Bachelier wasn't merely identifying new applications for existing mathematical frameworks, he was identifying the need for a mathematical apparatus that did not exist. So Bachelier proceeded to develop it and attempted to apply it to the problem of option pricing. As well as introducing the mathematical modelling of Brownian motion, he also developed the Chapman-Kolmogorov equation and made other important developments that showed how the mathematics of physics could be transferred to the emerging field of stochastic analysis. His work was not mathematically rigorous, but it was recognised by the leading stochastic theoreticians of the following half-century such as Kolmogorov as they developed the formal mathematics of probability theory.

Despite being well-known to probability theorists (and being cited in the most important stochastic analysis papers of the century), Bachelier's work, and in particular, his attempts at option valuation, went completely unnoticed by the economics profession for over 50 years. This changed sometime in the mid to late 1950s, when a leading American statistician of the time, Jim Savage, sent a note to a few leading economists asking if they had read Bachelier. Paul Samuelson, the MIT professor who can be viewed as the leader of the new breed of quantitative financial economists that emerged in the third quarter of the century, was one of the recipients of the note. Samuelson had not heard of Bachelier or his work, but he browsed the MIT library and found a copy of his 1900 thesis. Inspired by what he read, he circulated Bachelier's work amongst his peers and students. Bachelier had not solved option pricing, but he was able to make some accurate statements about option pricing behaviour—for example, showing that an at-the-money option price would increase in proportion to the square root of its term to maturity (this is not strictly true under more realistic stochastic processes for the underlying asset than the arithmetic Brownian motion assumed by Bachelier, but is nonetheless usually a good approximation). He was also able to derive many probability statements for the behaviour of the underlying asset such as the probability that it would reach a particular price over a given time period (again under the assumption that it followed an arithmetic Brownian motion).

Bachelier had done enough to show Samuelson that his application of continuous-time stochastic processes to derivative pricing opened up many possibilities for further research. And the work that had been done by stochastic mathematicians in the intervening fifty years to develop Bachelier's modelling intuitions into rigorous mathematics provided a ready-made modelling toolbox for a new generation of financial economists.

The next notable contribution to follow Bachelier on the application of Brownian motion as a description of the stochastic path behaviour of asset prices took a full 59 years to appear. M.F.M. Osborne, a researcher at the US Naval Research Laboratory in Washington DC, published a paper, 'Brownian Motion in the Stock Market',[27] which broke new ground in several ways. Osborne proposed that stock prices followed a geometric Brownian motion—that is, it is the natural logarithm of the asset price that is normally distributed, rather than the asset price itself. Osborne does not appear to have been aware of Bachelier, but the use of a geometric rather than arithmetic process was fairly intuitive: the possibility of negative prices was removed; and the process could be viewed as a multi-period generalisation of the single-period mean/variance description of security returns introduced by Markowitz a few years earlier. From this point on, geometric Brownian motion would be the standard modelling assumption for the stochastic process of an asset price. Secondly, Osborne performed some statistical tests of how well empirical asset price behaviour (US stock prices from various sources going as far back as 1831) conformed to the geometric Brownian motion model. Osborne tested for this in two ways: by considering the cross-sectional dispersion of many stock price changes on a given date; and by analysing how the variance of stock prices increased with the length of the measurement interval (up to ten years). Rubenstein has noted that Osborne was the first researcher to perform this second type of test on asset price data.[28] Osborne concluded that the empirical data accorded reasonably well with the geometric Brownian motion model (though he noted that the empirical data produced tails that were fatter than those predicted by the model).

In 1965, Paul Samuelson, now armed with Bachelier's option pricing thesis and Osborne's geometric Brownian motion proposal and supporting empirical stock price analysis, synthesised these ideas in a paper that examined option

[27] Osborne (1959).

[28] Rubinstein (2006), p. 135.

pricing under the assumption that the underlying asset follows a geometric Brownian motion.[29] Samuelson, with the assistance of the world-leading stochastic processes professor Henry McKean, who was also at MIT, was able to derive a call option pricing formula under the assumption that the option had some constant required return β and the underlying stock had a constant expected return of α (and followed a geometric Brownian motion with volatility σ). The theory of stochastic analysis was now sufficiently developed to have permitted a closed-form solution to be found for the discounted present value of the option pay-off. But, critically, Samuelson had no theory for how to determine the option discount rate β. He recognised that the call option was a 'leveraged' exposure to the underlying asset in the sense that a given change in the stock price would have a bigger proportional impact on the call price than the stock price. In the CAPM terminology, the call option's beta was bigger than that of the underlying asset. So he argued it must be the case that β was greater than α. But the size of the option's leverage (and hence beta) also evidently varied with the underlying asset price—it was a stochastic process itself. Samuelson had no strategy for allowing for that. He was close. He had perfected the application of stochastic processes to determine the option's expected pay-off, but he had no way of transforming it into an economically justified present value.

1973 saw the publication of two papers—the Black–Scholes paper[30] and a paper by Merton[31]—that used a crucial economic insight to determine how Samuelson's formula could be parameterised to deliver an arbitrage-free option pricing formula for the first time. Collectively, the publication of these papers represented a watershed moment: the era of modern option pricing theory starts here. Scholes and Merton were awarded the Nobel Prize in Economics in 1997, primarily for the contributions in these papers (Fisher Black died in 1995 and so could not receive the award).

The Black–Scholes paper arrived a few months before Merton's. Although the papers were written contemporaneously and cover similar ground, the Black–Scholes one is the natural starting point for the reader as it is somewhat more direct and less elaborate (at least for the non-mathematician). Merton is more general and more formal. The critical breakthrough in Black and Scholes's treatment of the option pricing problem came with the recognition that changes in the option value were entirely driven by changes in the stock price—over an instantaneous period, the option and the underlying

[29] Samuelson (1965b).

[30] Black and Scholes (1973).

[31] Merton (1973).

asset were therefore perfectly correlated assets. This implied that a continuously rebalanced portfolio of the option and the underlying asset could be constructed that would generate a *certain* rate of return. That is, a portfolio could be constructed that included one unit of the underlying asset and an appropriate short position in the call option that provided an equal and offsetting exposure to movements in the underlying asset. The short position would be determined by the rate of change of the call option value with respect to the underlying asset. In a footnote in their paper, Black and Scholes attribute this critical idea of the riskless dynamic portfolio to Robert Merton.

Mathematically, the portfolio would generate an instantaneous certain return if it was of the form:

$$\textit{Hedge Portfolio} = S - \frac{1}{\frac{\partial C}{\partial S}} C(S,t)$$

The certain return earned by the hedge portfolio must be the risk-free rate in order to avoid arbitrage—if it was greater (lower) than the risk-free rate, an investor could borrow (invest) at the risk-free rate and create unlimited profits by investing in (shorting) the hedge portfolio. Black and Scholes assumed the stock price followed a geometric Brownian motion. As the call option price is a function of the stock price, stochastic calculus (in particular, Ito's lemma) can be used to express a change in the call option price as a function of the derivatives of its value with respect to the underlying asset and time:

$$dC(S,t) = \left(\frac{\delta C}{\delta t} + \mu S \frac{\delta C}{\delta S} + \frac{\sigma^2 S^2}{2} \frac{\delta^2 C}{\delta S^2} \right) dt + \sigma s \frac{\delta C}{\delta S} dZ(t)$$

where μ and σ are the drift and variance terms of the underlying assets' geometric Brownian motion and dZ(t) is the underlying assets' Brownian motion or Weiner process.

From this and the assumption that the hedge portfolio must earn the risk-free rate, together with the elementary boundary conditions of the option pay-out at maturity, a partial differential equation can be derived which is familiarly known as the heat-transfer equation of physics (or simply the heat equation), and for which there was a well-established pre-existing solution.

Just like that, the arbitrage-free option pricing equation was obtained (under various technical assumptions such as investors' having the ability to continuously rebalance asset positions without transaction costs). The observation that dynamic hedging could replicate the option's pay-offs (or equivalently, generate a risk-free portfolio) was pivotal and had transformed the option pricing challenge into a strictly *relative* problem: it did not require a grand, unified explicit model of how the economy priced all risky securities; instead, it took the current value of the underlying asset as an exogenous input; that input, together with a description of the underlying asset price's stochastic process and the risk-free interest rate, was enough to define the arbitrage-free price (at least for their case where the stochastic process was 'well-behaved' and followed a geometric Brownian motion). Moreover, not all characteristics of the underlying asset's stochastic process had to be specified in their option pricing equation. The expected return of the underlying asset (α in Samuelson's notation) did not even feature in the Black–Scholes pricing equation.

Black and Scholes were not satisfied with providing a single route to their arbitrage-free pricing equation: they also provided a second, alternative approach to its derivation that was based on the CAPM. The CAPM was a single-period model. It had no direct answer to the problem of valuing a security whose beta changed continuously and stochastically. However, Black and Scholes showed that the CAPM could be used to determine the return required on the option at any given instant. They did this by first showing that the option's beta can be related to the underlying asset's beta by the equation:

$$\beta_C = \beta_S \frac{\partial C}{\partial S} \frac{S}{C}$$

This allowed the required return of the option to be written in terms of the underlying asset beta, and by equating this with the expected return implied by the Ito expansion of the change in the value of the call option as a function of the underlying asset and time, they were again able to obtain the heat equation that they obtained in the dynamic hedging derivation (and the various β terms cancel out). In the historical context of the development of financial economics of the previous 20 years, this CAPM approach to the option pricing formula would perhaps have been the most natural to a contemporary economist, but it was the dynamic hedging argument that proved to be most revolutionary in its broader applicability to derivative pricing and hedging.

Black and Scholes did not use the concept of risk-neutral valuation in either of their two derivations of the option pricing formula. They noted, almost in passing, that the option price does not depend on the expected return on the underlying asset (though the expected return on the option *does* depend on the expected return on the underlying asset through the above *beta* relationship). They did not note that their option pricing formula gave the risk-neutral expected option cashflow discounted at the risk-free rate (i.e. the Samuelson formula with $\alpha = \beta = r$).

It was Merton who generalised and extended the Black–Scholes result in his 1973 paper, and arguably he provided a more mathematically rigorous treatment of the derivation of the pricing formula. Merton highlighted that the Black–Scholes formula was equal to the risk-neutral cashflow expectation discounted at the risk-free rate, though this was not a central focus of his paper—the power of the generality of risk-neutral valuation was not yet appreciated or anticipated. Merton also generalised the Black–Scholes formula by introducing a stochastic (rather than fixed) interest rate into the model. He showed that in this case, the dynamic hedge portfolio needed to include a third asset alongside the underlying asset and call option: a zero-coupon risk-free bond that matures at the option maturity date. He also extended the formula to allow for the impact of dividends paid on the underlying asset during the term of the option. As a final flourish, he even included an option pricing formula for a 'down-and-out' exotic call option (where the option expires if the underlying asset falls to a specified level at any time before maturity).

With these two papers, the permanent and rigorous foundations of arbitrage-free option pricing theory had been laid. The vast research programme on option pricing theory that their work inspired over the following decade would focus on a few key themes: applying option pricing theory to the theory of capital structure; understanding the fundamental valuation ideas that could be identified in the arbitrage-free option pricing formula (in particular, risk-neutral valuation); fully developing the mathematical rigour of the arbitrage-free valuation so as to apply it to a broader range of problems (such as yield curve modelling). We discuss each of these threads of development briefly below (yield curve modelling is discussed in greater detail in the next section).

The financial economics academic profession around this time was almost obsessed with option pricing. Exchange-traded option contracts were, especially in the 1970s, relatively unimportant, niche financial instruments (though trading volumes increased significantly in the years following Black–Scholes–Merton). The academic interest in option pricing theory was primarily driven by the recognition that option-type pay-offs arose in many forms of more widely relevant contingent claims. Perhaps most importantly, the Modigliani–Miller capital structure proposition could be viewed as a form of put-call parity, where the equity of the firm was a call option on the value of the firm's assets, and the debt of the firm was a risk-free bond less a put option on the value of the firm's assets. For Black and Scholes, this was the ultimate area of interest, as is reflected in the title of their paper, which refers to the pricing of corporate liabilities. Modigliani and Miller showed that, in well-functioning markets, the sum of the claims on the firm always added up to the same total. But they did not say anything about how to value each of the different claims. Option pricing theory opened up the possibility of answering this question.

The application of the Black–Scholes option pricing formula to the valuation of corporate liabilities was further developed by Robert Merton in a 1974 paper.[32] The paper did not introduce significant new, fundamental theories. Rather, it formally clarified the conditions required for the use of the Black–Scholes formula in the valuation of corporate liabilities, and it tabulated and discussed the corporate bond prices, yields and spreads implied by the formula for various assumptions for firm asset volatilities, and the term and level of debt (all assuming the value of the underlying assets of the firm follows a geometric Brownian motion). This paper has resulted in the terminology of the 'Merton model' being used to refer to corporate bond pricing models that use the insight that a corporate bond is a risk-free bond less a put option on the value of the firm's assets. Such models have been widely used in modern risk management practice in banking and credit risk modelling.

Curiously, the analogous insight for equities—that equities can be viewed as a call option on the value of the firm's assets—has not been so widely used as the basis for the stochastic modelling of equity returns and the measurement of equity risk. Such a modelling approach can provide a logical economic explanation for the well-documented failings of modelling equities directly as a geometric Brownian motion. Since the 1970s, academics and practitioners have taken geometric Brownian motion as the starting point for equity modelling, and have then extended it in various complex ways to capture the fat tails and negative skew that are observable in empirical equity return data and that are implied by equity option market prices. By starting with the assumption that it is the value of the underlying assets of the firm that follow the geometric Brownian motion, and then recognising that equity is a call option on the value of the firm, similar dynamics can be produced and arguably in a more economically coherent way. Under this approach, an option on the equity of the firm becomes a 'compound' option—an option on an option. Robert Geske, a financial economics professor at UCLA, developed this logic and obtained the valuation formula for the compound option when the underlying firm asset value follows geometric Brownian motion in a

[32] Merton (1974).

paper published in 1979.[33] In his paper he argues that this approach to valuing equity options is a better theory that produces better empirical results:

> The introduction of these leverage effects adds a new dimension to theoretical option pricing. Any change in the stock price will cause a discrepancy between the compound option value and the Black–Scholes value [when applied directly to the equity of the firm]. The qualitative discrepancies between these two formulas correspond to what practitioners and empiricists observe in the market – namely, that the Black–Scholes formula underprices deep-out-the-money options and near-maturity options, and it overprices deep-in-the-money options.[34]

We now move onto the next thread of post-Black–Scholes–Merton research: the general application of risk-neutral valuation as a method of valuing options. Cox and Ross explored this topic in an important paper published in 1976,[35] where they argued:

> The fact that we use a hedging argument to derive [the option pricing equation] and the fact that *P*(*S*, *t*) [the price of the option] exists uniquely means that given *S* and *t* the value of the option, *P*, does not depend directly on the structure of investors' preferences. Investors' preferences and demand conditions in general enter the valuation problem only in so far as they determine the equilibrium parameter values. No matter what preferences are, as long as they determine the same relevant parameter values, they will also value the option identically .... A convenient choice of preferences for many problems (although one can envision problems where another preference structure might be more suitable) is risk neutrality. In such a world equilibrium requires that the expected returns on both stock and option must equal the risk free rate.[36]

By this time, the concept of risk-neutrality in valuation was not new. As we saw above, Merton noted in his 1973 paper that the Black–Scholes formula was equivalent to the risk-neutral expected cashflow of the option discounted at the risk-free rate. Furthermore, Samuelson and Merton had derived an early glimpse of the risk-neutral valuation equation in a paper published in 1969.[37] Nonetheless, Cox and Ross highlighted the generality of its applicability as

[33] Geske (1979).

[34] Geske (1979), p. 76.

[35] Cox and Ross (1976).

[36] Cox and Ross (1976), p. 153.

[37] Samuelson and Merton (1969). There are also other competing claims for the earliest derivation of risk-neutral valuation, but the Samuelson and Merton work was most closely related to the option valuation literature.

a computational technique for option valuation. They illustrated this point by showing how it could be used to value options under a range of assumed stochastic processes for the underlying asset that were alternatives to the now ubiquitous geometric Brownian motion. Perhaps most importantly, they established the link between risk-neutral valuation and arbitrage-free prices. As Merton himself has written, 'They [Cox and Ross] were the first to recognise that this relation [risk-neutral valuation] is a fundamental characteristic of "arbitrage-free" price systems in continuous-trading environments.'[38]

Cox and Ross collaborated with Mark Rubinstein to publish in 1979 another seminal paper in the history of option pricing theory: 'Option Pricing: A Simplified Approach'.[39] The paper attempted to cut through the increasingly sophisticated stochastic calculus of the option pricing theory of the previous few years and clarify the underlying economics. They did this by abandoning the use of continuous-time stochastic processes to describe the underlying asset behaviour. Instead, they considered option pricing theory in the presence of the simplest possible stochastic process for the underlying asset: a single discrete time-step where the underlying asset can take two possible end-period values with specified (non-zero) probabilities (often referred to as a binomial tree). They then extended this tree structure to multiple time-steps and used a backward recursive process (like that used by Pascal in 1654 in his solution of the Problem of Points discussed in Chap. 1) to find option values at every 'node' of the tree.

Cox, Ross and Rubinstein showed that the key option pricing theory results—risk-neutral valuation produces the arbitrage-free price and the same pricing function can be obtained by a dynamic hedge strategy; the option price is not a function of the underlying asset's expected return but does depend on its volatility—applied in this simpler setting. They then showed that the binomial tree option pricing formula converged with the Black–Scholes formula as the number of time-steps in the binomial tree increased towards infinity. This method was very useful for two quite different reasons: it provided a highly intuitive approach to teaching option pricing theory that allowed the fundamental economic logic to be accessed without resorting to advanced mathematics; and the backward recursive tree process provided an advanced analytical approach for valuing path-dependent option features.

A remarkable decade in the history of option pricing theory was rounded off by two papers co-authored by Michael Harrison, a Stanford Professor in

[38] Merton (1990), p. 335.

[39] Cox, Ross and Rubinstein (1979).

operations research.[40] These papers took the risk-neutral valuation logic of the 1976 Cox and Ross paper, and further developed it into a more general and mathematically rigorous valuation theory. These papers showed that an 'equivalent martingale measure' could be used to define the arbitrage-free prices of all forms of securities (under appropriate technical conditions). This meant that, in an arbitrage-free market, there existed a probability measure such that the discounted prices of all securities (stocks, bonds, options, other derivatives) must follow a martingale: that is, the expected value of the ratio of the security price to the risk-free account must be 'driftless' over time. The papers' mathematical rigour came at the expense of accessibility. As noted by Davis and Etheridge, 'This paper [Harrison and Pliska] has turned financial economics into mathematical finance'.[41] It marked the completion of an extraordinary stream of economic research that found immediate and extensive application in the financial sector.

In stark contrast to the binomial trees of Cox, Ross and Rubinstein, the complete and generalised mathematical description of dynamic hedging, replication, arbitrage, martingales and risk-neutral valuation was formidable. A quantitative revolution in finance was first heralded by Redington and Markowitz in 1952. Their work required a limited mathematical tool-kit to understand it—essentially little more than elementary differential calculus and, in Markowitz's case, also basic mathematical statistics. By 1981, work at the cutting-edge of financial theory required the post-doctoral quantitative skills of the rocket scientist. Unfortunately, this inaccessibility could get in the way of the 'big ideas' that were highly intuitive and yet still with profound implication for the management of financial risk. We will revisit this topic in later discussions of how actuarial thinking evolved alongside these developments in the second half of the twentieth century.

Option pricing theory was, and is, just that—a theory. The development of the arbitrage-free pricing models of Black, Scholes, Merton, Cox, Ross and the rest all rely on some assumptions that are quite evidently unrealistic. Of these, perhaps the most fundamentally important is that the underlying asset of the derivative follows a form of stochastic process that permits dynamic hedging to deliver perfect replication of the option's pay-off. If, for example, the assumption of geometric Brownian motion was relaxed to also permit discontinuous jumps to occur in the underlying asset price, it would not be possible to perfectly replicate the option pay-off through continuous rebalancing of a portfolio of the underlying asset and cash. The logic of arbitrage-free

[40] Harrison and Kreps (1979); Harrison and Pliska (1981).

[41] Davis and Etheridge (2006), p. 114.

pricing would therefore no longer hold. Given the empirical tendency of markets to suffer short-term jumps in price that are difficult to reconcile with a continuous form of stochastic process for the price, this has and should prompt caution in applying the model in 'real-life' applications (including actuarial ones).

As with any model, it is crucially important that the model user understands its assumptions and limitations. These limitations have generally been well-understood by market practitioners, arguably since the 1970s and particularly since the equity market crash of 1987. Market option prices are not fully explained by the Black–Scholes model—as is evidenced by the 'smile' in Black–Scholes' implied volatilities that has long been observed. Nonetheless, these implied volatilities are also the currency in which traders quote option prices—such is the ubiquity of the Black–Scholes model. The academic developments in option pricing theory of the 1970s have had a most profound impact on the practices of structuring, pricing and hedging complex financial market securities; and also in the theoretical valuation of forms of non-linear contingent claim that can occur in an immensely wide array of economics and business—from executive compensation to the valuation of oil fields. As we shall see later, eventually it played an important role in British actuarial thought and practice, particularly in the life sector.

Yield Curve Modelling

Economic theories of the term structure of interest rates have been part of economic thought since around the start of the twentieth century. Irving Fisher, the American economist whose eponymous separation theorem we discussed earlier, was one of the early theorists on what determines the shape of the interest rate yield curve. In 1896 he developed what became known as the Expectations Hypothesis. This says that long-term interest rates simply reflect the market's expectations for the future path of the short-term interest rate.[42] Under this hypothesis, the expected return on risk-free bonds is the same for all terms of bond.

The Keynesian revolution in macroeconomics that emerged in the 1930s as a response to the Great Depression generated, amongst other things, new ideas in the theory of interest rates. One such idea was the Liquidity Preference Hypothesis. This was mainly developed by the British economists John Maynard Keynes and John Hicks as part of broader economic models

[42] Fisher (1896).

of the behaviour of saving, investment and money.[43] The Liquidity Preference Hypothesis said that the expected future path for the short-term interest rate was not the only factor that would impact on the prices of long-term risk-free bonds: a risk premium would need to be offered to investors to induce them to bear the additional price volatility associated with holding longer-term bonds. Under this hypothesis, the risk premium offered by longer-term bonds would always be positive and would increase with the bond term.

A third economic theory of the determination of the risk-free yield curve was developed in US academia in the 1950s and 1960s, most notably by Modigliani and Sutch (yes, *that* Modigliani).[44] This theory, which became known as the Market Segmentation Hypothesis or Preferred Habitat Hypothesis, could be viewed as a generalisation of the Liquidity Preference Hypothesis. It argued that different investors had different investment horizons and this horizon determined investors' bond maturity preferences. Investors would require a risk premium to induce them to hold bonds with terms different to their maturity preference. The Liquidity Preference Hypothesis can be viewed as a special case of this theory where all investors have an instantaneously short investment horizon. Under the Market Segmentation Hypothesis, the term premiums embedded in risk-free bond prices could be positive or negative at different points on the term structure, depending on investors' investment horizons and the supply of bonds.

In a world of deterministic interest rates, a simple arbitrage argument shows that the Expectations Hypothesis must hold true. So theories of how yield curves are determined are really about how interest rate *uncertainty* impacts on bond prices. This is a complex and quantitative problem to tackle rigorously—to make meaningful progress, a mathematical description of the nature of interest rate uncertainty is inevitably required; and, intuitively, any theory must recognise that risk-free bonds of similar durations are close substitutes for each other and must be priced in a consistent way that does not create any trivial free lunches (i.e. the bond prices must be arbitrage-free). These two ingredients—a stochastic process for an economic variable and its use in the determination of arbitrage-free asset prices—were the keys to Black–Scholes–Merton's option pricing breakthrough. So it is unsurprising to find that the breakthrough in how interest rate uncertainty impacts on bond prices was directly inspired by the Black–Scholes dynamic replication and no-arbitrage argument.

---

[43] Keynes (1936); Hicks (1939).

[44] Modigliani and Sutch (1966).

This breakthrough was achieved in a seminal paper by Oldrich Vasicek that was published in 1977.[45] Vasicek was a Czech probability theorist who migrated to the USA in the late 1960s and began a career in investment banking in California. In his role as a researcher at a bank, he attended seminars by Fisher Black and Myron Scholes in the early 1970s where they presented their early ideas on option pricing. Vasicek saw that a risk-free bond of a given term could be viewed as a form of derivative security where the short-term interest rate was the underlying asset. This opened up the possibility of applying the dynamic hedging ideas of Black–Scholes to find a formula for the arbitrage-free behaviour of the entire yield curve.

Vasicek first specified a general mathematical structure for the uncertainty in future interest rate behaviour by defining a stochastic modelling framework for the short-term interest rate. He assumed that the instantaneous short-term interest rate (which we will refer to as the short rate) followed a continuous-time stochastic process that had a few key properties: it was continuous (the short rate could not jump); it was Markovian (its future value depended only on its current value and not on previous values); and it had a single source of randomness (what would now be called a single-factor model). He then supposed that the yield on a risk-free bond of any term could be expressed as the expectation for the short rate over the term of the bond plus a risk premium component. Of course, if there are no restrictions on the how the risk premium component behaves as a function of the term of the bonds, this statement is just a tautology—the risk premium at any point in the term structure is merely the difference between the market bond yield and the expected short rate path over the bond term. One of Vasicek's key results was to show that in order for the yield curve to be arbitrage-free, there were well-specified constraints on what form the risk premium component could take across the term structure of risk-free bonds.

The assumption that the short rate process had a single source of randomness (and that the risk premium component of bond prices did not have any additional source of randomness) implied that all variation in bond prices of all terms must be driven by the same single source of randomness. This, in turn, implied that all risk-free bond returns would be instantaneously *perfectly correlated* (though bond expected returns and volatilities would vary with term). This perfect correlation implied that some combination of any two bonds of different maturities could be held that would (instantaneously) produce a certain return (this would involve being long one bond of a given term, and short another bond of another given term). In other words, the

[45] Vasicek (1977).

short-term interest account could be *replicated* by a dynamic combination of any two risk-free bonds. Like Black–Scholes, Vasicek argued that as this dynamic portfolio was instantaneously riskless, its rate of return must be the short rate in order to avoid arbitrage.

This argument was very general—for *any* two bonds of two different terms, portfolio weights could be found such that the hedge portfolio would be instantaneously riskless. The relative portfolio weights would be a function of the relative sizes of the bond volatilities at the two different bond terms, and these in turn would be determined by the short rate's stochastic process. Vasicek had not yet considered a specific form of stochastic process—the dynamic replication argument held for any short rate process that satisfied his general modelling assumptions. From this construction, and in very similar style to Black–Scholes, Vasicek used the dynamic hedge portfolio argument to derive a partial differential equation for the bond price of any term *t*, which was a function of the current short rate, its drift and variance processes and the risk premium assumed to apply to the bond of term *t*.

This analysis uncovered the key arbitrage-free property of risk-free bond prices in Vasicek's framework. It showed that every risk-free bond's instantaneous expected return in excess of the short rate must vary in linear proportion to the bond price's instantaneous volatility. So at any given moment, a ratio of the excess expected return to volatility existed that was constant across all terms of risk-free bond. The risk premium component of all bond yields was determined by this ratio, which Vasicek called the 'market price of risk'.

Mathematically, this market price of risk could be written as:

$$q(t) = \frac{\mu(t,T) - r(t)}{\sigma(t,T)}$$

where $\mu(t,T)$ is the instantaneous drift rate at time t of the price of a bond maturing at time T; $r(t)$ is the short rate at time t, and $\sigma(t,T)$ is the instantaneous volatility at time t of the price of a bond maturing at time T. In Vasicek's framework, no-arbitrage implied that $q(t)$, the market price of risk at time t, did not vary as a function of the bond term T.

In this setting, the risk premium component could be positive or negative, and it could vary in size over time, but at any given instant, it could only take one sign across all points on the yield curve and it varied across the yield curve in this prescribed linear way. If a bond had a risk premium different to that implied by the market price of risk and the bond's volatility, it would be possible to generate arbitrage profits by constructing a riskless portfolio that generated returns in excess of the short rate.

This restriction on the term structure of the risk premium had significant implications for the economic theories of the term structure. The Expectations Hypothesis (market price of risk is zero) and the Liquidity Premium Hypothesis (market price of risk is always positive) could both be accommodated by Vasicek's no-arbitrage relationship, but it placed significant limits on the broader Preferred Habitat Hypothesis. In essence, Vasicek had shown that investors' ability to reconstruct bond maturities through dynamic portfolios of bonds of other maturities meant that there was a limit to the way in which maturity preferences could impact on the relative differences in arbitrage-free bond prices. For example, in Vasicek's model, if the ten-year bond offered a negative risk premium and the 20-year bond offered a positive risk premium, investors could perfectly replicate the ten-year bond pay-off through a dynamic combination of the 20-year bond and cash that would cost less than the ten-year bond price.

To complete his analysis of this general arbitrage-free bond pricing framework, Vasicek then noted that the bond pricing equation that he derived through the dynamic hedging process was equal to the bond price implied by the expected short rate when the market price of risk is assumed to be zero (though this was not a central object of his analysis, it was essentially an incidental by-product of his 'real-world' modelling presentation). In the same way that Black–Scholes–Merton showed that the risk-neutral stock process determined the value of any option on the stock, Vasicek had shown that the risk-neutral short rate process determined the arbitrage-free price of risk-free bonds of all terms.

Having established the above framework and its general properties, Vasicek then proceeded to illustrate it with the use of a simple example: he assumed that the short rate followed a mean-reverting process with a normally distributed volatility process (sometimes referred to as an Orstein-Uhlenbeck process):

$$dr = \alpha\left(\mu - r(t)\right)dt + \sigma dZ(t)$$

He also assumed the market price of risk was some constant q (i.e. it did not vary over time). These model dynamics were never meant to be taken as a recommendation of a realistic description of interest rates, but rather to provide a simple and tractable illustration of Vasicek's pricing framework. Nonetheless, this short rate model is universally referred to as the 'Vasicek model'. With this modelling specification, the arbitrage-free bond pricing equation could be solved to provide an analytical formula for the bond price as a function of term. And the expected return and volatility of every risk-free bond were well-defined functions of the parameters of the short rate process and the market price of risk (as well as the bond's term).

Vasicek's demonstration that the Black–Scholes–Merton idea of arbitrage-free pricing by dynamic replication could be used to price the risk-free term structure was the permanent foundation for a new branch of quantitative finance. A vast amount of yield curve modelling research followed in the next two decades that further developed Vasicek's big idea. One of the most significant developments to arise in the years following the Vasicek paper came from a pair of papers published by Cox, Ingersoll and Ross (CIR) in 1985.[46] These papers were far less accessible than Vasicek's paper. Vasicek's paper had been very focused in its ambitions: he showed that, under certain assumptions, a given stochastic process for the short rate together with a market price of risk could determine the arbitrage-free pricing of risk-free bonds of all terms. Cox, Ingersoll and Ross broadened the setting. Rather than specifying a framework where the stochastic process for the short rate is specified exogenously, they developed a broader stochastic model of the wider economy (production possibilities, wealth, individual's utility) and all the assets that traded in it, and then derived the short rate stochastic process that was implied by the specified economy system.

Before specifically considering interest rate behaviour, their first paper derived the by now ubiquitous risk-neutral valuation property for their economic system. It is interesting to note that Vasicek did not focus the presentation of his bond pricing mathematics in a risk-neutral setting. He used a 'real-world' probability measure, and showed that the bond pricing equation could use these real-world probabilities together with the market price of risk assumption to obtain the bond price (and this could be presented as a risk-neutral valuation where the market price of risk transforms the real-world probability measure into a risk-neutral one). After Vasicek, the biggest application of the arbitrage-free yield curve modelling that he pioneered was increasingly in interest rate derivatives pricing and hedging rather than in macroeconomic descriptions of how interest rate uncertainty impacted on yield curve behaviour. These applications were most efficiently addressed by working in risk-neutral probability measures, and the use of real-world probabilities increasingly disappeared from the yield curve modelling literature. Cox, Ingersoll and Ross's grand economic framework seamlessly moved between real-world probabilities and risk-neutral probabilities.

[46] Cox, Ingersoll and Ross (1985), 1 and 2.

The CIR paper, 'A Theory of the Term Structure of Interest Rates', broke new ground beyond Vasicek in a number of important ways. Like Vasicek, they derived the arbitrage-free bond pricing equation via a stated stochastic process for the short rate. Their short rate model (below) assumed that the volatility of the short rate moved in proportion to the square root of the short rate:

$$dr = \alpha\left(\mu - r(t)\right)dt + \sigma\sqrt{r(t)}dZ(t)$$

This was a slightly more sophisticated and realistic model of interest rate behaviour than that used by Vasicek. Most notably, it precluded the possibility of negative interest rates (this has traditionally been seen as a desirable feature of interest rate models, though that view has been subject to some revision in the 2010s!). Like Vasicek, they derived an analytical solution for the risk-free bond pricing equation implied by the short rate process. Cox, Ingersoll and Ross also went a step further and considered how the same valuation framework could be used to value bond derivatives and, as an illustrative example, derived the arbitrage-free pricing equation for a call option on a risk-free bond. This was an early illustration of the potential applicability of stochastic yield curve models to pricing and hedging interest rate derivatives. A huge expansion in the types and volumes of such securities occurred in the final two decades of the twentieth century. This was facilitated by the advanced mathematics of these models and motivated further advancements in the theory and its implementation.

Finally, Cox, Ingersoll and Ross also highlighted the limitations of the single-factor models that had been exclusively considered by Vasicek and themselves up to this point. The single-factor model structure had severe limitations—most obviously, it implied that the entire term structure was perfectly correlated. From a derivatives pricing perspective, it would prove difficult to recover the market prices of derivatives that depended on both bond price volatilities and correlations with such models. From a risk management perspective, these models implied that a ten-year liability could be perfectly hedged by dynamically rebalancing long and short positions in a three-year and two-year risk-free bonds. These limitations had been recognised by Brennan and Schwartz in the immediate aftermath of the Vasicek paper, and they published a specific two-factor arbitrage-free interest rate model in 1979.[47] Cox, Ingersoll and Ross developed a more general multi-factor modelling setting within the arbitrage-free pricing framework first

[47] Brennan and Schwartz (1979a).

pioneered by Vasicek. You may recall that the perfect correlation of risk-free bond returns was an important step in Vasicek's bond pricing derivation. Cox, Ingersoll and Ross showed that Vasicek's dynamic replication argument could still work in this multi-factor, decorrelated yield curve world—more factors just meant more risk-free bond holdings of different maturities were required to construct the dynamically riskless portfolio.

These three themes beyond Vasicek that Cox, Ingersoll and Ross developed—more realistic volatility behaviour for the short rate; the application of the model to interest rate derivative pricing; and multi-factor short rate modelling—were themes that were further developed by academics and practitioners over the following few years. Hull and White,[48] Black, Derman and Toy[49] and Black–Karasinski[50] were important contributions that pursued these themes in an extremely active research programme that was at least partly motivated by the huge growth in the trading of interest rate derivatives in securities markets.

A notable departure from these threads of development arose in a paper by Heath, Jarrow and Morton (HJM) published in 1992.[51] The HJM approach reconsidered how to describe the stochastic evolution of the yield curve. As we have seen, short rate models explicitly specified a stochastic process for the short-term interest rate, and then deduced the arbitrage-free pricing of all risk-free bonds from that process. So one explicit equation (the short rate process) determined the behaviour of every point on the yield curve. This was very parsimonious, but it also restricted modelling flexibility and meant that there were limited degrees of freedom to fit to calibration targets such as volatilities of different points of the yield curve (or to simultaneously fit to the prices of many different types of interest rate derivatives).

Instead of only explicitly specifying a stochastic process for the short rate, HJM suggested explicitly specifying a stochastic process for every point on the yield curve. This general framework provided almost unlimited freedom in specifying the volatility and correlation structure of the different points on the yield curve. This flexibility was very powerful in the context of interest rate derivative pricing (which was the intended purpose of their framework). HJM then found the no-arbitrage conditions that specified the risk-neutral drifts that must be generated by each forward rate of the yield curve in order to avoid arbitrage.

[48] Hull and White (1990).

[49] Black, Derman and Toy (1990).

[50] Black and Karasinski (1991).

[51] Heath, Jarrow and Morton (1992).

Their paper was closely related to the quantitative work of the Harrison option pricing papers discussed above. HJM showed that their interest rate framework could be embedded as the interest rate modelling piece of the broader mathematical economic framework of Harrison and Pliska. They were therefore able to make use of the 'equivalent martingale measure' concept of Harrison and Pliska to determine the drift processes for any forward rate on the yield curve. Where Vasicek's yield curve modelling was inspired by Black–Scholes, the modelling framework developed by HJM was inspired by Harrison and Pliska: with this paper, yield curve modelling had fully caught up with and become an integral part of option pricing theory.

Market Efficiency

Our discussion of the emergence of key ideas in financial economics has focused mainly on the development of economic *theories*. That is, a range of theoretical results have been discussed (for example, the Capital Asset Pricing Model) that have been developed deductively from a set of starting axioms (investor risk aversion and non-satiation, and so forth). In all cases, these results were subject to various forms of empirical testing, and such testing has consistently formed a substantial part of financial economics' research output. But this final section is somewhat different in that it is related to a stream of work that is intrinsically empirical: it is focused first and foremost on how well 'real-life' financial markets work—not in theory, but in practice. In particular, this stream of financial economics considers the *informational efficiency* of financial markets' prices. Pricing efficiency in this context refers to how well market prices reflect relevant information and how quickly prices react to new information. Its empirical nature and its implications for the possible lack of usefulness of large swathes of financial services practitioners have made it one of the most contentious areas of financial economics. This was true many decades ago and it remains true today, particularly as later research has painted a more complex and nuanced picture of real-life market behaviour than that implied by market efficiency's major research results of the 1960s and early 1970s.

We noted some detailed empirical studies of stock price behaviour in the discussion of option pricing theory—in particular, Osborne's 1959 research that provided an empirical basis for the use of geometric Brownian motion as a reasonable model of stock price behaviour. There are also some earlier examples of empirical research that date back to the first half of the twentieth century. But improvements in the collation of security price data and growing computing power stimulated a new wave of empirical analysis of security prices in the 1950s.

Besides Osborne, another important example of this empirical work was provided by Maurice Kendall, who was director of research techniques at the London School of Economics. Kendall presented a detailed empirical study of the time series behaviour of financial market prices to the Royal Statistical Society in 1952.[52] The study provided the most detailed statistical analysis to date of the time series behaviour of stock prices. Kendall considered UK equity market behaviour over the ten-year period between 1928 and 1938, and wheat prices on the Chicago Board of Trade between 1883 and 1934. In both cases he could find little evidence of statistically significant serial correlations at any time-lag. As a trained economist with a faith in rational market responses to the business cycle, this lack of trend or apparent signal in the price process confused and alarmed Kendall:

> At first sight, the implications are disturbing … it seems that the change in price from one week to the next is practically independent of the change from that week to the week after. This alone is enough to show that it is impossible to predict the price from week to week from the series itself … The series looks like a "wandering" one, almost as if once a week the Demon of Chance drew a random number from a symmetrical population of fixed dispersion and added it to the current price to determine the next week's price.[53]

Professor R.G.D. Allen, in his vote of thanks, shared Kendall's despondency, noting that the paper's results were 'a very depressing kind of conclusion to the economist'.[54] However, some speakers from the floor had a different economic interpretation for the lack of trend or predictability in the price data. A Professor Champernowne commented that 'the low serial correlation coefficients found in this particular series may reflect the success with which the professionals are doing their job',[55] whilst Professor Paish elaborated: 'It seems inevitable that where prices are based on expectations markets are as likely to go down as up. If the markets thought they were more likely to go up they would have gone up already.'[56]

A decade after Kendall's empirical study, Paul Samuelson published a theoretical paper, 'Proof that Properly Anticipated Prices Fluctuate Randomly'.[57]

---

[52] Kendall (1953).
[53] Kendall (1953), p. 13.
[54] Kendall (1953), p. 26.
[55] Kendall (1953), p. 27.
[56] Kendall (1953), p. 30.
[57] Samuelson (1965).

This paper provided some mathematical formality to the intuitions of Professors Champernowne and Paish: in well-functioning markets, the no-serial-correlation results of Kendall were exactly what *should* happen. But Samuelson was not the first to argue that well-behaved prices should exhibit random fluctuations. Louis Bachelier's arithmetic Brownian motion process was almost certainly the first no-serial-correlation model of the stochastic paths of financial market prices. In his (translated) words: 'the mathematical expectations of the buyer and the seller are zero'.[58]

In 1965, Eugene Fama, then a young assistant professor at the University of Chicago Business School, published a research paper, 'The Behaviour of Stock Market Prices'.[59] As an empirical study of the statistical properties of stock market prices, this covered some similar ground to Kendall, but was more comprehensive and wide-ranging. Like Kendall, Fama found that serial correlations in daily stock returns were generally very low (Fama tested the returns of US stocks in the Dow Jones Industrial Average over the period 1957–1962). He also performed some other forms of statistical tests such as runs tests to provide further evidence of statistical independence of returns through time.

Fama's paper covered a couple of other important topics that would become increasingly relevant in future years. First, Fama analysed the *shape* of the distribution of returns. He concluded that there was strong statistical evidence that daily stock returns had fatter tails than those implied by a normal distribution. Second, as well as considering the behaviour of individual stocks, he also analysed empirical data on the returns of mutual funds (he considered 39 mutual funds over the period from 1950 to 1960). His analysis led him to two conclusions: mutual funds, as a whole asset class, did not beat the equity market over the period; and no mutual fund consistently outperformed the others year-on-year through the ten-year period.

In his 1965 paper, Fama argued that the observations of statistical independence of returns through time and the inability of any mutual funds to consistently outperform the market or each other were both forms of evidence in support of what had become known as the Random Walk Hypothesis—the idea that market prices varied randomly and unpredictably from one period to the next. In a further paper, published in 1970, he developed these

[58] David and Etheridge (2006), p. 28.

[59] Fama (1965).

ideas further. This paper, 'Efficient Capital Markets: A Review of Theory and Empirical Work',[60] is one of the most famous and influential financial economics papers ever published. As its name suggests, it was a review of the by-then abundant empirical analysis of stock market behaviour that had accumulated over the previous 15 years. But it was more than a review. Fama took those various threads of analysis and wove them into a clear body of evidence in support of the notion of *efficient markets*, which he defined as where 'security prices at any time fully reflect all available information'.[61]

The theory of market efficiency was concerned with how prices responded to information. In Fama's crystallisation of efficient markets, he proposed three levels of market efficiency that corresponded to three different information sets: weak-form efficiency, where efficiency meant prices fully reflected all information in historical price movements; semi-strong efficiency, where efficiency meant prices fully reflected all publically available information (a semi-strong-form efficient market therefore must also be weak-form efficient as historical prices were public information); and strong-form efficiency, where efficiency meant prices fully reflected all information, both public and private (so a strong–form efficient market was also semi-strong and weak-form efficient). His paper reviewed the empirical evidence that had been published in relation to each form of informational efficiency.

The evidence for weak-form efficiency was naturally found in the statistical testing of historical price data. These tests took two broad forms: testing for statistical independence of returns through time (mainly by serial correlation testing such as that done by Kendall and Fama); and testing the profitability of mechanical trading rules (the idea being that any 'excess' profitability of such rules would not be consistent with efficient markets). Fama's 1965 paper included some analysis of these mechanical trading tests, and he published a paper in 1966 with Marshall Blume with further testing of such rules.[62] Fama's review of the evidence relating to weak-form efficiency allowed him to conclude that 'the results are strongly in support'.[63]

The empirical evidence for semi-strong market efficiency was largely based on analyses of how market stock prices reacted to major public announcements of relevant information such as earnings statements and stock splits. The basic idea was that if the market was efficient the price impact of these announcements would be immediate, and subsequent expected returns would

[60] Fama (1970).
[61] Fama (1970), p. 383.
[62] Fama and Blume (1966).
[63] Fama (1970), p. 414.

therefore be unaffected by the announcement. Fama again concluded that these studies invariably provided support for the semi-strong form of the efficient market hypothesis. The implications of the semi-strong hypothesis are the most provocative to investment professionals as it implies that active fund management cannot be expected to outperform the market except by luck. This is consistent with Fama's survey of mutual fund performance in his 1965 paper. Such studies also have a longer history—in 1933, the US economist Alfred Cowles published a paper which showed that buying-and-holding would tend to outperform the recommendations of stock market forecasters.[64] Evidence in support of the semi-strong hypothesis can provide an intellectual basis for market indexing or passive investment management—a form of investing that has rapidly grown in popularity since the 1980s.

In considering the strong form of market efficiency, Fama conceded that there was evidence that corporate insiders have monopolistic access to information that is not in the share price. But even in this case he argued that the investment community is unable to access and use such information to outperform the market. Overall, Fama's conclusion was emphatic: 'In short, the evidence in support of the efficient markets model is extensive and (somewhat uniquely in economics) contradictory evidence is sparse.'[65]

Inevitably perhaps, however, empirical evidence contradicting the efficient markets hypothesis quietly started to accumulate in the decade following Fama's emphatic declaration of efficient markets victory. The watershed moment arrived in 1978 when Professor Michael Jensen, a leading financial economist of the period, edited a special edition of the *Journal of Financial Economics* that was dedicated to reviewing this stream of research.[66] Amongst other studies, this research included several analyses of the returns on a diverse range of mechanical trading strategies (including securities such as investment trusts and exchange-traded stock options). If strategies were identified that earned statistically significant excess risk-adjusted returns (after trading costs), this would be regarded as evidence inconsistent with weak-form market efficiency. Unlike in Fama's mechanical trading tests of the 1960s, the edition tentatively concluded that several such strategies could deliver excess risk-adjusted returns. But this type of study raised an interesting question: did excess risk-adjusted returns look good for these strategies because markets were mispricing assets or because the theoretical models for assessing the risk-adjusted required returns were wrong (or were missing some features that are important to these

---

[64] Cowles (1933).

[65] Fama (1970), p. 416.

[66] Jensen (1978).

complex strategies)? This left some ambiguity in the conclusions which helped to shape the future direction of financial economics research.

Jensen's special edition helped to create an environment within the financial economics profession where challenge to the accepted wisdom of perfectly functioning financial markets was an accepted part of academic orthodoxy. It ushered in a new era of empirical research in financial economics where the identification of potentially irrational market behaviour was suddenly highly in vogue.

In 1981, Robert J. Shiller published a provocative paper where he argued that the volatility of stock market returns was much, much higher than could be explained by changes in rational expectations for levels of future dividend pay-outs.[67] Using a dividend discount model for equity market valuation, he showed how, under some assumptions about the stochastic properties of the dividend pay-out process, a relationship between the year-on-year volatility of dividend pay-outs and year-on-year volatility of equity price changes could be established. Shiller's long-term empirical analysis of dividend pay-outs and stock market volatility in the USA implied that market volatility was 'five to thirteen times too high to be attributed to new information about future real dividends'.[68] In deriving this analysis, the dividend discount model he used assumed a constant real required return. He inverted the analysis and considered how volatile the real discount rate would need to be to generate the observed level of market volatility. He found it would need to have an annual standard deviation of 4–7 %, which he dismissed as economically unfeasible.

Shiller's work generated considerable academic controversy and prompted a notable response from Robert Merton, one of the financial economics profession's established leaders of the period. In a paper with Terry Marsh published in 1986,[69] the authors argued that Shiller's conclusions were highly dependent on his assumed form of stochastic process for dividends. Marsh and Merton's key point was that firms' managers liked to smooth dividend pay-outs as much as possible. But, as was shown by Modigliani and Miller decades earlier, the rational or intrinsic value of the firm should be determined by the performance of the firm's underlying assets, and not by its dividend policy. If investors understood that managers preferred to smooth dividends over time, then they would be more sensitive to changes in dividend pay-outs (if the firm still had to reduce dividends even though management prefer to pay stable dividends, this signalled things must be pretty bad). Their general point was that inferring rational levels of return volatility from observed dividend policy

[67] Shiller (1981).

[68] Shiller (1981), p. 434.

[69] Marsh and Merton (1986).

was very difficult because dividend policy did not necessarily have a direct relationship with the true value of the firm. To prove this, they showed that the opposite statistical conclusion could be reached from Shiller's data when they specified an alternative form of stochastic process for dividend pay-outs (which they argued fitted better to empirical dividend pay-out behaviour).

Despite Merton's protestations, the genie was out of the bottle. Other leading financial economists followed Shiller's lead and produced further analysis to support the argument that volatility in stock market returns was inexplicably high. Richard Roll, in his presidential address to the American Finance Association in 1987,[70] presented an empirical analysis that argued that, even with the benefit of hindsight, 60 % of US equity stock market daily price volatility was inexplicable (in the sense that the price variation in a firm's stock could not be explained by observable new information relating to the firm, its industry or general economic and market impacts). This was not the order of magnitude of excess volatility that Shiller had reported, but it was perhaps all the more plausible for that. Roll's address opened the door to possible behavioural explanations: 'Several authors have suggested that volatility of asset prices can be better explained by psychological factors, fads, etc., than by information. The results above are actually consistent with such a view'.[71]

If Shiller and/or Roll were right that short-term equity volatility was inexplicably higher than could be justified by changes in fundamentals, what did that imply about long-term equity behaviour? If 'extra' volatility was continuously feeding into stock returns without any form of self-correction, equity prices would become infinitely dislocated from underlying economic reality. Eugene Fama, the economist more associated with efficient markets than any other, worked with another Chicago economist, Kenneth French, to provide some further insights into the empirical behaviour of longer-term equity returns. Fama and French published two significant papers on this subject in 1988.[72] The first paper, 'Permanent and Temporary Components of Stock Prices', identified statistically significant mean-reversion (negative serial correlation) in historical (1926–1985) US stock market returns over three- to five-year horizons. Previous tests of serial correlation in stock market returns such as Kendall's had used equity data series of a more limited size (Kendall used a total equity data horizon of ten years). Fama and French's more comprehensive data analysis suggested there was a noteworthy cumulative effect which was highly significant over longer holding periods.

[70] Roll (1988).

[71] Roll (1988), p. 565.

[72] Fama and French (1988a); Fama and French (1988b).

Seen alongside the work of Shiller, Ross and others, Fama and French's research suggested that short-term equity market volatility was excessively high, and that some of this 'excess' or 'temporary' volatility was removed over time by a form of correction mechanism in equity market prices (which manifested itself statistically as a material mean-reverting component in the price process). Their second paper of 1988, 'Dividend Yields and Expected Stock Returns', took this analysis further: if mean-reversion was an important element of long-term equity market behaviour, was it possible to observe at any given point in time whether this mean-reverting component of returns was above or below its mean level? Fama and French suggested this was possible and indeed trivially easy: dividend yields appeared to be meaningful predictors of long-term equity performance. High dividend yields predicted strong returns over the following two to five years, low dividend yields predicted the opposite.

From a market efficiency perspective, this was a profound challenge to even the weak-form of the efficient market hypothesis. But there were some caveats. First, whilst long-term expected returns did vary with the starting level of the dividend yield, it was unclear whether this was mispricing resulting from fads, bubbles or some other form of irrational behaviour, or whether this reflected rational changes in required returns due to time-variation in the riskiness of equities or in investor risk appetite. Furthermore, an inevitable consequence of analysing longer-term empirical behaviour is that there is a smaller sample size to observe. As the leading twenty-first-century financial economist John Cochrane has pointed out, when dealing with such long-term trends, we may really only have a few observable data points:

> What we really know is that low [stock] prices relative to dividends and earnings in the 1950s preceded the boom market of the early 1960s; that the high price/dividend ratios of the mid-1960s preceded the poor returns of the 1970s; that the low price ratios of the mid-1970s preceded the current boom.[73]

This limited volume of empirical data constrained the degree of consensus reached within the financial economics profession on the topic of long-term security price behaviour, and it continues to do so today. Robert Shiller and Eugene Fama received the 2013 Nobel Prize in Economics for their work in this field (along with Peter Hansen). Their acceptance speeches featured an exchange of views that highlighted how much work remained to be done to find a consensus explanation for their empirical findings.

[73] Cochrane (2005).

Whilst it took until the 1980s for the notion of mean-reversion and 'time diversification' to gain academic credence, it has arguably been part of investor intuition for as long as equity markets have existed. As custodians of long-term liabilities, mean-reversion in long-term returns doubtless played a role in first attracting life offices and their actuaries to equities as an asset class in the 1930s. For example, in a letter to F.C. Scott, the managing director of the Provincial Insurance Company, in June 1938, John Maynard Keynes wrote:

> A valuation at the bottom of the slump tends to bring out an unduly unfavourable result as against an investment policy which on the whole avoids equities; since it allows nothing for the nest egg in hand arising out of the fact that such a valuation is assuming in effect that one has purchased a large volume of equities at bottom prices ... *Investment policy which is successful in averaging through time will produce the same good results as insurance policy which is successful in averaging through place* [emphasis added].[74]

Time diversification can only arise if a component of the price change process is temporary. As we shall see in Chap. 5, this idea was embedded in how actuaries modelled and measured equity risk in the context of long-term liability business in the late twentieth century. This was clearly inconsistent with the financial economics of the 1960s and 1970s. It was not as inconsistent with the financial economics of the 1980s and beyond as actuaries have sometimes been led to believe.

[74] Keynes (1983), p. 67.

5

Life Offices After the Second World War: The Underwriting and Management of Financial Market Risk (1952–2004)

The history of actuarial thought in the British life assurance sector over the second half of the twentieth century is tumultuous. This was increasingly the case as the century wore on, reaching something approaching a crisis by the century's end, from which the profession started to reorientate itself in the early 2000s. The root causes of the late twentieth-century challenges initially emerged earlier in the century and have already been briefly noted. The most important of these was the increasing trend, started in the 1920s and 1930s, of abandoning the strictly risk-averse investment disciplines of nineteenth century Bailey in order to pursue greater investment in equities and other risky asset types.

This apparent increase in risk appetite could be viewed in the context of a broader long-term trend in with-profits policy design towards it being unambiguously configured as a long-term savings vehicle as opposed to a more balanced mix of savings and life assurance protection. As a long-term savings product provider, with-profits funds were exposed to increasing competition from other sectors of the financial services industry such as banks and asset managers. The stress of competition doubtless placed commercial pressures on actuaries, whose role was to determine the charging, reserving requirements and bonus policies of life products.

From an intellectual perspective, the radical developments in relevant related disciplines that occurred over the second half of the century created opportunities and threats for the actuarial profession. Financial economics, the application of advanced quantitative techniques to finance and new modelling possibilities created by advances in computing technology each

C. Turnbull, *A History of British Actuarial Thought*,
DOI 10.1007/978-3-319-33183-6_5

contributed to a revolution in financial and risk management practices in other parts of the financial and corporate sectors. Would these developments make traditional actuarial skills obsolete? Could the profession embrace these new ideas and capabilities and add them to their professional toolkit? This represented a formidable challenge for the British profession. As we have seen, actuarial concepts and their applications evolved significantly over the 100 years between 1850 and 1950. The core technical skills required in actuarial work, however, had not changed fundamentally over that period. The tried and tested actuarial syllabus of the mid-twentieth century was not easily adapted to the radical and technical developments in finance, quantitative techniques and computer modelling that would emerge in the following decades.

The challenge was cultural as well as intellectual. The British actuary of this era stood resplendent in his status as a purveyor of techniques too complex and opaque to be understood by anyone else; his professional judgement was beyond challenge; his track record was without blemish. He was not minded to be told that his methods were antiquated or that his thinking was flawed. An intellectual and cultural clash was afoot.

As ever, this is a story replete with remarkable actors, and it is perhaps all too easy for hindsight to blind us to the real challenges and conflicts they faced as they made the progress they did.

Equities and With-Profits: Smoothing and Guarantees (1952–1976)

Chapter 3 noted how actuarial thinking on life office investment strategy began a journey in the 1920s, away from the highly conservative nineteenth-century tenets of Bailey. By the late 1940s, the Dalton 'cheap money' era had further prompted British life offices to move materially into equity and property investments in an effort to maintain their investment yields in excess of those that had been guaranteed in with-profit premium rates. This asset allocation trend continued unabated into the 1950s and 1960s. By 1951, 20–25 % of life office assets were invested in equities and property;[1] by 1961 this allocation had grown to over 30 %.[2] Life offices had a mixture of both with-profit and without-profit liabilities. The without-profit business was effectively backed by bonds, and so the with-profit business's equity/property exposures were significantly higher than the above figures. By 1981, Frank

[1] Dodds (1979), p. 176.

[2] Dodds (1979), p. 50.

Redington noted that for some British with-profit funds the equity/property allocation was 100 %.[3] In the decades following the Second World War, British with-profits business had transformed into something different, and the actuarial thought of British life actuaries had to catch up.

There were a couple of key (and related) questions that required actuarial answers. How did substantial equity/property investment impact on life office solvency assessment? And how should with-profit bonuses be distributed in the presence of significant equity/property investment? Most actuarial energy was initially invested in the second of these questions. In the 20 years following the end of the Second World War, equity markets performed remarkably well (especially in nominal terms). This was driven by both a high rate of dividend growth and a fall in equity market dividend yields. Between 1950 and 1960, dividend pay-outs more than doubled and market values increased by a greater amount. Long-term interest rates steadily increased during this era (from a low of 2.6 % in 1946 to 6.8 % by 1966). Life office solvency was therefore generally not felt to be a major concern. Conversely, there was a pressing need to deal with the embarrassment of riches that with-profit funds had found themselves holding. This was made particularly pertinent by the increasing competition from other forms of savings vehicle such as unit trusts.

The crux of the bonus distribution challenge was that the reversionary bonus system did not recognise any equity/property capital gains until they were crystallised in asset book values, and the writing-up of book values prior to the sale of the asset was generally viewed as actuarially distasteful. In the competitive environment, this was a fundamental issue. The only way equity returns naturally fed into the assessment of distributable surplus was through the dividends received. By 1960, the 'reverse yield gap' had emerged: equities generated a lower (initial) income than risk-free bonds. Distributable surplus would therefore be reduced by increasing equity allocations, even following a period when the equity holdings had materially outperformed bonds in terms of market value growth. Moreover, during this time, market practice was such that actuaries were under significant pressure not to reduce bonus rates other than in the most exceptional circumstances. This further incentivised the actuary against releasing unsustainable rates of equity market value increase into distributable surplus. As a result, with-profit pay-outs started to significantly lag comparable unit trust pay-outs, and life office estates increasingly bulged with undistributed and unrecognised equity market value gains.

This situation was unsustainable—the with-profit bonus system was not fit for the purpose of distributing surplus from equity-dominated asset

[3] Redington (1981), p. 378.

allocations. By the mid to late 1950s, British life actuaries started to recognise that a more flexible bonus system would be necessary in order to distribute equity capital gains equitably amongst the policyholders whose premiums had generated them. It was crucial that this was done in a way that did not create the expectation that such distributions could be predictably delivered every year. This required two new devices: a special or terminal bonus that was distinct from the reversionary bonus and that did not have the same policyholder expectations for stability attached to it; and an acceptable method for writing up book values of equity/property assets to market values in order to transfer some gains into distributable surplus. It is difficult to overstate how counter to traditional British actuarial principles these developments were. But several papers were published between 1959 and 1976 that progressively established the new orthodoxy.

An Institute paper presented to Staple Inn in 1959 by the Equity and Law actuary Norman Benz was an important early landmark in establishing these radical ideas in mainstream actuarial thought.[4] Those ideas were given further support in another Institute paper written by P.E. Moody in 1964.[5] These papers reflected the growing recognition that intergenerational policyholder equity and competition in the long-term savings market both demanded that policyholders receive the capital gains generated by their investments. This naturally led to an increasing embrace of the contribution principle of bonus distribution that had been well-established in the USA for so many decades. For example, in an influential 1968 paper,[6] Skerman argued that 'the distributed profits should be allocated between policyholders in relation to the contribution which their policies have made to them'.

However, in the 1950s and 1960s (and beyond), there was still a prevalent view amongst British actuaries that with-profit funds should deliver something more stable than what was directly generated by the market returns of their invested premiums. Both Moody and Benz advocated a *smoothed* distribution of equity gains to policyholders. The concept of using the with-profit fund's estate to deliver a smoothed equity return to policyholders would remain a permanent central tenet of British actuarial thought. The 1976 report of the Institute of Actuaries' Working Party on Bonus Distribution with High Equity Backing stated that 'the general view of the working party was that conventional with-profits business should smooth out the fluctuations in

[4] Benz (1960).

[5] Moody (1964).

[6] Skelman (1968).

investment return' and that 'the estate should be used to ... level out fluctuations in experience'.[7]

Benz and Moody explored *how* to smooth the distribution of equity capital gains to with-profit policyholders. In both papers there was a particular wariness about distributing equity capital gains that arose through falls in yields rather than through growth in dividend pay-out levels. Both Benz and Moody advocated the idea of using an actuarial estimate of the future dividend growth rate to determine a steady writing up of asset book values and hence release to distributable surplus, irrespective of how the market yield was behaving. In his paper, Skerman agreed that market value changes arising due to changes in equity dividend yields should not enter into distributable surplus. In the Staple Inn discussion of Skerman's paper, A.C. Stalker advocated the use of the historical average dividend yield to determine equity book value write-ups. Whilst such approaches could be substantially dislocated from market value movements, the implications of the Fama and French empirical equity studies that would emerge some 25 years later arguably offer some degree of intellectual support for these ideas. But such an approach was not without practical difficulty: did with-profit funds have the capital to underwrite larger distributions than those implied by market value gains when equity market yields were unusually high? Did policyholders understand that they may not get a material portion of 'their' equity capital gain when markets performed very well?

Whilst Benz and Moody advocated a quantitative method for developing a smoothed distribution of the equity capital gains, Skerman did not welcome such a mechanical approach to valuing assets. Instead, he argued that the release of equity gains into distributable surplus should simply be a matter of actuarial judgement as part of the actuary's role in the 'proper steering'[8] of the with-profit fund. Such an approach placed a huge amount of discretion in the hands of the actuary, even by British actuarial standards. In such a system, an actuarial discipline that balanced policyholder security against intergenerational equity and prudence against competitive marketing pressures would be critical to the long-term success of with-profits as a popular and sustainable long-term savings vehicle.

Life office practices evolved concurrently. Leading British with-profit funds started to use terminal bonus as an additional degree of freedom to deliver more equitable pay-outs to with-profit policyholders in the late 1950s. The Prudential, with Frank Redington as its chief actuary, was a leader in this innovation, introducing terminal bonuses in 1956. The terminal bonus was

[7] Kennedy et al. (1976), p. 12.

[8] Skelman (1968), p. 72.

an almost universally adopted mechanism amongst British with-profit offices by the end of the 1960s. However, such was the success of equity markets during this era that a real risk emerged of the terminal bonus becoming established in the minds of policyholders and salesmen as just as stable and predictable as the reversionary bonus. The 'reasonable expectations of policyholders' were enshrined in UK legislation by the Insurance Companies Act of 1974, and this further concentrated actuarial minds on bonus policy. But the extreme equity market experience of 1973 and 1974 prompted several offices to reduce terminal bonuses and one or two offices even reduced them to zero.[9] This suggests the actuary was able to use the terminal bonus mechanism broadly as intended.

What of actuarial thought on the appropriate level of equity allocation in with-profit funds and its implications for solvency? When equity investment was first introduced to UK with-profit funds in the 1920s and 1930s, this topic received little consideration. Equity allocations were small and solvency buffers were usually large, and there was little discussion of what upper limit should be placed on the equity allocation. Raynes's influential papers of 1928 and 1937 said virtually nothing on this question. In the closing remarks in the discussion of Raynes's 1937 paper, something is finally mentioned in relation to liabilities when George Recknell, the actuary at Keynes's National Mutual, suggested:

> Concerning the percentage of ordinary shares which should be purchased there was something to be said, as a very rough rule, for investing on more or less conventional lines to the extent of the funds needed to support the contractual liabilities, while leaving the surplus free to invest in ordinary shares and real estate.[10]

Haynes and Kirton's 1952 paper had similarly advocated Recknell's form of asset-liability matching: fixed liabilities (sums assured, including accrued reversionary bonuses) should be matched by bonds, and the excess funds could be invested in equities and property. A more ambitious discussion of the equity asset allocation in a with-profit fund was developed in 1957 in an influential Institute paper written by J.L. Anderson and J.D. Binns,[11] who were, respectively, the actuary and investment secretary of Scottish Widows at the time. Anderson and Binns suggested that with-profit funds should invest more in equities than the maximum level implied by the above principle that guaranteed benefits were matched by bonds:

[9] Kennedy et al. (1976).

[10] G.H. Recknell, in Discussion, Raynes (1937), p. 505.

[11] Anderson and Binns (1957).

> Suppose that k is the maximum depreciation on present market values which is envisaged. Estimate the value of a function we shall call the "remainder" R, i.e. the excess of the total assets at market value over the liabilities on a gross premium basis ... It can then be argued that R/k can safely be invested in equities provided the balance of the fund is reasonably matched.[12]

The full-matching approach advocated by Recknell, Haynes and Kirton was equivalent to setting $k = 1$, i.e. assuming equities could fall to zero. By setting k to less than 1, the with-profit fund opened up the possibility that poor equity market performance would result in the office failing to meet its guaranteed liabilities: of course, the probability of this possibility depended on the assumed size for their k parameter (and on how quickly the fund could shift out of equities if it wished to as market values fell). Their logic could be viewed as a precursor to a modern 'portfolio insurance' strategy, which is typically couched in the logic of Black-Scholes's dynamic hedging (though Anderson and Binns gave no indication that it was their intention for with-profit funds to reduce their probability of insolvency through very frequent rebalancing of the equity allocation to the new level implied by market value changes).

Anderson and Binns noted that the Dow Jones Industrial Index had fallen by over 80 % during 1929–1932, but dismissed this period as exceptional, and suggested using a k parameter value of 0.6. In the discussion of the paper that followed at the Institute meeting, their proposal excited little alarm, except for one lone voice, S.H. Cooper, who noted:

> If he had correctly understood the ... remainder, it followed that any depreciation in excess of the limit envisaged would absorb the whole of the free reserves of the fund ... and would encroach upon the cover required for the basic liabilities of the fund. In that context it seemed to him that they should be concerned with possibilities rather than probabilities, and he had been rather surprised to see the authors dismiss the American experience of 1929–32 as exceptional.[13]

Despite such occasional protestations, life office equity/property allocations continued their inexorable rise in the following years. By 1976, allocations to equity and property were around 50 % of life office assets.[14] This excited little actuarial concern. Conventional actuarial wisdom embraced equities as an inflation-proof long-term asset which life offices were uniquely equipped to hold on behalf of policyholders due to their large estates and the ability it gave

[12] Anderson and Binns (1957), p. 125.

[13] S.H. Cooper in Discussion, Anderson and Binns (1957), p. 143.

[14] Dodds (1979), p. 50.

them to protect policyholders from equity markets' irrational short-term market movements. In concluding his history of the first 200 years of Equitable Life, Ogborn in 1962 mused:

> What seems important at the time of writing is the greater freedom of investment which the life offices have felt both necessary and desirable, in the current economic climate, for the purpose of giving their policyholders some share of profits from the expansion of industry and some protection against the possible effects of further inflation, should it recur.[15]

By the 1960s, British life actuaries had learned to feel quite sanguine about the financial market risk exposures that their businesses were underwriting in ever-increasing scale. This contrasted with the technical sophistication that was developing in the pricing and management of financial market risk outside the actuarial world. An interesting next 40 years would ensue for the profession.

Equity-Linked Maturity Guarantees: Towards Risk-Based Solvency (1971–1982)

Unit trusts and unit-linked business emerged as a major competitor to with-profit policies in the UK long-term savings market in the 1960s. By the early 1970s, unit-linked business had become a significant business line for many British life offices. These products were generally much simpler than with-profit policies, and their financial management required less actuarial input. An interesting complication arose however, when life offices attempted to differentiate from the competition by providing maturity guarantees with their unit-linked funds—typically these would provide a minimum guaranteed pay-out of a return of the premium(s) paid at the specified maturity date. The maturity date typically varied between ten and 25 years. These guarantees were provided on underlying funds that were typically invested entirely in equities.

The pricing and reserving for these guarantees was not initially subject to any actuarial sophistication. In some cases, the reserve was merely retrospective, i.e. an accumulation of the (small) additional premiums charged for the provision of the guarantee. Life actuaries found themselves having to catch up with their products, and by the second half of the 1970s they were further prompted into action by the industry regulator's stated intention to impose

[15] Ogborn (1962), p. 256.

some prescriptive measures on reserving for these guarantees. Somewhat perversely, the relative simplicity of the unit-linked product created an actuarial risk management challenge. With-profit funds' discretionary levers of investment policy, bonus policy and policyholder cross-subsidy were not available to the actuary to mitigate the guarantee risk embedded in unit-linked business. This meant that the office bore the full exposure of the risks that they wrote in the policy; and the resultant transparency of those risks meant that their measurement and modelling were a lot more straightforward than was the case for with-profit business.

In 1968, at the Staple Inn discussion of Skerman's with-profit bonus paper, Jim Pegler (who at the time was only months from becoming the Institute President) raised the general topic of how to balance security with surplus distribution. Pegler raised the possibility of using a *probabilistic* basis for setting reserves. This is notable as perhaps the earliest such reference in a British actuarial paper, though Pegler recognised it would be challenging to implement and stopped short of advocating such an approach:

> The author drew attention to the over-riding need to ensure security for policyholders ... they had to find a way of arriving at a reasonable degree of security, to which absolute priority should be given ... He imagined that theoretically they could approach the problem from the point of view of the Theory of Risk and estimate a probability of ruin. But even supposing that a reasonably accurate numerical figure could be obtained, he doubted whether it was helpful in practice to consider whether a probability of, say, 1 in 1,000 or 1 in 10,000 should be aimed at; he was choosing figures more or less at random, as he had little idea of the order of magnitude which was relevant.[16]

The application of the probability of ruin concept to a real-life with-profits business and all its complexities was a daunting challenge in 1968. But for unit-linked business, it was much more feasible. As actuarial thinking caught up with unit-linked maturity guarantee practices, it was natural that a probability of ruin approach to setting adequate maturity guarantee reserves should at least be considered. The Prudential actuary Sidney Benjamin pioneered the development of this train of thought with a paper that was first discussed at Staple Inn in 1971. The meeting was described by one of the actuaries present as 'by far the stormiest I have ever attended'.[17] Indeed, it was so stormy that Skerman, as President, closed the meeting and classed the discussion as confidential so that

[16] Pegler in Discussion, Skelman (1968), p. 94.

[17] P. Smith in Discussion, Corby (1977), p. 274.

its content would never be published.[18] The paper was never published by the Institute. It was only formally presented and published five years later in 1976 at the twentieth International Actuarial Congress in Tokyo.

What was it about Benjamin's work that provoked such a convulsive reaction within the actuarial profession? In stark terms, the British actuarial profession had not been trained or educated to think about risk. Or, to put it more kindly, they had been trained to assume that an insurance office could *diversify away* risk. This worked well, up to a point, for mortality and other insurance risks. But it did *not* work for financial market risk, which was inherently non-diversifiable—all policyholders had guarantees on the performance of funds that would behave similarly. There was an elegant equivalence in the retrospective and prospective reserves produced by charging for expected costs and accumulating net premiums. But this was of no use when a contingency reserve was required for a non-diversifiable risk that could not be fully funded by the premiums received (at least not without a highly sophisticated dynamic investment strategy which was far from the actuarial minds of the time). With non-diversifiable risk, an adequate prudential reserve may be an order of magnitude greater than the value of the premiums accumulated from charging for the risk. Benjamin's paper put the actuarial profession face-to-face with the realities of non-diversifiable risk, taking it into uncharted waters. It was a seminal moment in the history of the British profession. With the benefit of hindsight, it marked the start of a 30-year struggle to modernise its thinking on the measurement and management of financial market risk.

In the framework of Benjamin's paper, the required maturity guarantee reserve was calculated as the discounted present value of the maturity guarantee shortfall (i.e. the shortfall in the final underlying fund value relative to the guaranteed maturity proceeds) assessed at a specified percentile level (the accepted probability of ruin). This calculation required two key inputs: an assumption for the acceptable probability of ruin; and a probabilistic model of the behaviour of the underlying equity funds that would allow the probability distribution of the guarantee shortfall to be assessed. This reserve could be funded by the discounted present value of the premiums that would be charged for providing the guarantees. If the reserving requirement exceeded this present value, this excess would have to be funded using the office's capital.

Benjamin developed his probabilistic equity model by considering the history of annual UK equity returns for the 51 years from 1919 to 1970. He conducted various statistical tests on the time series and concluded that there was not statistically significant evidence to reject the null hypothesis that the annual

[18] S. Benjamin in Discussion, Corby (1977), p. 277.

returns were independent. Given this result, he argued that a probability distribution for the *n*-year equity index could be generated by randomly sampling (with replacement) *n* points from the 51 historical annual return data points. He then suggested that the probability of ruin be set at 2 %, and estimated the reserve required by generating 50 simulation paths and using the one that produced the largest reserve.

This was all undoubtedly methodologically quick and dirty, but it captured the essence of the problem and it produced insightful results. Benjamin's calculations implied that the *mean* ten-year guarantee shortfall for annual premium business paying a premium of one and with a maturity guarantee of ten was 0.11 (i.e. 1.1 % of the guarantee), and that the second percentile was 4.5 (i.e. 45 % of the guarantee). Discounting at a risk-free yield of 2.5 % implied a starting reserve of 39 % of the maturity guarantee would be required. Some of this reserve could be funded by accumulating the premiums charged for granting the guarantee, but, at typical charging rates, this still implied that the office would need to fund an additional reserve of over 30 % of the maturity guarantee. Benjamin himself wrote that this reserve was 'unexpectedly high' and suggested that it meant 'the contract is probably not a commercial proposition'.[19]

There was something of the spirit of Bachelier in Benjamin's research, at least during this stage in his career. He had a talent for taking ideas such as game theory or risk theory and envisioning how they could be newly applied to actuarial problems. But his work needed further development and refinement to be rigorously implemented and made accessible to the wider profession. A couple of papers quickly followed the Tokyo Congress that provided this development: a substantial paper by W.F. Scott[20] and a shorter note by David Wilkie[21] that were both published in 1977.

Scott followed Benjamin's general probabilistic risk-based framework but his implementation of the reserving assessment differed in some key ways. Most significantly, Scott's analysis of historical equity returns led him to reject the assumption that annual equity market returns were independent over time. Scott used the same data as Benjamin—UK equity market returns from 1919 to 1970 as calculated from the de Zoete equity index—but Scott identified statistically significant serial correlation of -0.30 at a two-year lag. Benjamin never explicitly tested serial correlation in his analysis of the independence of historical returns but instead used runs tests and a few other forms of statistical test of independence. Scott tested lags of one to four years,

[19] Benjamin (1976a), p. 25.
[20] Scott (1977).
[21] Wilkie (1977).

and whilst the serial correlation was negative for each of two, three and four years, it was only the two-year lag correlation that was statistically significant from zero. Nonetheless, Scott concluded this was sufficient evidence to reject the independence hypothesis. This analysis is noteworthy for its early documentation of evidence of statistically significant negative serial correlation in long-term empirical equity market returns. It anticipated Fama and French's work by more than a decade (and there is nothing to suggest Fama and French were aware of this work). Scott's explanation of this negative serial correlation also anticipated, at least in a heuristic way, Shiller and Roll's excess short-term volatility arguments. Scott argued that the extraordinary UK equity market volatility of 1973 to 1976 (during which time dividend yields moved from 3 % to 12 % to 5 %) could only be explained by concluding 'the market panicked and then corrected itself'[22] (though he was not the first actuary or market participant to suggest financial markets could behave in such a way).

The stochastic modelling skills of the profession at this time were still embryonic, and Scott had a preference for a simple model that would not require Monte Carlo simulation methods in its implementation. So, rather than developing a stochastic model of equities with negative serial correlation in returns, he instead used a standard random walk model but with an annual volatility that was lowered to reflect the long-term volatility that would arise in the presence of the assumed negative serial correlation. Specifically, he used an annual equity return volatility assumption of 10 % rather than the 19 % that he found in the annual historical returns data. He did not provide any statistical basis for the substantial size of this adjustment.

Scott's rejection of the independence of equity returns through time and the consequent assumption that long-term returns would therefore be materially less volatile naturally resulted in significantly reduced reserving requirements for long-term maturity guarantees. Scott's modelling implied that the ten-year annual premium return-of-premium maturity guarantee required an initial reserve of 15 % of the guarantee rather than the 30 %-or-greater estimate produced by Benjamin. And that was with a 0.5 % probability of ruin instead of Benjamin's less ambitious 2 % assumption.

Scott's paper was primarily focused on reserving, but he also considered how to set the premium that should be charged for provision of the guarantee. He argued the guarantee premium should be based on the expected guarantee shortfall discounted at the risk-free yield, appealing to the 'equivalence principle'. This implied very low premiums for the guarantee—for example, the expected cost of the ten-year guarantee was estimated at 0.5 % of the guarantee

[22] Scott (1977), pp. 374–75.

(a Black-Scholes put option price would have implied 4–5 %). He also discussed how the reserving requirement could be mitigated and considered several approaches. None of them, however, recognised that the guarantee was a put option and that a body of theoretical work and trading practice had recently been developed on how to price and hedge such options.

Wilkie's paper further refined Scott's analysis. Again, the most significant development was in the assumptions underlying the stochastic modelling of the equity market return. Wilkie confidently grasped the nettle that Scott had avoided: he developed an explicitly auto-regressive equity model that was fitted to the serial correlations observed in the historical returns and that was implemented using simulation. This modelling suggested Scott's downward adjustment of annualised volatility from 19 % to 10 % was too great. Wilkie also updated the historical dataset to include returns up to 1977, thereby including the exceptional volatility period of 1973–1976. These resulted in reserving estimates that were closer to Benjamin's than Scott's.

Wilkie also criticised Scott's argument that the premium charged for the guarantee should be calculated on the basis of expected cost, arguing that shareholders would require a return on the capital that they needed to use to support the business (Benjamin's paper had made a similar argument). Wilkie suggested that shareholders would require a return of 2 % in excess of the risk-free rate on their capital on the grounds that the riskiness of their exposure was comparable to reasonable quality corporate bonds. He argued this cost should be added to the expected guarantee shortfall. Whilst such an approach was more sophisticated than Scott's, it was still far from utilising the available economic insights on option pricing that were published a few years earlier by Black, Scholes and Merton.

Benjamin, Scott and Wilkie differed in their recommended modelling assumptions, but they all shared a substantial common ground that was a radical departure from traditional actuarial thought: maturity guarantees on unit-linked business ought to require a contingency reserve that should be assessed by specifying a probability of ruin, γ, and then using a stochastic model of the underlying unit funds to determine how much reserve is required to fund guarantee shortfalls with probability $(1-\gamma)$.

This rapid actuarial research output of 1976 and 1977 had one curious institutional feature: none of it was published by the Institute of Actuaries. The papers of Scott and Wilkie were both Faculty papers, whilst, as we have seen, Benjamin had been published under the auspices of the twentieth International Actuarial Congress. The Institute had not touched maturity guarantees since the Benjamin Staple Inn debacle of 1971. This was to be addressed by an Institute working party led by F.B. Corby, but it never published its findings due to 'problems in resolution of fundamental points'.[23] So Corby went ahead and published his own paper with his personal views in 1977.[24]

Corby rejected the probability-of-ruin-and-stochastic-model reserving approach advocated by each of Benjamin, Scott and Wilkie. He argued that 'it was unlikely that it [his working party] would be able to derive a model of stock market behaviour which would be satisfactory for extrapolation into the future and which would be generally acceptable as a basis for reserving' and that 'in the United Kingdom … little attention has been paid to a ruin probability approach'.[25]

Corby explained his proposed alternative: 'The approach followed is to assume a trend line for the performance of the relevant index together with a spread about that line. To obtain the reserve at inception of a single contract, it is assumed that all purchases are made at the top of the range and all sales (i.e. maturities) at the bottom of the range.'[26]

Thus the reserve for an annual premium product was calculated by assuming that the guarantee shortfall at maturity was:

$$G - \frac{1-k}{1+k} \sum_{t=1}^{n} (1+r)^{t}$$

where r is the assumed long-term expected equity return and k is some measure of the extent to which the equity market can diverge from this trend.

This naturally led to the question of how to set the parameter values for k and r. Corby did not have a clear answer for that. He tabulated results for a range of values for k (0.2, 0.3, 0.4) and r (5 %, 7.5 %, 10 %) and noted that values of $k = 0.4$ and $r = 7.5$ % produced results similar to Benjamin's. So his approach was simpler and more transparent, especially to actuaries who did not have a familiarity with stochastic modelling, but it was somewhat arbitrary and Corby's only means of implementing it was to select parameters that were consistent with the stochastic modelling analysis developed elsewhere. His approach still demanded that a choice of stochastic model and calibration be made, even though he believed that it was not possible to make such a choice in a way that would be satisfactory or acceptable.

[23] Corby (1977), p. 261.
[24] Corby (1977).
[25] Corby (1977), p. 262.
[26] Corby (1977), p. 264.

The Corby paper was also the institute's first explicit engagement with financial economics' developments in option pricing theory. Corby noted the existence not of the original Black-Scholes-Merton papers, but of a number of papers that had been written more recently by academics at the University of British Columbia on the application of option pricing theory specifically to unit-linked maturity guarantees.[27] These papers by Brennan, Schwartz and the actuary Phelim Boyle are notable as the earliest work on applying option pricing theory to the pricing of the financial guarantees found in life assurance policies. They directly applied risk-neutral valuation to the guarantees and showed that they could be hedged using portfolios with dynamically rebalanced holdings of risk-free bonds and underlying units. Corby relegated discussion of these concepts to an appendix and delegated the writing of the discussion to another actuary, P.J. Nowell.

Nowell dismissed the option pricing work as a theoretical irrelevance, concluding: 'Here practice and theory are irreconcilable … the investment procedure [dynamic hedging] is a perfectly reasonable theoretical concept but as a practical proposition it is one which contains risks greater than the risk which it is designed to eliminate'.[28] Corby and Nowell threw the baby out with the bath water. There was no analytical basis for their assertion that real-life dynamic hedging would result in an overall *increase* in the life office's risk position. And the fundamental economic insight of the guarantee replication argument and its implications for pricing appeared to be lost on them. However, it is interesting to note that a couple of speakers at the Staple Inn discussion of the paper raised the prospect of 'immunising' the office from guarantee losses by holding a negative exposure to the underlying units in the contingency reserve.[29] These actuaries did not refer to option pricing theory or dynamic hedging. To them, delta hedging was an intuitive extension of Redington's immunisation theory.

Whilst Corby and Nowell's discussion of option pricing theory and dynamic hedging was rather perfunctory and dismissive, the topic and its application to unit-linked maturity guarantees received further attention from the profession in the following years. In particular, T.P. Collins published an Institute paper in 1982 which discussed in detail the concept of investing the guarantee premiums in a dynamic hedging strategy.[30] Collins's analysis was intelligent and open-minded. He provided a thorough examination of the challenges of

[27] Brennan and Schwartz (1976b), Boyle and Schwartz (1977). Also see Brennan and Schwartz (1979b; 2).
[28] Corby (1977), p. 273.
[29] Fagan, p. 282, and Seymour, p.285, in Discussion, Corby (1977).
[30] Collins (1982).

a 'real-life' implementation of a dynamic hedging strategy, such as transaction costs and discrete-time rebalancing frequencies. He highlighted that the hedging strategy is exposed to large short-term movements (up or down) in the underlying asset value and that large switches in the hedge portfolio may be required as the guarantee nears maturity when the underlying unit value is close to the guaranteed amount.

Collins modelled what the historical performance of hedging and 'conventional' investment strategies would have been over the period of 1930–1978 in the UK. He concluded that the dynamic hedging strategy 'compares unfavourably with the conventional strategy'.[31] However, he caveated this conclusion with the observation that 'The conclusion … depends critically on whether the [historical] price series used in the investigations is representative of the underlying distribution … if the price had not risen sharply after the end of 1974 the immunization [dynamic hedging] strategy would have proved superior'.[32]

Collins correctly identified the couple of key features of the historical return series that drove his conclusion. One was the negative serial correlation (mean-reversion) in the returns. This effect reduced the risk of the conventional strategy, but did not reduce the cost of the hedging strategy (which is driven by the short-term volatility experience and is not affected by serial correlation). Secondly, the hedge strategy did not mitigate the impact of exceptionally large short-term movements that could occur between hedge rebalancing actions. This was partly a reflection of the limited sophistication of his modelled hedge strategy implementation, but nonetheless reflected the genuine 'gap' risk that is present in a dynamic hedge program. The paper was a perceptive contribution to thinking on the applicability of dynamic hedging of long-term guarantees, though its empirical emphasis missed the pricing insights at the core of option pricing theory.

A working party report on unit-linked maturity guarantees was published in 1980[33] (by which point Collins's paper had already been submitted but not yet published). The working party tried to establish an actuarial orthodoxy on the vexed topic of how to reserve for these guarantees. Benjamin and Wilkie were both members of the working party, whilst Corby was not. Its conclusions were therefore unsurprisingly broadly consistent with the reserving approach advocated by Benjamin, Scott and Wilkie and did not consider Corby's approach. The working party recommended that a probability of ruin

[31] Collins (1982), p. 280.
[32] Collins (1982), p. 281.
[33] Ford et al. (1980).

approach to reserving should be used, and that a stochastic model of equity returns be used to assess the reserve.

Wilkie developed a new version of his stochastic model of equity returns for use by the working party in its reserving analysis. Its notable new feature relative to the model he used in his 1977 paper was that the equity total return index was not modelled directly, but instead as the ratio of dividend pay-outs to dividend yield. Explicit stochastic processes were developed for dividends (its logarithm followed a normally distributed random walk with upward drift) and dividend yields (a mean-reverting lognormal process). The equity price was the ratio of the two. This structure provided the flexibility to capture the volatility patterns implied by Wilkie's view of material mean-reversion in long-term equity returns—it could be fitted to views for both short-term and long-term equity return volatility that were consistent with significant serial correlation in equity prices. It produced similar results to Wilkie's 1977 model. That is, it produced a significant long-term mean-reversion effect that resulted in lower long-term price and return volatility than a 'random walk' model with the same year-to-year volatility assumption. The reserves estimated in the 1980 working party report were broadly similar to those produced in Wilkie's 1977 paper.

The working party report discussed dynamic hedging and, whilst noting it was a subject which merited further investigation, they considered that 'there is no basis for reducing maturity guarantees reserves because a company follows some form of immunization strategy and, in fact, a company that follows such a strategy without fully appreciating the difficulties could well require greater maturity guarantee reserves than would otherwise be the case'.[34]

The working party report brought the British actuarial profession's ten-year saga with unit-linked maturity guarantees to a close. The quantum of reserves recommended by the working party was not substantially different to the rough-and-ready estimates produced by Benjamin a decade earlier. The working party's two fundamental recommendations—that appropriate risk-based reserves should be held for unhedged market risk, and that no reduction in reserves can be obtained by hedging the risk—made the maturity guarantee feature of unit-linked policies commercially untenable and it was no longer sold in volume by British life offices. In 1981, Duncan Ferguson, who would go on to be Institute President in 1996, was able to state at a Staple Inn sessional meeting: 'The considerable expertise which actuaries have now developed to handle linked policies has forced them to accept the logic that capital guarantees are too expensive to be given ... and that in the end their offices

[34] Ford et al. (1980), p. 112.

can afford to do little more than provide policies under which the customer takes nearly all the risks.'[35]

This story reflects the stresses that arose when life offices underwrote, in a very transparent way, significant volumes of non-diversifiable financial market risk rather than (largely) diversifiable mortality risk. As we have seen, this trend towards taking more market risk had been underway for the previous 50 years. But up until the unit-linked maturity guarantee product, this risk-taking had been cloaked in the complexities of with-profits business. The simplicity of the unit-linked maturity guarantee made this risk-taking highly transparent, and it caught the profession unprepared.

Nonetheless, over the ten years following Benjamin's stormy evening at Staple Inn—the sessional meeting that never was—the profession made real progress, not least because of the precocious young Scots actuary David Wilkie. The profession had accepted the principle that life offices should hold contingency reserves beyond the retrospective accumulation of premiums to support financial market risk exposures. They had adopted a stochastic modelling approach that could be used to provide reasonable estimates of the required reserves. The working party and Collins had produced intelligent and unprejudiced appraisals of the merits of dynamic hedging. Whilst they had significant reservations, they had indicated a desire to constructively engage with the new ideas of financial economics and option pricing theory to further actuarial thought and practices. Significant and permanent developments in actuarial thought on the management of financial risk had been achieved.

The Wilkie Model (1984–1995)

The Wilkie Model was the ubiquitous stochastic asset model for British actuarial use from the mid-1980s to the end of the 1990s—and, to a lesser degree, beyond. Its genesis can be found in the stochastic equity modelling that Wilkie developed and applied as a member of the Maturity Guarantees Working Party. Wilkie further developed this work into a broader stochastic asset model—which quickly became known as the Wilkie model—in a paper presented to the Faculty of Actuaries in 1984 and published in the *Transactions of the Faculty of Actuaries* in 1986.[36]

---

[35] D.G.R. Ferguson, in Discussion, Redington (1981), p. 389.

[36] Wilkie (1986a).

The Wilkie model shared many stylistic and methodological similarities with the equity modelling that Wilkie had developed for the Maturity Guarantees Working Party. Equity prices were again modelled as the ratio of dividend pay-outs to dividend yields using a time series model structure. The model was expanded beyond the working party's asset modelling in two key dimensions: stochastic models for the long-term government bond (consol) yield and inflation were developed; and a 'cascade' structure was introduced such that the time series models of dividends, dividend yields and consol yields included dependencies on past and present inflation as well as past values of themselves. The modelling approach was heavily based on time series analysis and postulated some quite convoluted statistical relationships in the joint behaviour of equities, consols and inflation. The model structure and calibrations were driven directly by the relationships Wilkie found in the historical data (based on UK series from 1919 to 1982). He was not concerned with imposing many economic restrictions or null hypotheses on the projected asset behaviour. These characteristics were again evident when Wilkie published an updated version of the model in 1995.[37] This update extended the asset coverage of the model, most notably by including models of short-term government bond yields and property.

Wilkie's preference for modelling historical behaviour as he found it, without imposing strong economic prior beliefs, led to some striking implications for the risk/reward profile of different asset classes and, especially, for what could be generated by dynamic asset strategies. The structure and calibration of the Wilkie model implied that most variation in equity dividend yield was due to changes in the expected future return of equities (without any corresponding change in risk) rather than being a signal of a significant change in dividend growth expectations. As Wilkie himself noted in a 1986 paper,[38] this meant that dynamic asset strategies that switched between equities and bonds using simple rules based on their relative yield levels could generate returns in excess of those produced by simply buying and holding equities.

This begged an interesting quasi-philosophical modelling question: should actuaries' risk and capital assessments include 'credit' for investment strategies that take account of projected market 'inefficiencies'? Should these inefficiencies be assumed to prevail into the future? Are these strategies gaming the asset model or are they reflective of a fundamental reality that ought to be incorporated into the assessment of the risk profile of the business? Such issues remain at large in twenty-first-century principle-based reserving.

---

[37] Wilkie (1995).

[38] Wilkie (1986; 2).

These questions generated very little discussion in the 1980s actuarial papers and sessional meetings. They did, however, attract greater attention (and discomfort) amongst actuaries in the 1990s, particularly from those in the vanguard of modernising actuarial thought through the firmer embrace of financial economic principles. Arguably, this is perhaps ironic as Fama and French's 1988 dividend yield and expected return empirical analysis that emerged between the first publication of the Wilkie model and the profession's eventual interest in financial economics could provide some intellectual support from financial economics for Wilkie's approach.

Risk and return anomalies also existed in the relative behaviour of other asset classes—in particular, between equities and property. Wilkie had a fairly limited volume of available property data (28 years from 1967 to 1994) and it implied average property returns significantly in excess of the average return generated by the equity calibration, yet with a very similar risk profile. Such a feature might possibly be rationally explained by illiquidity premiums or related transaction costs, but it was never Wilkie's objective to try to provide an economic explanation for market behaviour: he simply calibrated to what he found.

The other notable historical feature of the Wilkie model was its approach to modelling the yield curve. In his original 1984/1986 paper, Wilkie only attempted to model a single yield (the consol yield), so the issue of how to consistently model all points on the yield curve never arose. But Wilkie chose a model structure for the consol yield that was quite convoluted. The yield was a function of its value in the previous three years (and hence was non-Markovian). It was unnecessarily complicated—Wilkie himself noted in his original paper that a first-order autoregressive process could produce very similar results and in the 1995 update of the model he implemented this change. Writing later, Wilkie acknowledged that the original consol yield model may have been an example of 'over-parameterisation'.[39]

Wilkie never really grasped the nettle of producing a full stochastic yield curve model. In his 1995 paper, he developed a short-term interest rate model to go alongside the consol yield, but he left it to the user to decide how to join the two dots. He referenced polynomial functions in the 1995 paper, and in a later paper of 2003 he suggested a specific polynomial functional form to interpolate between the two yields.[40] He did not consider arbitrage constraints or advocate the use of the arbitrage-free yield curve models that had been developed since Vasicek's 1977 paper. (Interestingly, Brennan and Schwartz derived the arbitrage-free yield curve pricing function for a two-factor model

[39] Wilkie et al. (2003), p. 319.

[40] Wilkie et al. (2003), pp. 356–58.

where the first factor is the short rate and the second factor is the consol yield in 1982[41]). Indeed, Wilkie seemed uninterested in Vasicek's arbitrage-free logic and its insights for constraints on the term premiums generated across the yield curve. He was fairly dismissive of the entire stochastic yield curve modelling stream of financial economic research, writing in 1995:

> Many alternative yield curve models have been proposed in the academic literature ... Unfortunately, to my mind, they are usually based on an assumption about how yield curves ought to behave rather than being based on how they actually do behave.[42]

This comment highlights Wilkie's deeply empirical modelling philosophy. And there was some truth in his observations: the arbitrage-free yield curve structures of the financial economics literature had features that were deduced from the arbitrage-free condition rather than purely from empirical observation; moreover, most of these models were developed for the purposes of derivative pricing rather than to describe the fullest description of 'real-world' yield curve dynamics. Equally, however, Wilkie offered no clear rationale for why these approaches would not be the most appropriate way of jointly modelling the behaviour of, say, a ten-year bond yield alongside his cash and consol yield; and he did not raise or address the potential limitations of using a yield curve model that could permit bond arbitrage and that provided no means of controlling the relative size of the term premiums available across the term structure. Of course, the importance of this point depended on the use to which the model was put. In the 1980s and 1990s, the model was mainly used in the analysis of long-term asset allocation choices between asset classes such as equities, property and long-term government bonds in the context of funding long-term liability cashflows. For these purposes, the absence of a robust yield curve model was relatively unimportant. But for later actuarial applications such as the analysis of alternative derivative hedging strategies for guaranteed annuity options, the model was wholly inadequate. Of course, Wilkie never envisaged or advocated the use of the model in this type of application when he developed it.

For context, it is worth noting, however, that some actuarial research was published that did try to embrace the new yield curve modelling technology of Vasicek et al. and showed how it could be made use of in actuarial practice. In particular, Phelim Boyle, who we met earlier as one of the pioneers

---

[41] Brennan and Schwartz (1982).

[42] Wilkie (1995), p. 874.

of the application of option pricing theory to life assurance guarantees, published a paper in the *Journal of the Institute of Actuaries* as early as 1978 that showed how Redington's duration calculations could be generalised to allow for the volatility term structure implied by modern yield curve methods.[43] Redington's duration approach had assumed parallel yield curve movements, which implied that long rates were as volatile as short rates. Empirical evidence, and arbitrage-free yield curve models, suggested long rate volatility was lower than short rate volatility. This meant that if matching the duration of a set of liability cashflows with a long bond and a short bond, more would really be required in the long bond than was implied by Redington's duration calculation. This was a fundamentally important point for asset-liability duration management and Boyle showed how the new financial economic modelling approaches could cast fresh light on it.

The Wilkie model's public domain and wide application meant that it was subject to considerable peer scrutiny—most notably, a working party was established in 1989 to review the model and it published a paper with its findings in 1992.[44] But it focused mainly on the statistical quality of the time series fits, especially for the inflation model, rather than on arguably the more fundamental question of whether such a statistically orientated and empirical approach was desirable and what its economic limitations would be.

The Wilkie model was an important step forward for the British actuarial profession in embracing the probabilistic analysis of the financial market risks that increasingly dominated the balance sheets that actuaries oversaw. More generally, it was also important in motivating the profession to develop its quantitative modelling skills. But it was also a product of the profession as it existed. In particular, Wilkie's reticence to incorporate any of the insights that could be delivered from the (then-recent) developments in financial economics into his model could be viewed in hindsight as a missed opportunity for the further development of the profession's financial risk management thinking.

With-Profits: Reserving and Financial Economics (1987–2004)

Whilst an actuarial consensus on the uneconomical nature of capital guarantees as it pertained to unit-linked business had been established by 1981, this view would typically not extend to the capital guarantees provided

[43] Boyle (1978).

[44] Geoghegan et al. (1992).

under with-profit policies. With-profits business and its guarantees had some undoubtedly significant differences with unit-linked business and its maturity guarantees. In the context of the risks and costs to the life office of providing guarantees, linked and with-profit business had two differences of the utmost importance. First, the office had significant control over the investment strategy of the with-profit fund. If assets performed poorly, they could, in theory, be switched into bonds that matched the guarantees before the surplus funds were exhausted; second, in with-profits, the cost of the guarantees could be borne by other generations of policyholders (rather than the office or estate) through reductions in their future bonuses. If one generation of policyholders did not have accumulated asset values sufficient to meet the minimum guarantee attaching to their policy, this could be funded by reducing the bonuses that would be distributed to other policyholders. These broad powers of actuarial discretion provided risk management levers that could, in principle, substantially mitigate the financial market risks to the office that were created by with-profit guarantees.

Actuaries, and indeed the UK life office regulator, were therefore significantly more sanguine in the 1980s about with-profit guarantees than those found in their unit-linked cousins. There was no body of opinion advocating the extension of the stochastic modelling and probability of ruin approach developed for unit-linked maturity guarantee reserving—the 'considerable expertise' that Duncan Ferguson referred to—into the domain of with-profit reserving. This was partly because it would be viewed as unnecessary as the office did not have the same degree of risk exposure, and partly because it would be much harder to model the complexities of with-profit business and actually demonstrate that these risk management levers could be adequately operated by life office actuaries to manage the guarantee risks. This scepticism around the benefits of using stochastic modelling in the actuarial management of with-profit business is well-reflected by Frank Redington's comments in 1976:

> He [Redington] expressed uneasiness at leaning upon computers to answer these unanswerable questions. If there was a beam in the eye of the question, why worry about a mote in the eye of the answer? It was easy to fail on the back of an envelope as with a computer, but a great deal more instructive. This was a countryside to explore on foot and not by fast car.[45]

Some British life offices did develop and internally use stochastic 'model offices' in the 1980s and 1990s, but the application of stochastic models and probability of ruin approaches to British with-profit statutory reserving

[45] Redington in Discussion, Kennedy et al. (1976), p. 48.

was deferred for over 20 years until the beginning of the twenty-first century. What about the use of option pricing theory in charging and reserving for with-profit guarantees? We saw above that the (unit-linked) Maturity Guarantees Working Party did not actively make use of option pricing theory in its work on the pricing, reserving and risk management of unit-linked guarantees. However, in the following years the British profession, and in particular David Wilkie, started to explore the potential applicability of option pricing in actuarial work. In 1987, Wilkie published a notable paper on the use of option pricing in the analysis of with-profit guarantees and bonuses with his Institute paper, 'An Option Pricing Approach to Bonus Policy'.[46]

Wilkie showed that with-profit style pay-outs could be replicated by a mixture of investments in unit-linked funds and put option contracts. The recognition that with-profit guarantees could be considered as a form of put option was not remarkable and, as we saw above, this was first considered by Phelim Boyle some ten years earlier. Wilkie broke new ground, however, by considering how different reversionary bonus strategies could impact on the option costs associated with the guarantees. Wilkie also tentatively suggested that the cost of the options that correspond to the reversionary bonus strategy could be used as a reference point for a deduction from with-profit terminal bonuses as a form of charge for the provision of the guarantee (when the fund performance is good enough for a terminal bonus to be paid).

Wilkie's approach resulted in relatively high estimates of the guarantee cost as it did not allow for the cost-mitigating with-profit features of office-controlled investment strategy or inter-generational cross-subsidy. A 20-year single-premium contract's guarantee cost inclusive of reversionary bonuses was estimated to require a deduction from the final proceeds of the policy of, on average, around 15 % to 25 % across a range of stochastic scenarios.[47] It was, however, the natural first step in applying option pricing ideas to with-profit guarantee charging and Wilkie, now established as the British actuarial technical thought-leader of his generation, was the natural actuary to lead the way. The paper also helped to more generally introduce option pricing ideas to the actuarial profession. Collins's excellent work back in 1982 was never presented at a Staple Inn sessional meeting, and before the presentation of Wilkie's paper in June 1987, option pricing theory had never been discussed at an Institute sessional meeting. In opening the discussion of Wilkie's paper, J.M. Maud started with the telling statement:

---

[46] Wilkie (1987).

[47] Wilkie (1987), p. 55.

> Until a few months ago I was almost entirely ignorant about option pricing theory. Having been introduced to the subject, I was astonished that a theory with so many obvious uses in actuarial work is so little known to many of us.[48]

However, British actuarial thinking on the use of option pricing theory in with-profits reserving and risk management lay largely dormant for the next decade. At the start of the 1990s, more pressing matters were at hand. The actuarial profession and its practices in the with-profits sector were under scrutiny like never before. Similar to the experience with unit-linked maturity products 20 years earlier, there was a sense that actuarial methods were failing to keep up with a changing world. With-profit products were under great competitive pressure from the alternative investment vehicles of asset managers and banks. Life offices had evolved their with-profit offerings to compete: 'conventional' regular premium with-profit business was on the wane, and was being superseded by 'unitised' with-profits, which was a more flexible, recurrent single premium product design. The net premium valuation method was obviously less applicable to the solvency assessment of single premium products, and, in any case, there was concern amongst actuaries at the start of the 1990s that the net premium statutory valuation requirements had become unreasonably stringent.

There was also increasing regulatory pressure to take explicit consideration of Policyholders' Reasonable Expectations (PRE) in the management of with-profits, and for these considerations to be reflected in with-profit reserving methods. This was not easy in the net premium valuation method, where no explicit allowance was made for the future bonuses that policyholders could reasonable expect to receive. Meanwhile, Canada and Australia had recently enacted new solvency methods that dispensed with net premium valuations. These changes were largely driven by pressure from accounting standard setters to produce valuations that could conform to generally accepted accounting principles. Working parties were established by the Institute and Faculty in 1993 to investigate a modernisation of with-profit reserving methods. This started a process that meandered for a decade as actuarial leaders tried to reconcile their traditional methods and training with the demands of a fast-changing external environment.

One of the first deliverables from this process was the 1996 paper by a working party, led by P.G. Scott, which had been established with the remit to investigate alternatives to the net premium valuation for statutory solvency

[48] J.M. Maud in Discussion, Wilkie (1987), p. 78.

assessment.[49] This paper proposed two fundamental changes to the statutory solvency valuation method for with-profit business: the net premium valuation should be replaced with a gross premium bonus reserve valuation (which should include future reversionary bonuses but not terminal bonuses); and a second value should also be reported, which was intended to be a measure of the 'realistic' value of the policy having reference to policyholders' reasonable expectations (and which was loosely interpreted as the policyholder's asset share, i.e. the accumulated value of their invested premiums).

The arguments presented in the paper and the profession's discussion of them rather resembled the actuarial discussions of the late nineteenth century. Net premium liability valuations taken alongside asset market values produced spurious variation in free assets; it did not have regard to the reasonable (bonus) expectations of policyholders; but if a gross premium bonus reserve valuation was used, what level of future bonus should be reserved for? And why did the working party recommend reserving for future reversionary bonuses but not terminal bonuses? (A certain logic for this position existed. European Union legislation required the liabilities to be discounted based on the yield of the assets backing the liabilities. The working party argued that in the scenario that equities did not generate any capital growth, no terminal bonuses would be paid, and hence they could be ignored in a statutory valuation that only took account of equity dividend yields.)

Risk margins were incorporated into the valuation under both the net premium method then in place and in the gross premium approach proposed by Scott's working party, but there was no rigorous basis for determining what these margins should be. There was no underlying concept of what level of security the solvency valuation should aim to deliver—there was no probability-of-ruin objective. Furthermore, there was no consideration of option pricing ideas or how financial market risk could be managed in with-profits and how reserving methods could be aligned with risk mitigation (i.e. *risk-based*).

The two key recommendations of the Scott working party—that the net premium valuation be replaced by a gross premium bonus reserve valuation for conventional with-profits, and that a second valuation be submitted that reflected the terminal bonus expectations of policyholders—were broadly rejected by the actuarial profession. So another working party was established in 1996 with terms of reference to determine whether the ideas put forward by Scott could be developed into something better. This new working party, led by P.W. Wright, published a paper reporting their findings in 1998.[50]

[49] Scott et al. (1996).

[50] Wright et al. (1998).

It rejected the key recommendations of the Scott working party and recommended that the net premium method be retained for conventional with-profits. However, part of their reasoning for this recommendation was not that the Scott working party's findings were necessarily flawed, but rather that conventional with-profits business was now being written in such low volumes that a major change in its valuation method was hard to justify. As Frankland put it at the Institute discussion:

> At the start of this century prophetic actuaries spoke of the 20th century being that in which a replacement would finally be found for the net premium valuation method. Today's prophets appear to accept that, within a decade, the replacement of the net premium valuation method for conventional with-profits business will cease to be a matter of materiality in the valuation of a with-profits life office.[51]

Like the Scott working party, the Wright working party recognised that solvency and PRE must be taken into account in reserving for unitised with-profits business. But unlike Scott, the Wright working party advocated achieving this through a single reserving figure which would be somewhat akin to the greater of the two calculations suggested by Scott. The first would be a 'bonus reserve test' which would be the present value of guaranteed maturity values. Like in the with-profit regulations of the time, this present value would be calculated in a 'resilience' scenario that featured a prescribed series of stresses to assumptions. This scenario was not explicitly assessed using stochastic models and probabilities of ruin as per the unit-linked maturity guarantee reserving practices, but its intention was fundamentally similar (the effect was not so severe because the tests were weaker than the tails produced by the stochastic equity modelling of the Maturity Guarantees Working Party and because asset allocations were less risky for with-profit business). The important point that the Wright working party advanced was that it was not sufficient for this reserve to be based only on the *current* level of the guarantee—some explicit allowance must be made for future reversionary bonuses that would be consistent with PRE. The allowance should be consistent with the reversionary bonuses that would be paid in the circumstances assumed in the statutory valuation basis (including those used in the resilience test). Like the Scott working party, the Wright working party argued that no allowance should be made for terminal bonus in this leg of the calculation (this was a principle

[51] Frankland, in Discussion, Wright et al. (1998), p. 1036.

that was much cherished by British life offices and its actuaries and that was only permitted in EU legislation after a great deal of political horse-trading).

The second leg of the reserving calculation was the surrender value of the policy that was consistent with PRE. This would generally be closely related to the asset share of the policy (i.e. the accumulated value of the premiums at experienced investment return, net of charges and expenses). A requirement for the reserve for a with-profit policy to have a floor that was equal to or close to asset share implied that the policy could not provide substantial capital support to the business, even when the level of guarantee was very low. PRE implied that terminal bonuses had to be reserved for after all—even if it was only the terminal bonus that had 'accrued' within current surrender values. The Institute discussion of these proposals was very mixed. To some, the principle of reserving for terminal bonus was anathema, and they argued that PRE did not mean a 'guarantee' of a surrender value that was closely related to the policy's asset share. To others, the proposals were merely a codification of then-prevailing best practices. But whether they liked it or not, some 40 years after the first terminal bonuses were paid on British with-profit policies, it appeared inevitable that regulatory pressure on the explicit treatment of PRE would result in their inclusion in statutory solvency assessment.

During this episode of statutory reserving navel-gazing and prevarication, there was very little research published by the profession on the application of option pricing concepts to reserving or charging for with-profit guarantees. That changed, however, in the late 1990s. Over the period between 1997 and 2004 several important papers were published on this topic. By the mid-1990s, the profession's internal debates on if and how to use financial economic ideas had become increasingly strained and adversarial. This perhaps reflected a more general malaise that was impacting on the profession and its traditional areas of focus. Actuaries were facing an unprecedented degree of scrutiny from others in the financial, regulatory and even legal sectors. Financial economics was increasingly being viewed by some actuaries as a threat to their core doctrines rather than as a tool they could utilise. Criticism emerged, especially in the pensions field (discussed below in Chap. 6), that suggested some actuarial struggles were arising from their economic illiteracy. Some actuarial 'traditionalists' now rejected financial economic thinking, not on the grounds that it was a nice theory with no practical utility (Corby and Nowell's position), or because it needed to be considered more thoroughly in order to make use of it (Collins and Wilkie's position), but because it was simply wrong and ought to be wholly rejected.

This latter view was the position of Robert Clarkson, who made his arguments in a *Journal* paper, 'An Actuarial Theory of Option Pricing', published in 1997.[52] This paper must surely rank highly on the list of the most eccentric papers ever published by an actuarial journal. Clarkson took the reader on a journey that pondered the ideas of many great thinkers such as Einstein, Keynes, Bernoulli, Newton, Adam Smith, Mandelbrot, Halley, Hayek and himself before concluding that 'we have to abandon this [Black-Scholes] methodology completely'[53] and 'it seems unscientific in the extreme not to conclude that a completely new paradigm of option pricing is urgently required'.[54] This rejection was based on the argument that 'its formal mathematical derivation is completely detached from reality'.[55]

He rejected the most basic and general insights of well-established option pricing theory. He argued that it was 'commonsense' that a call (put) option price must increase (decrease) with the expected return on the underlying asset.[56] This whimsical rejection of the risk-neutral valuation concept would have been plausibly breathtaking to other financial professionals. Clarkson proposed a 'new approach' to option pricing. But it was fundamentally the same one that had been advocated for unit-linked maturity guarantee pricing back in the 1970s by Benjamin and Wilkie: calculate the guarantee price as the real-world expected cashflow, discounted at a rate consistent with a static asset mix, and add a loading for the cost of the capital that would be required to support the risk associated with the (unhedged) option. Clarkson also argued that financial markets exhibit 'systematic over-reaction' and that the expected cashflows and capital should therefore be calculated on the basis of a mean-reverting stochastic process, which again was consistent with the Maturity Guarantees Working Party. Beyond the hyperbole, his proposed grand new idea was not new at all. If anything, it was an anachronism. It failed to produce prices consistent with put-call parity—the most basic (and model-independent) fundamental property of put and call option pricing. Clarkson did have many sensible practical observations to make, and his general critique of the limitations of economic theory to practical financial practice was not without some basis, but his overall position was so extreme and out-of-touch with the modern financial sector that it was an embarrassing reflection on the British actuarial profession that it provided a platform for such views in the late 1990s.

[52] Clarkson (1997).

[53] Clarkson (1997), p. 333.

[54] Clarkson (1997), p. 335.

[55] Clarkson (1997), p. 367.

[56] Clarkson (1997), p.335.

The Faculty discussion of the paper highlighted that by this point in time an impressive generation of younger British actuaries such as Kemp, Cairns, Exley, Smith, Mehta, Macdonald, Speed and Bowie had emerged who had independently developed expertise in financial economics and who worried that the British actuarial profession was dangerously behind other financial professionals in their positive utilisation of these ideas. Bowie gave a particularly impassioned speech where he argued that 'it is nothing short of misplaced arrogance to repudiate a perfectly respectable and successful science on the basis of a criticism [use of abstract or unrealistic models] to which our own profession is also subject'.[57] This ultimately bode well for the 'catching-up' that the profession would go on to undertake in the 2000s.

In the late 1990s with-profit funds were experiencing a period of difficult publicity that required life offices and the actuarial profession to be clearer about how the product worked and, in particular, which policyholders were paying for what. Improving the transparency around guarantee costs and how they were charged for was increasingly another important factor—along with PRE and financial reporting requirements—driving change in actuarial methods in with-profits. As an appointed actuary commented at an Institute sessional meeting in April 2000:

> Recent events, including guaranteed annuity options, potential mortgage endowment shortfalls, and a public suspicion that life offices have "squirreled away" policyholder money over the years in the form of orphan estates, have made the public and consumer press less inclined to trust life companies and the "black box" processes of the traditional with-profits fund.[58]

A flurry of papers was produced between 1999 and 2004 that significantly enhanced the actuarial profession's thinking on the application of option pricing ideas to the assessment of the costs and risks associated with with-profit guarantees. These papers provided new insights into potential transparent guarantee charging and risk management approaches for with-profits. The first of these papers, 'A Market-Based Approach to Pricing With-Profit Guarantees',[59] was published in 1999 by the Faculty of Actuaries' Bonus and Valuation Research Group. The group was led by David Hare, who would go on to become President of the Institute and Faculty in 2013. The paper can be viewed as a natural, if somewhat belated, development of Wilkie's 1987

[57] Bowie, in Discussion, Clarkson (1997), p. 385.

[58] Saunders in Discussion, Hare et al. (2000), p. 722.

[59] Hare et al. (2000).

with-profits and options paper and the 1980 Maturity Guarantees Working Party paper. The British actuarial profession's 1990s issues with quantitative techniques and financial economics were referred to by Hare in his introduction of the paper at the Faculty sessional meeting: 'Mention of the word "stochastic" can cause some actuarial eyes to glaze over, and the inclusion of equations like the Black-Scholes pricing formula can prove major deterrents to a wide readership of a paper'.[60]

The Hare paper applied the Maturity Guarantees Working Party's approach to reserving for guarantees—a probability-of-ruin approach using a 1 % probability and a stochastic equity model with mean-reverting properties—to determine the reserves required for with-profit guarantees at various durations and with various levels of equity backing ratio (which were assumed to be static). Unsurprisingly given the similarity in methodology, these results were consistent with the Maturity Guarantees Working Party results. The paper also considered a guarantee charging approach that had two components: the expected (real-world) guarantee shortfall plus a loading for the cost associated with the capital that would need to be held to maintain a 1 % probability of ruin. As discussed above, this was consistent with the 1970s thinking of Benjamin and Wilkie, and also with the more recent writings of Clarkson. The paper then went on to also consider a 'market-based' approach to pricing the guarantee. That is, put options (and their market prices) were used to match the guarantees (for a given equity asset allocation). They introduced an interesting wrinkle to the use of options—they argued that in order to compare with the capital-based approach, the strategy should not pay out on guarantee shortfalls beyond the 1 % probability of ruin level, and so the strategy involved a put spread where the office sold a put option with a strike at the equity index value at the 1 % probability of ruin level. They concluded that the two approaches to pricing the guarantees produced broadly comparable results (though this would naturally be a function of subjective parameter choices for the expected-shortfall-plus-cost-of-capital-loading approach). Like in the Wilkie 1987 paper, these with-profit guarantee costs were generally higher than was intuitive to with-profit actuaries because it made no allowance for the cost-mitigation levers that were available to the with-profit actuary (such as reducing the fund's equity asset allocation level when surplus assets were eroded).

At the start of the 2000s, significant changes in global insurance financial reporting and UK insurance regulation were afoot. In 1997, the International Accounting Standards Board started work on developing a new accounting

[60] Hare in Discussion, Hare et al. (2000), p. 197.

standard for insurance. In the following years its intention became clear: the principle driving the Standard would be that a balance sheet-driven approach should be used for insurance firms' financial reporting; profit should be measured as the change in the value of assets and liabilities; and those valuations should be done at fair value (which, for deep and liquid markets, meant market value). The Faculty and Institute Life Board established a working party in 1999 to consider what a 'fair value' approach to asset and liability valuation 'might offer for the development of an improved approach to reporting for prudential supervisory purposes'.[61] The terms of reference of the working party provided the motivation:

> It is recognised that the existing methods of actuarial valuation for long-term insurance, particularly for supervisory purposes, are inadequate for their purpose under certain circumstances and economic conditions.

This would be the third actuarial working party in ten years to make an attempt at improving actuarial methods for life office statutory reserving. It would be easy to suspect it was a task that was beyond the profession. But this time was rather different: now the profession merely had to follow the lead given to it from outside. Historically, financial reporting was based on adjustments to the actuary's statutory solvency valuations (the 'modified statutory basis'). Now the roles would be reversed. Financial reporting standards would lead the specification of the profit reporting requirements, and the job at hand for the actuarial profession was to consider how these developments could help the profession make their necessary improvements in statutory valuations.

This Fair Value Working Party, chaired by C.J. Hairs, presented its findings at Institute and Faculty meetings in November 2001 and their paper was published in the *British Actuarial Journal* in 2002.[62] Its findings were fairly revolutionary in comparison to the progress of the Scott and Wright working parties. This working party supported the fair value approach to financial reporting. More fundamentally, it concluded that prudential solvency assessment should also be based on the fair value approach. They advocated a risk-based capital system for with-profits that made explicit use of probabilistic models and a defined probability of ruin; the ruin event could be defined not simply as failing to fund all liability cashflows as they fell due, but as fair value insolvency over some specified time horizon that should be related to the time taken to close out risk positions. They recognised that with-profit liabilities

[61] Hare et al. (2000), p. 204.
[62] Hairs et al. (2002).

would require market-consistent stochastic models to estimate fair values, and these models would likely need to be complex in order to adequately capture management actions. There was no mention of net premium valuations.

All this was quite revolutionary and it is fascinating to see that relatively little dissension arose in the Faculty and Institute discussions of the paper. Thirty years after Benjamin's aborted unit-linked maturity guarantee paper on stochastic reserving approaches, the actuarial profession had now come to accept that such principles should be implemented in with-profit funds, the traditionally undisputed home of actuarial judgement and discretion. But the profession was now dancing to someone else's tune. The UK financial regulator, the Financial Services Authority (FSA), the International Association of Insurance Supervisors and the International Association of Actuaries had all recently suggested a move to a more risk-based capital system that could reduce the scale of regulatory arbitrage that prevailed between countries and between insurers and banks. The emerging New Basel Capital Accord for banks would use probabilistic models and even encouraged institutions to develop their own internal models for capital assessment.

Meanwhile, the UK's FSA published consultations in 2002 and 2003 that would ultimately enshrine these concepts in the statutory solvency system.[63] It required market-based measures of with-profit liabilities and guarantee costs. Whilst the actuarial profession may have been fairly sanguine about its reserving methods for conventional with-profits business, the UK regulator was not. In particular, the regulator was concerned with 'the risk that a firm using the net premium method might fail to keep enough reserves to meet a policyholder's reasonable expectations that bonuses would increase (if markets improve, and asset values increase)'.[64] The FSA proposed a 'twin peaks' approach where the firm would need to hold the greater of two values: the net premium valuation currently undertaken in accordance with EU legislation; and a 'realistic present value of expected future contractual liabilities plus projected fair discretionary bonus payments'.[65]

What did this 'realistic present value' imply? The first task was to codify the 'projected discretionary bonus payments' in a way consistent with PRE. For this, the with-profit fund would be required by the FSA to set out its with-profit bonus, smoothing and investment policies to policyholders in a document called 'Principles and Practices of Financial Management'. The realistic present value had to be assessed using assumptions consistent with those

[63] Financial Services Authority (2002), Financial Services Authority (2003).

[64] Financial Services Authority (2003), p. 11.

[65] Financial Services Authority (2003), p. 11.

documented policies. Once the pattern of these discretionary payments had been specified, they then needed to be valued. The consultations were unambiguous in their intentions in this regard: 'For the purposes of valuing the contracts and the embedded options and guarantees … the methods require market consistency i.e. the use of model assumptions, or option prices that replicate the costs of hedging such risks in the market'.[66]

The FSA was the first insurance regulator in the world to attempt to implement such a reserving method for complex long-term life business such as with-profits. But the initiative did not come completely out of the blue. It can be seen as the natural conclusion of a process, led from outside the actuarial profession, to put in place objective, transparent, measures of risk and capital in life business that reduced the reliance on actuarial judgement and discretion. An alignment of regulatory reserving methods with financial reporting standards was also an increasingly important driver (and was a prime motivation of the abandonment of the net premium valuation method in Canada and Australia). The FSA proposals aligned with emerging International Accounting Standards Board's proposals around fair value accounting.

The FSA was also not intent on allowing any undue delay to the implementation of their radical proposals. For more than a decade, the British actuarial profession had prevaricated over reserving methods for with-profit business. The FSA's proposals of 2002 and 2003, which were far more radical and created quite significant new intellectual and technological demands, were to be implemented in 2004. The actuarial profession had some work to do and three papers were published in the *British Actuarial Journal* in 2003 and 2004 that proposed how these market-consistent or realistic valuations of with-profit liabilities could be undertaken.[67] Up until this point in time, no stochastic methods had been used in with-profit statutory reserving. The FSA proposals didn't merely try to bring with-profits reserving in line with unit-linked maturity guarantee reserving as was discussed by Hare et al. It went much further, defining a fundamental shift in how to quantify a reserving requirement—the reserving requirement was to be based on the 99.5th percentile of the one-year change in the fair value balance sheet rather than on a percentile of the ultimate cashflow shortfall that could emerge over the full run-off of the liability outgo (i.e. over 30 or 40 years or more). As Hibbert and Turnbull wrote in the introduction to their paper:

[66] Financial Services Authority (2003), p. 23.

[67] Hibbert and Turnbull (2003), Dullaway and Needleman (2004), Sheldon and Smith (2004).

> Actuarial philosophy towards the valuation of liabilities has traditionally been based on the notion of funding ... the funding approach will tell us what reserves are required to meet the liabilities [cashflows] with a given level of confidence. By contrast, the thinking behind the economic valuation of a liability is very different. The economic value is defined as the sum of money required to establish a portfolio of assets that – provided that they are invested in a particular way – will replicate the liability as closely as possible. This special portfolio is called the hedge portfolio.[68]

Smith and Sheldon's introduction provides further background on the industry and regulatory context:

> The introduction of the realistic balance sheet is, in part, a response to the difficulties that un-hedged guarantees have caused the life industry in recent years. Reliance on long-term solvency tests runs the risk that we overlook more imminent problems, compounded by the use of over-optimistic assumptions and models used to determine capital needs. While there has been some criticism of the transfer of banking techniques to life assurance, the rate of deterioration in life offices' finances over the last three years demands a greater focus on the short term.[69]

Hibbert and Turnbull's (2003) paper presented an approach to economic valuation of with-profit liabilities that could be viewed as an evolutionary next step beyond the approaches of Wilkie in 1987 and Hare et al. in 2000. As was noted above, the Wilkie and Hare papers did not attempt to identify how the unique features of with-profits—the forms of discretion that the life office has in managing the product—could impact on the assessed cost of with-profit guarantees. As a result, they produced counterintuitively high measures for the guarantee cost. This failing had become well-understood. The Fair Value Working Party's 2002 paper had noted 'stochastic models of with-profits funds often show alarmingly high ruin probabilities because of inadequate modelling of management actions in adverse scenarios'.[70]

Hibbert and Turnbull suggested that a stochastic simulation approach—with an asset model calibrated to observable market prices—could be used to fully capture the impact on guarantee costs of the with-profit fund's ability to dynamically revise investment policy and bonus policy. Naturally, any results would be highly sensitive to the form that this dynamism was assumed to take, but this point had already been addressed by the FSA's requirement for the 'Principles and Practices of Financial Management' documentation.

[68] Hibbert and Turnbull (2003), p. 726.

[69] Sheldon and Smith (2004), p. 547.

[70] Hairs et al. (2002), p. 233.

One of Hibbert and Turnbull's points of emphasis was that the market-consistent valuation methodology could also be used to identify the hedging required in the estate to mitigate the risks left behind after the available (PRE-compliant) management actions had been implemented. Simulation modelling together with a codification of the with-profit funds' actuarial management levers could allow the costs and risks of with-profit guarantees to be analysed like any other derivative contract, albeit a complex one.

The Hibbert and Turnbull case study results suggested that the economic cost of guarantees could be as much as halved by allowing for the life office's ability to dynamically manage investment policy and bonus policy under assumptions that were typical of the time. Dullaway and Needleman's 2004 paper provided additional illustrative results of the impact on liability valuation of dynamic rules for the discretionary management features of with-profits business, again highlighting that these effects were material and could be captured by a simulation model.

The paper by Sheldon and Smith further discussed the technical challenges of market-consistent valuation of life business—particularly for the calibration of market-consistent asset models (choice of risk-free asset, extrapolation of implied volatilities, and so on). It highlighted the important point that market-consistent valuation of very long-term, complex liabilities could never be completely objective, even if that was the desire of the regulator. The paper also showed how some formulations of dynamic management actions could be captured in closed-form guarantee pricing formulae, thereby avoiding the need for the 'brute-force' method of Monte Carlo simulation.

Collectively, these three papers provided a technical grounding for the actuarial implementation of market-consistent valuation of long-term life assurance liabilities. Such techniques could be used in solvency reserving, guarantee pricing and hedging of the market risk created by the guarantees. The FSA's 2004 timetable was successfully met. The use of market-consistent valuation as a basis for life assurance reserving went on to be adopted more widely—for example, in the European Union's Solvency II system. The experience demonstrated that whilst the British actuarial profession had arguably lost its ability to be a master of its own destiny, it was capable of technical excellence in implementation when it was told what needed to be done.

Guaranteed Annuity Options (1997–2003)

Guaranteed annuity options (GAOs) —options to convert the cash proceeds of a life policy into an annuity at a guaranteed minimum rate—represent a particularly difficult episode of post-war British life office history. The GAO story contains a number of the threads that have already been identified in our above discussions of post-war British life offices: long-term financial guarantees provided under the stress of competition with little initial actuarial consideration for charging and reserving requirements; a lack of actuarial interest in 'modern' thinking on economic risk measurement and hedging; an over-reliance on the assumption that the unique features of with-profits could mitigate whatever economic risk was created by guarantees; ultimately, in the face of actuarial prevarication, the intervention of external parties (regulators, and, in this case, the courts) to compel a resolution.

Forms of guaranteed annuity option were written by British life offices from the end of the Second World War, and, to a lesser degree, earlier. Long-term interest rates rose almost continuously between 1950 and 1973 (consol yields increased from 4 % to 17 % over this period) (Fig. 5.1). GAOs were valuable to the policyholder in *low* interest rate environments and they therefore did not create any financial stress for life offices over the third quarter of the twentieth century. From the outset, however, there was some awareness amongst actuarial thought-leaders that such long-term guarantees were inherently risky for the life office. A particularly notable critic was Dick Gwilt, who was President of the Faculty of Actuaries from 1952 to 1954 and the principal executive of Scottish Widows from 1946 to 1960. Speaking at a faculty meeting in 1948, Gwilt voiced his concerns:

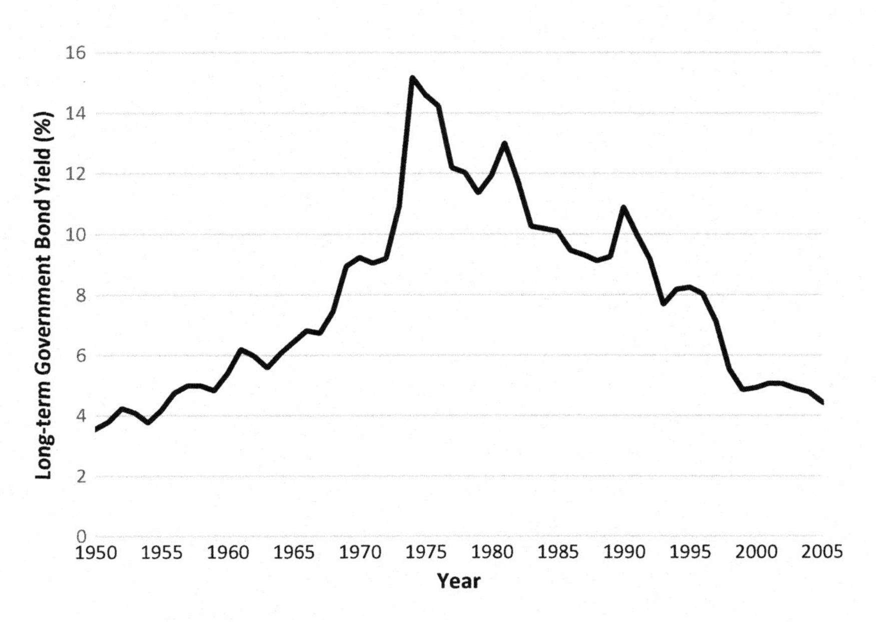

Fig. 5.1 Consol yields (1970–2005)

> Some of the guarantees given by offices at the present time [1948] appear to me to be based on somewhat optimistic assumptions about the level of interest rates many years hence. I cannot see any justification for giving options on a basis which may involve offices in serious loss if interest rates are low but do not give a chance of profit if interest rates are high for, in that event, the policyholders take the cash and secure the benefit of the high rates. I feel we ought to keep this matter prominently before us.[71]

A few years later at the 1952 Faculty meeting where Haynes and Kirton's interest rate risk management paper was presented, Gwilt again voiced his concerns about these long-term options:

> I think there is no doubt that many actuaries pay lip-service to the dangers of options, but, unfortunately the stress of competition leads them to grant terms which in the long run may prove to be a serious embarrassment and source of loss. I would go so far as to predict that the time will come when offices will bitterly regret some of the options which have been so light-heartedly granted in recent years [1952].[72]

Gwilt was not the only actuary with such worries about these product features at this time. In the Staple Inn discussion of Redington's seminal immunisation paper, Haynes commented:

> Reserves ought to be set up immediately to meet the chance that the options might become onerous and that the actuary's assessment of current surplus should be reduced – in some cases very extensively reduced – by reason of the options having been granted ... to my mind, the fundamental answer was to restrict the granting of options to an absolute minimum.[73]

These comments are particularly striking in the context of the life offices' experience of guaranteed annuity options in the 1990s and 2000s, particularly as such options were written up until the mid-1980s and no reserves were held for them until the 1990s. But they also reflect a more general phenomenon that would recur in post-war British life actuarial experience: actuaries were struggling to keep up with the sales and product development functions of life offices. Whether it was guaranteed annuity options on with-profit policies, or maturity guarantees on unit-linked savings products, or perhaps even

[71] Gwilt in Discussion, MacLean (1948), p. 318.

[72] Gwilt in Discussion, Haynes and Kirton (1952), p. 213.

[73] Haynes, in Discussion, Redington (1952), p. 331.

with-profit equity asset allocations, the actuarial profession found it difficult to adequately influence their firms' decision-making in the face of the 'stress of competition'. In all these cases, some actuarial thought-leadership was evident that could identify concerns and issues with the levels of associated risk being generated, but actuaries' ability to influence and lead life offices' corporate decision-making had started to wane.

It was the guaranteed annuity options written in the 1970s and first half of the 1980s that ultimately proved Dick Gwilt correct. At the time they were originally written, these options were generally far out-the-money. Indeed, they were viewed as so out-the-money that they need not be charged for or reserved for in any way. Even in the 1950s, actuaries were aware that out-the-money options had some value to the policyholder. Haynes and Kirton noted in 1952:

> In practice, guaranteed options are always "valuable" to the policyholder, even though a margin is retained between the current rate of interest and the interest rate adopted in calculating the option.[74]

This observation, however, did not influence the business practices of British life offices—GAOs were considered sufficiently worthless to the policyholder to be given without charge, yet also to be a product feature that was powerful in driving sales. Guaranteed annuity rates varied, but typically GAOs written in the 1970s and early 1980s provided a guaranteed annuity rate of around 11 % for a level annuity on a male aged 65. The a(55) mortality table typically in use in 1970s life offices implied a break-even long-term interest rate of 6–7 % for the GAO. Consol yields in 1975 were around 15 % and hence the options were deeply out-the-money. However, long-term interest rates had been below 6 % in the early 1960s so there was little reason for the possibility of rates again falling to these levels over a 25-year horizon to be considered completely outlandish.

Two unrelated phenomena emerged over the 1980s and 1990s that transformed GAOS from a deep out-the-money guarantee into a highly material liability. The heady rises in long-term interest rates of the third quarter of the twentieth century were reversed over the final quarter. Long rates fell inexorably from the early 1980s through to the end of the century (and beyond): the UK long-term interest rate fell from 13 % in 1981 to below 5 % by 1999. Secondly, the mortality assumptions of the a(55) table, which was designed for immediate annuitants in 1955, proved wholly inadequate for pricing annuities at the time these contracts matured. Pensioner longevity improved

[74] Haynes and Kirton (1952), p. 185.

at an exceptional rate over the period from the 1960s to 1990s. For example, the one-year mortality rate for a 65-year-old male in the Permanent Male Assurances (PMA) 92 mortality table was less than half of that in the PMA 68 table. This was largely unanticipated by British actuaries. It had a material impact on GAO liabilities—the break-even interest rate of a 11 % guaranteed annuity rate increased from 6–7 % to 8–9 % when a(55) was replaced with a typical 1990s longevity basis.

What actions were taken by life offices and their actuaries as the double-whammy of lower long-term interest rates and lower pensioner mortality rates emerged? Most offices stopped writing GAOs on new business during the late 1980s. By this time long-term government bond yields had fallen below 10 % and the prevailing market annuity rate approached the guaranteed annuity rate. Ceasing to write new GAO business could not, however, entirely stop the offices from accepting further GAO liabilities on to their books: the terms of the existing business applied the guarantees to future regular premiums and sometimes to one-off additional single premiums that the policyholder could choose to pay.

By 1993, the combination of further falls in long-term interest rates (by then below 8 %) and lighter pensioner longevity assumptions brought significant volumes of GAOs into-the-money for the first time. Some temporary respite was provided by the short-lived increase in long-term interest rates during 1994–1996 but by 1997 rates had fallen below their 1993 levels. UK insurance regulators were curiously inactive on the subject of GAOs over this period. The scale of life offices' GAO exposures was perhaps not well-understood by regulators or indeed the life offices and their actuaries at this time. In January 1997, the Life Board of the Institute and Faculty of Actuaries established an Annuity Guarantees Working Party to establish the scale and nature of the potential GAO problem and how life offices were managing or planning to manage it. The terms of reference of the working party noted:

> Currently there is no accepted practice for reserving for these guarantees and there is no published research to guide Appointed Actuaries in setting reserves. The DTI [Department of Trade and Industry, the UK insurance solvency regulator at the time] have not published any guidance or regulations specific to annuity guarantees.[75]

The working party published their report in November 1997.[76] It provided an insightful survey and intelligent commentary on the GAO situation at

[75] Bolton et al. (1997), Appendix 1.

[76] Bolton et al. (1997).

British life offices and the prevailing actuarial practices around it in the key areas of reserving, hedging and bonus policy (including implications for PRE). But it stopped short of providing any recommendations. It was never published in the *British Actuarial Journal* or presented at an Institute or Faculty sessional meeting. With hindsight it seems a curiously understated contribution from the actuarial profession.

The working party found that around half of British life offices with GAO liabilities were ignoring them entirely in their statutory reserving at the time of writing. The other half reserved for the greater of the cash maturity value and guaranteed annuity value, as implied under the statutory reserving basis. This reserving adjustment was straightforward in the context of GAOs on unit-linked business, but the majority of GAOs were written on with-profits business, and incorporating the GAO into the net premium valuation was less straightforward. As the working party noted:

> 'Adding in allowance for annuity guarantee reserves is another twist on top of a series of potential adjustments to standard net premium reserves.'[77]

The statutory net premium valuation included a resilience test that required firms to calculate the capital required after some specified stresses to valuation assumptions such as changes in equity values and interest rates. But, as was noted in Chap. 3, the net premium valuation has a perverse sensitivity to interest rate changes. In particular, it understates the economic balance sheet impact of a fall in interest rates as a lower interest rate basis generates a higher net premium. As a result of this, many life offices found they had greater exposure to interest rate rises than falls under the net premium valuation. In those cases, no resilience reserve was required for the contingency that GAO costs would increase due to further falls in long-term interest rates (though that only arose because they were required to hold reserves that incorporated a loading for the imaginary fall in future regular premiums that would arise if interest rates increased). This was an archaic actuarial feature that reflected poorly on the profession's ability to understand and manage financial risk. The originators of the net premium valuation never intended it to be used in solvency resilience testing, and it is an obviously flawed tool for analysing exposure to future changes in interest rates.

UK long-term interest rates continued to fall substantially over the late 1990s—from 8.0 % in 1996 to 4.9 % in 2000. GAOS were by this time unambiguously in-the-money options. Dick Gwilt's prediction that such

[77] Bolton et al. (1997), Section 3.

options would become a 'serious embarrassment and source of loss' had, half a century later, finally crystallised. GAOs costs were no longer a possible contingency. They had arrived and something had to be done. With-profit funds' estates could meet some or possibly all of the losses but the general first preference of life offices in the late 1990s was to manage the GAO losses through reductions in with-profit terminal bonus pay-outs. This led to the much vexed question of *whose* terminal bonuses should be reduced to fund the GAO cost, and how this stacked up against PRE. The simplest approach would be for all current policyholders (both those with and those without GAOs) to receive the same reduction in terminal bonus that applied to the (pre-GAO) cash maturity value. This reduction could be calculated to fund all GAO costs, or the cost could be shared between current policyholders and future policyholders (i.e. the estate). Such an approach may or may not have been deemed consistent with PRE, perhaps depending on the specific constitutional arrangements of the given with-profit fund and the policyholder communications it had made.

However, some life offices, most notably Equitable Life, the venerable institution whose early history we discussed in Chap. 2, pursued a different approach. Equitable Life had become something of an outlier in terms of many life office practices in the second half of the twentieth century. Unlike virtually all other life offices, it reported on a gross premium rather than net premium valuation method; it did not sell any policies through independent advisors but only through a tied salesforce; its appointed actuary was also its chief executive through the 1990s until 1997. Most strikingly, and most importantly for our discussion, it had a policy of not retaining any estate. The idea was that all policyholders would receive an asset share-based pay-out, and hence there would be no need for the office to hold any additional assets beyond the assets accrued from the investment of existing policyholders' premiums (notwithstanding that the with-profit policies included contractual guarantees whose ultimate costs could exceed the policyholders' asset shares).

The Equitable believed that they had communicated their asset share-based bonus policy such that 'the policyholder had effectively and knowingly mandated to the directors absolute discretion over each terminal bonus addition'.[78] The Equitable used this 'absolute discretion' to make each individual policyholder bear the final cost of the GAO that they had been granted—that is, at an individual policyholder level, the terminal bonus was reduced so that the post-GAO value of the policy would be equal to the cash maturity value of the with-profit policy if it did not include a GAO. This meant the GAO

[78] Corley et al. (2001), p. 12.

was irrelevant to the value of the policy pay-out (except in the circumstance where the GAO cost implied the required equalising terminal bonus was negative). This was a fairly striking practice. Importantly from a PRE perspective, it was not a bonus policy that had ever been explicitly communicated to policyholders when the policies were sold, or indeed at any time prior to its implementation.

This bonus policy was questioned by some GAO policyholders through the pensions ombudsman and the Equitable sought clarification from the courts. The case went all the way to the House of Lords, the highest court of appeal in England at the time. In 2000, the law lords found that the Equitable's 'differential' terminal bonus policy was unlawful. The Equitable immediately cut bonuses on all policies to zero in order to try to meet their GAO liabilities. But the Equitable had sailed too close to the wind and was now floundering. It put itself up for sale and closed to all new business in December 2000.

The House of Lords' judgement and the closure of the Equitable to new business sent shockwaves through the British actuarial profession. Within weeks of the Equitable closing to new business, the Faculty and Institute established a committee of inquiry to review its implications for the role of actuaries in life offices and their actuarial professional guidance. It published its findings in September 2001.[79] It called for a rewriting of key aspects of actuarial professional guidance, and for external peer review of the work of the appointed actuary.

The hedging of guaranteed annuity options through the purchase of derivatives from investment banks started in the late 1990s and received a fillip as the Lords judgement removed other management options. Such hedges were generally financed by with-profit funds' estates, and they secured the financial safety of the leading British life offices. The hedges cost much more than they would have had they been purchased in 1997, but nonetheless have proven valuable investments in the seemingly interminable environment of falling long-term interest rates of the 2000s.

With the benefit of hindsight, how does the GAO story reflect upon the British actuarial profession? In 2003, David Wilkie co-authored a paper with Heriot-Watt Professor Howard Waters and the doctoral candidate S. Yang that developed some retrospective analysis.[80] They considered how actuaries ought to have thought about charging and reserving for GAOs in 1985, given the state of actuarial thinking at that time. Specifically, they argued that the Maturity Guarantees Working Party paper of 1980, and the Wilkie

[79] Corley et al. (2001).

[80] Wilkie et al. (2003).

model presented to the Faculty in 1984, collectively provided actuaries with an adequate conceptual and analytical framework to charge and reserve for GAOs in an appropriate way. Their analysis showed that in 1985 the Wilkie model would have implied that there was a c. 20 % probability of a GAO attached to a 25-year endowment maturing in-the-money. A probability-of-ruin reserving approach would have implied that a reserve of around 10 % of a single premium would be required at a 95 % probability level. Based on the guarantee charging approach used in the Maturity Guarantees Working Party (not an option price but the expected cost plus charge-for-cost-of-capital approach), a charge for the GAO of around 5 % of the single premium would have been justified. There can be little doubt that had such reserving and charging parameters been in place, the result for GAOs would have been similar to that of unit-linked maturity guarantees: a lot less would have been written.

Wilkie, Waters and Yang were, of course, factually correct—a stochastic modelling approach to pricing and reserving for guarantees had been well-established in British actuarial thought by the mid-1980s. However, at this time, these concepts had only been rigorously applied within the domain of unit-linked business. In the 1980s there was no serious backing from the actuarial profession to apply these concepts to reserving for any of the guarantees in with-profits business. Some leading British life offices were undertaking internal stochastic analysis of their with-profit business in the 1980s, but few, if any, were suggesting that statutory reserving for with-profits business be undertaken in the way that had been implemented for unit-linked maturity guarantees.

In short, in the 1980s and 1990s there existed a misplaced confidence within the actuarial profession that with-profits business provided enough risk management levers to the actuary to make stochastic analysis unnecessary, difficult to apply and of limited value. Redington's 1976 statement that the actuarial analysis of with-profits 'was a countryside to explore on foot and not by fast car' is the most colourful expression of this perspective. There is doubtless some truth in Redington's view: it is easy to substitute computer modelling output for hard thought, and for insight to be lost amongst a plethora of ill-understood random numbers. But, by the 1980s and 1990s, was the actuarial profession doing enough walking of the with-profit countryside? And was the profession guilty, through hubris or fear, of rejecting a whole body of thinking that had emerged in the previous decades which could provide the key to the measurement and management of the increasing financial risks that it was charged with managing? Boyle and Hardy published a GAO actuarial retrospective in 2003[81] which forcefully concluded:

[81] Boyle and Hardy (2003).

> This entire episode should provide salutary lessons for the actuarial profession. It is now clear that the profession could have benefited from greater exposure to the paradigms of modern financial economics, to the difference between diversifiable and non-diversifiable risk, and to the application of stochastic simulation in asset-liability management ... [this] would have enabled insurers to predict, monitor and manage the exposure under the guarantee.[82]

It is evident that, particularly during the 1980s and 1990s, there was a strong reticence on the part of the British actuarial profession to adopt or make use of financial economics in their thinking and best practices. With hindsight, these were partly wasted decades for the profession in this respect. On a positive note, after taking their direction from regulators, financial reporting standard-setters and judges, real progress was made in the 2000s in actuarial education, research and practices that started to rectify this damage.

[82] Boyle and Hardy (2003), p. 150.

6

British Actuarial Thought in Defined Benefit Pensions (1905–1997)

The recognisable British private staff pension scheme, funded by a mixture of employer and employee contributions based on percentage of salary, paid over the period of service of the employee and invested in a trust fund, with benefits defined with reference to the salary history and years of service of the employee, first emerged over the second half of the nineteenth century. Ad hoc unfunded pension arrangements arose amongst some large employers earlier in that century. The occasional historical precedent of earlier funded arrangements for post-employment benefits can also be found: the earliest funded scheme for the provision of widows' annuities to 'employees' is thought to be the Scottish Ministers' Fund, which was established in the mid-eighteenth century.[1]

Actuaries have provided technical and professional advice on the financial management of defined benefit pension funds since around 1875. The British profession has been actively engaged in this role continuously from then to the present day. Whilst the task of funding long-term life-contingent pensions had many parallels with the actuarial work in the more-established life assurance sector, pensions work has also always had its own unique requirements and challenges: typically, pension benefits were even longer-term than the liabilities that arose in the life sector; and the ultimate size and cost of the benefits were a function of a wider array of uncertain variables than the liabilities of life offices. Whilst rates of interest and mortality tables were similarly vital to pensions work, the cost of the pension benefit would also depend on other variables such as long-term salary inflation and the rates at which employees would leave their employer prior to retirement. The trust arrangement that

[1] Crabbe and Poyser (1953), p. 1.

C. Turnbull, *A History of British Actuarial Thought*,
DOI 10.1007/978-3-319-33183-6_6

formed the bedrock of funded pension arrangement also implied a different legal and commercial context to that of a mutual life office. This raised subtle issues about the *purpose* of funding which had important implications for the objectives that the actuary's advice should attempt to meet.

Amidst these new challenges, there was a plethora of external drivers of change in pensions actuarial thought that will be familiar from the history of actuarial thought in the life sector: economic and political shocks; increasing government regulation in both solvency and consumer protection that gradually saw statutory responsibilities displace unbridled actuarial judgement and discretion; financial reporting metrics that demanded alignment with accounting principles; the emergence of financial economics and the new insights it could provide into financial management and the difficulty in reconciling them with traditional actuarial approaches.

This chapter identifies some of the key developments in pension actuarial thought that arose over the long history of actuarial involvement in the field. Their historical context can help us understand the ways that thinking in this field meandered, circled and occasionally underwent revolutions. A final word on terminology—the terms final salary pension scheme, defined benefit pension fund and so on are used interchangeably in this history, and none generally provide a perfect description of how these pension arrangements worked. Most of the schemes had pensions based on final salary, but a minority were based on salary levels over a longer period of employment. Pension benefits were generally not always prescriptively defined: for significant periods of the history early leaver benefits and in-force pension increases included significant discretionary elements that would tend to be greater when funding levels were strong.

Early Thought on Defined Benefit Pension Funds (1905–1921)

Early actuarial thought on defined benefit pension funds focused on the development of the algebra and substantial mechanical computation necessary for the valuation of a scheme's liabilities and the assessment of a required contribution rate under a given set of deterministic assumptions for the scheme's future development. In these early years, there was considerable discussion of how to set the decrement bases for the liability projection, but relatively little focus at this time on how to set the valuation interest rate. Asset valuation was rarely, if ever, addressed. Investment was predominantly in long-term government bonds, and in the late Victorian and Edwardian eras, long-term interest rates were remarkably stable by later standards: between 1875 and 1910 the long-term government bond yield moved within the range of 2.0–3.2 %. Valuing assets at the lower of book value and market value seemed simple, prudent and adequate.

This actuarial work was fundamentally similar in spirit to the reserving calculations made in life offices, but the calculations were even more convoluted and computationally intensive. Ralph Hardy developed the first defined benefit pension fund valuations around 1875. In doing so, he originated the commutation factor as a vital computational aid for this task.[2] Hardy's work, however, was never formally published. The earliest British actuarial professional papers on funded pension schemes were published during the Edwardian era by disciples of Hardy such as George King (in 1905) and Henry Manly (in 1911).[3] Actuarial papers on funded pension schemes were nonetheless sparse—prior to the conclusion of Second World War, only a handful of notable papers on pension schemes were published in the British professional actuarial journals.

King's paper presented the actuarial calculations that were required for pension fund valuations and the assessment of the contribution rate that would fund the ultimate benefits—that is, to find a contribution rate expressed as a constant proportion of salary that had a present value equal to the present value of the liabilities. It was the first paper to set out the detailed calculations that were necessary to allow for treatment of the various decrements (death, pre-retirement withdrawals and retirement) and assumptions (mortality rates, interest rate, withdrawal rates, salary growth) in the valuation of the liabilities (pensions, death benefits, withdrawal benefits) and the setting of the contribution rate required to fund the projected liabilities.

King suggested that the withdrawal, mortality and retirement rates should be estimated by each scheme based on their own recorded experience. These could be suitably graduated using methods such as those developed by Woolhouse for life office mortality experience tables. He emphasised how these assumptions should vary across pension funds and railed against the use of assumptions based on pooled pension fund data. He recognised, however, that scheme-specific data would seldom be available for pensioner mortality at the older ages and he advocated the use of one of the published population mortality tables such as English Life Table, No. 3 for this purpose. He suggested a salary scale should be specified (i.e. how salaries vary by age) again based on the actual

[2] See King (1905), pp. 129 and 175.

[3] King (1905), Manly (1911).

experience of the scheme membership. Inflation was a relatively subdued economic phenomenon in this era—the UK inflation index was at the same level in 1904 as it had been in 1860; so the salary scale was intended to represent the growth in salary due to promotion and experience. In the examples he gave, he assumed that salaries would increase by about 80 % between the ages of 20 and 30, but by only 12 % between 50 and 60.

In considering the setting of the contribution rate, King only considered the calculation at the inception of the scheme, and hence avoided addressing the valuation of any pre-existing scheme assets. Nor did he discuss how the valuation interest rate should be set. It was simply a parameter that was subsumed into Hardy's commutation functions.

Manly's 1911 paper further discussed the estimation of the member decrement assumptions, and he also provided some analysis of the sensitivity of the valuation to those assumptions. Manly was more sanguine about pooling the experience of different pension funds for the purpose of estimating decrement rates, providing the pension funds reflected a similar profile of member. The paper included a 'Life and Service Table'—the withdrawal, mortality and retirement rates and salary scale based on scheme experience—for the aggregated experience of several large railway schemes. He noted that selection effects would produce pension scheme mortality that would typically be lighter than that observed on assured lives. Members who were seriously ill at pre-retirement ages would tend to withdraw from the scheme as they would no longer be able to work. The surrender of a life assurance policy, on the other hand, would tend to be made by a person in good health.

Manly highlighted the sensitivity of the projection of the pension scheme, and the contributions required to fund the liabilities, to these assumptions by comparing the results obtained with those produced by assumptions published in other early studies. He showed that his life and service table implied that a 5 % salary contribution would fund a pension of 1.55 % of average salary for every year of membership whereas alternative previously published assumptions would support 2.38 % of average salary for every year of membership.[4] Manly also provided detailed year-by-year projections of the pension scheme asset values, contributions and benefit cashflows that would be generated over a 90-year projection using his assumptions and the alternative. This highlighted how the solvency of the pension scheme could go awry if experience deviated significantly from the basis assumed in setting the contribution rate, thus highlighting the need for careful and regular monitoring of the pension fund's experience by the actuary.

[4] Manly (1911), p. 157.

The Staple Inn discussion of Manly's paper generated some interesting debate on the *purpose* of advance funding of pension liabilities that illuminated the actuarial views of the time. One speaker, Thomas Tinner, noted that a primary reason for advanced funding was to provide security to the pension fund member:

> Without a fund, or a guarantee from the employer, there was no security that when a man who had been paying all his period of service for the pensions of others came to retire his own pension would be provided by a younger generation. This was especially the case if the employer was a private firm.[5]

He also argued that the interest earned on advance funding of the pension benefits would reduce its cost:

> The effect of accumulating a fund was to reduce the cost of the pensions, because the interest earned would help to pay the outgoings.[6]

Meanwhile, W.O. Nash argued that the main aim of the pension scheme was to recognise a smooth accrual of the cost of the pension provision:

> What they were aiming at in starting a Pension Funds was an equalization of the annual expense for the pension.[7]

As we shall see below, these three reasons for advanced funding would be revisited and debated numerous times by actuaries, accountants and economists over the remainder of the twentieth century. The relative importance attached to each of these reasons would influence the choice and development of actuarial methods used in pension fund valuation.

The economic consequences of the First World War represented the first great economic shock to test British defined benefit pension schemes. It created considerable stress for the financial health and solvency of the pension funds. It also created new challenges for pension actuaries, particularly with respect to the treatment and management of inflation. Over the *Pax Britannica* period—the 100 years between the end of the Napoleonic War and the start of the First World War—inflation rates had at times fluctuated markedly both up and down but there was no discernible consistent upward trend in UK consumer prices. Indeed, the UK Consumer Price Index stood

[5] Tinner, in Discussion, Manly (1911), p. 219.
[6] Tinner, in Discussion, Manly (1911), p. 219.
[7] Nash, in Discussion, Manly (1911), p. 222.

some 20 % lower at the outbreak of the First World War in 1914 than it had been at the end of the Napoleonic Wars in 1815. Thus, the risks associated with future price inflation did not feature prominently in actuarial thought on final salary pension schemes prior to 1914. No explicit assumptions for price inflation or real salary growth featured in the work of King or Manly. Their age-dependent salary scales were intended to capture the typical path of career progressions, not economic inflation.

In the years 1915–1920 inclusive, the UK CPI index more than doubled. Salaries and wages increased at a similar rate. This substantially increased the value of pension liabilities without generating an offsetting increase in the value of pension fund assets (which, as noted above, were predominantly invested in long-term government bonds at this time). This scenario was radically different to anything within living memory of actuaries. The possibility of such a scenario and its consequences for pension funds had, naturally enough, not been seriously considered by pension actuaries prior to the war.

In 1921, G.S.W. Epps wrote an important, and at the time controversial, paper[8] on the implications of this experience for pension fund valuation, noting in his introduction that '[due to] enhanced scales of salaries or wages which have resulted from the change in the value of money … it is to be feared that in many funds a very serious position has to be faced'.[9] Epps used case studies to highlight that, for older, established pension schemes, the impact on the valuation of accrued service benefits and the resulting solvency levels was very substantial. Furthermore, at this time pension funds were not required to inflation-link pensions in payment, and Epps's valuations made no allowance for such increases. But significant pressure bore on the schemes to provide some increases to pensions in payment in order to mitigate the significant reduction in real value of pensions which had been inflicted on pensioners over the previous decade.

This exceptional economic environment also naturally had an impact on the asset side of the pension fund balance sheet. Between 1900 and 1920, long-term government bond yields more than doubled from 2.5 % to 5.3 %. The implications of this interest rate rise for the valuation of the assets and liabilities of pension funds was a topic of considerable controversy within the actuarial profession at the time. Epps recognised that the net impact of such interest rate rises may not necessarily be adverse for a pension fund:

[8] Epps (1921).

[9] Epps (1921), p. 405.

> It is, of course, true that materially higher rates than those assumed in past valuations can be looked for in the case of the accruing funds to be invested for many years to come; but the question as to how far it is prudent to meet past depreciation by raising the valuation rate of interest is one of extreme difficulty, the solution of which throws a great responsibility on the actuary concerned.[10]

The topics of asset valuation and depreciation and the related treatment of the liability discount rate received much attention at the Staple Inn discussion of Epps's paper. Some actuaries were reluctant to recognise the market value impact on bond prices, even if not to do so was evidently imprudent. For example, R.G. Maudling commented:

> As to the general treatment of depreciation, he might be accused of heresy, inasmuch as he could not bring himself to write securities down to their present market value ... He preferred to make a reserve for possible loss on realization, and to give considerable weight to the fact that securities might ultimately recover. He was not inclined to give much credit to present market values because they were more or less fictitious, being the values which a man who was compelled to realize was prepared to take, and he thought it would be extremely unwise to write everything down to that level.[11]

J. Bacon expressed a quite similar view to Maudling:

> He would certainly not write down investments to their present market price, but would simply set aside an investment reserve and write off any deficiency there might be by means of an annuity, believing as he did that in five or ten years' time, when they got to a state of equilibrium, the annuity set apart would probably not be required, and by that means he would have succeeded in giving a degree of safety to the fund without unduly crippling the employer.[12]

So, the above actuaries advocated valuing assets above market value, though with perhaps a partial write-down from their book value. A reserve would be included on the liability for future realisation of investment losses, but this would be smaller than the difference between the assets' actuarial valuation and their market value, reflecting the actuary's confidence that the assets would subsequently increase in value as rates reverted back to historical norms. No change in the liability discount rate would be made using this

[10] Epps (1921), p. 410.
[11] Maudling, in Discussion, Epps (1921), p. 444.
[12] Bacon, in Discussion, Epps (1921), p. 448.

approach. The current market yield was treated as an aberration that could be largely assumed away. However, S.G. Warner, an influential actuary who was Institute President during the years 1916–1918, suggested a quite different approach where assets were fully marked down to their current market values whilst the liability discount rate was increased consistently to reflect those market yields:

> Depreciation should be severely dealt with, but that there should be little hesitation about using something nearly approaching the rate of interest which that depreciation revealed as having been secured.[13]

This is perhaps the earliest actuarial advocacy of a market-based pension fund valuation. As we shall see below, the topic of consistent pension fund asset and liability valuation in the presence of significant changes in market interest rates would occupy pension actuaries' thoughts for much of the remainder of the twentieth century.

Consistent Asset and Liability Valuation (1948–1963)

The exceptional inflation experience of the years immediately following the First World War proved to be a transitory phenomenon. Indeed, over the course of the 1920s, a significant portion of the price increases of the post-war years were reversed by a decade of deflation. By 1933, the UK CPI stood at a lower level than at the end of the First World War. The Second World War inevitably generated a further bout of inflation, but in its immediate aftermath the political stance—in particular, that of Hugh Dalton, the then Chancellor of the Exchequer—was to seek very low long-term interest rates. In 1946, long-term government bond yields stood at 2.6 % (they were 4.6 % 20 years earlier). The position pension actuaries faced was therefore reversed from the time of Epps's 1921 paper: pension funds' bond assets now showed a considerable market value appreciation, whilst the prospective future available yield on new money was considerably lower than had been assumed in pension valuations.

C.E. Puckridge published a paper in the *Journal of Institute of Actuaries* in 1948 which analysed these circumstances and their actuarial consequences.[14]

[13] Warner, in Discussion, Epps (1921), p. 449.

[14] Puckridge (1948).

In the late 1940s, as 40 years earlier, standard actuarial practice was to value assets at book value or, if lower, market value (or perhaps somewhere in between), with liabilities valued using a broadly static valuation interest rate assumption. The liability valuation rate could be set with reference to the investment yield implied by the book value of assets, with some allowance made for how that yield was expected to evolve in the future. Puckridge, however, argued for an alternative approach that entailed a more systematic, consistent approach to the valuation of pension fund assets alongside liabilities:

> Value assets (including existing investments) and liabilities at the rate of interest which it is anticipated can be earned on future investments.[15]

Puckridge did not provide much specific guidance on how to estimate this interest rate, but he did note that the rate will be

> 'determined by considering the average rate at which investments could be made at the present time, taking into account the range of investments authorized for the particular fund and the proportion likely to be invested in the various classes; adjustments will have been made to allow for any interest margin that it is desired to incorporate and for the view taken on the probable trend of interest rates over a long period'.[16]

Puckridge's future investment rate was an inherently long-term and subjective parameter. He likely envisaged that it would result in more stable asset and liability values than the market-based approach suggested by Warner back in the discussion of Epps's paper in 1921. But the vital point of Puckridge's method was that the actuarial assessment of the financial health of pension funds required greater and more transparent consistency between the valuation of assets and liabilities than was found in the then standard actuarial practice. Whilst this may appear obvious in hindsight, it was not a particular point of emphasis in the previous half-century of actuarial thinking on defined benefit pension funds. Though Puckridge preferred to reject the direct use of market prices, he did recognise the difficulty in placing a value on assets that exceeded their current market value:

> It would in practice be difficult to adopt any value in excess of the total market value, and the balance would therefore be retained as a margin which would fall into surplus in the future as the excess interest is received.[17]

[15] Puckridge (1948), p. 2.

[16] Puckridge (1948), p. 12.

[17] Puckridge (1948), p. 12.

Puckridge's idea of valuing assets by discounting their future income at an interest rate that is consistent with the rate applied to the liabilities gained considerable ground within the actuarial profession over the following 15 or so years. Eventually, a form of this 'discounted income' approach to the valuation of all assets in a defined benefit pension fund would become standard actuarial practice. Of course, S.G. Warner's advocacy back in 1921 of using market values of assets along with a market-based liability discount rate could be viewed as one example of consistent asset and liability discounting. But pension actuaries were generally reluctant to shift their long-term liability valuation assumptions by the degree that could be implied by market movements. After all, in the context of open, expanding defined benefit pension funds, there would never be a need to sell assets, and actuarial judgement could apparently form a better view of yields available for future investments than current market prices. Puckridge's embryonic notion of setting a stable long-term liability discount rate and valuing the projected cashflows of both assets and liabilities at that rate emerged as the pension actuary's solution to consistent asset and liability valuation that would provide valuation stability through time.

Three actuarial papers published between the years of 1958 and 1963 were particularly important in further developing Puckridge's idea and establishing its practice in British pension scheme valuation. Once again, economic conditions provided a spur to this progress. Following the post-war Dalton era of politically induced low interest rates, economic reality had reasserted itself. Long-term bond yields reverted from 2.6 % in 1946 to 5.0 % in 1958. Pension actuaries were once again wrestling with the consequences of substantial bond asset depreciation.

The first of these three papers was written by D.F. Gilley and D. Funnell.[18] Gilley and Funnell, like Puckridge a decade before them, argued that both asset and liability cashflows should be discounted at the 'average yield on future investments'. However, in doing so, they also noted some of the perverse consequences that this may have. In particular, if this approach resulted in a deficit being identified in the scheme, the amount of money required to extinguish this deficit would not necessarily be equal to the assessed size of the deficit. Under the discounted income approach, actuaries were essentially operating in a parallel currency—actuarial valuation pounds had a different value to 'real-life' pounds. But they argued that their approach of valuing both sides of the balance sheet with a single discount rate was preferable to attempting to find consistency between liability valuation and book asset valuation by

[18] Gilley and Funnell (1958).

creating a discount rate that was an ill-defined blend of the current book yield and the assumed average yield on future investments.

Gilley and Funnell moved beyond Puckridge's work when they considered the valuation of *equity assets* within their discounted cashflow approach. As we will discuss further below, equities became an increasingly important asset class for pension schemes during the 1950s. Any asset valuation method for British pension funds therefore needed to be directly applicable to equities. The simultaneous valuation of equities and bonds immediately gave rise to the question of whether and how the *asset mix* of the pension fund should impact on the actuarial valuation of assets and liabilities. Should a portfolio of equities and a portfolio of bonds, each with the same market value, command the same actuarial asset value? Should a change in the asset mix result in a different valuation of liabilities through a different choice of liability discount rate? In a nutshell, should the higher expected return of equities be capitalised in the current assessment of the pension fund surplus/deficit?

Gilley and Funnell were unambiguous in their view: 'it would not be prudent for the actuary to reduce the contribution rate or weaken the basis of valuation by assuming a higher valuation rate of interest than the expected, long-term, gilt-edged rate'.[19]

So Gilley and Funnell were clear that the asset allocation choices of the pension scheme should not impact on the valuation discount rate applied to its projected asset and liability cashflows. However, that did not mean that they advocated the use of the market value of the equity holdings. They suggested that the market value of the equity portfolio be scaled by the ratio of the actuarial valuation of irredeemable consols (as implied by the valuation interest rate) to the market value of those consols. Or, equivalently, the equity market value should be scaled by the ratio of the market consol yield to the valuation interest rate, i. In this approach, the actuarial value of total assets would be the same for any split of assets between equities and consols. Thus, the asset allocation choice would not impact on the overall actuarial valuation of assets or liabilities. However, they added a caveat: the equity valuation may be reduced by a 'margin of arbitrary amount' if the actuary held a strong view that they were 'over-valued in the market'.[20]

These asset valuation ideas were quite controversial at the time. Traditionally, actuaries viewed the writing-up of asset values above their book values as a dubious and imprudent practice. Gilley and Funnell's approach could see assets valued significantly in excess of not just book values but *market values*.

[19] Gilley and Funnell (1958), p. 50.

[20] Gilley and Funnell (1958), p. 53.

The minutes of the discussion of their paper highlights the suspicion and hesitancy with which conventional actuarial thought welcomed these ideas. Gilley and Funnell's essential point, however, was that subjective assumptions about future investment conditions must arise in pension fund valuations, and their method provided a more coherent and transparent approach to setting those assumptions.

Heywood and Lander's 1961 paper, 'Pension Fund Valuations in Modern Conditions',[21] was the next significant contribution to British actuarial thought on the consistent valuation of pension fund assets and liabilities. Like Puckridge and Gilley and Funnell before them, Heywood and Lander again argued that the valuation discount rate should be set as the long-term interest rate that is expected to be earned on new money; and that assets should be valued consistently with that rate. However, Heywood and Lander differed from Gilley and Funnell in two fundamental respects. Firstly, whereas Gilley and Funnell had argued that the discount rate should reflect the expected future long-term gilt yield irrespective of the actual asset allocation of the pension scheme, Heywood and Lander believed that the expected return on the *actual* asset allocation of the pension scheme should be used. This meant that the liability valuation would be reduced by increasing the fund's allocation to equities. They suggested that a 100 % gilts allocation would merit a 3.5 % discount rate, whilst a 50/50 equity/gilts allocation would justify a discount rate of '4 % or possibly even higher'.[22]

The second, and related, point on which Heywood and Lander diverged from Gilley and Funnell was on the valuation of equity assets. Equity investment was becoming ever more material for UK pension schemes. Heywood and Lander noted that a typical approach at the time of writing was to invest 50 % of assets in equities and 50 % in fixed income. Equities had proved attractive to pension funds both as an inflation hedge and as a high-yielding asset class (though the reverse yield gap was just starting to emerge at this time). Like Gilley and Funnell, they noted that the practice of valuing assets at the lower of book value and market value was still prevalent. And they expressed strong unease with the use of market values in general on the grounds of their volatility:

> It seems very difficult to justify a method of valuing assets which places a value upon the fund which may alter substantially if the valuation date were varied by only some two or three months.[23]

[21] Heywood and Lander (1961).

[22] Heywood and Lander (1961), p. 323.

[23] Heywood and Lander (1961), p. 327.

They also noted that there was no reason to expect book values to provide an asset valuation that was consistent with the liability valuation basis. Like Gilley and Funnell, they concluded that a discounted income approach to asset valuation, using a discount rate consistent with that applied to the liability outgo, was the only way to obtain a consistent asset and liability valuation. For bond valuation, Heywood and Lander concurred with Gilley and Funnell: the projected cashflows of the bond could simply be discounted at the liability discount rate. For equities, they argued that equity dividends should be explicitly projected using a dividend growth assumption before discounting at the valuation rate. Thus, equities could be valued as a perpetuity discounted at the rate $(i - \theta')$ where i was the liability discount rate, and θ' was the assumed dividend growth. The dash was used because θ was defined as the assumed rate of salary inflation that should be applied in the liability valuation. Salary inflation was typically not explicitly allowed for in liability valuation at this time, but Heywood and Lander argued that credit for inflationary dividend growth on the asset side of the balance sheet should only be taken if consistent treatment of inflation was applied on the liability side of the balance sheet. Moreover, they specified that θ' should be less than θ (though they did not provide any specific rationale for this).

So, under Heywood and Lander's approach, the relationship between the actuarial valuations of equity and bond holdings were not directly constrained by their relation to their respective market values. This meant that a change in the equity/bond mix could change the actuarial valuation of assets as well as the actuarial valuation of liabilities. Shifting from bonds to equities would typically improve their assessment of the pension fund solvency position. However, their stipulation that, for reasons of consistency, this improvement should only be permitted if the liability valuation recognised a level of salary growth consistent (and not less than) the assumed equity dividend growth, meant that their approach would also be accompanied by a general strengthening of the liability valuation basis.

Finally, perhaps to hedge their bets, Heywood and Lander permitted a scalar parameter λ to be applied to the equity valuation as an 'arbitrary multiplier'. This was intended to allow the actuary to make any special adjustments that he deemed appropriate in the context of the valuation exercise. Whilst this might appear as something of a fudge, it prevented their methods from being cast as mechanical rules to be mindlessly applied by the actuary. It emphasised the application of actuarial judgement that was valued by the profession. In the Staple Inn discussion of the paper, Gilley suggested:

> Truly, the actuary valuing the fund was in a better position to assess the relative values to the fund of ordinary shares and irredeemable gilt-edged securities than was the market.[24]

Not all participants in the discussion held to the view however that actuaries were better equipped to value equities than the market. J.G. Day, whose writings we shall discuss below, commented:

> I am disturbed by the idea expressed in the paper that an actuary should be prepared to value ordinary shares in a way quite independent of market value, and that it should be presumed that his value would be superior to that of the market.[25]

The 1963 paper,[26] 'The Treatment of Assets in the Actuarial Valuation of a Pension Fund', by Day and McKelvey represented the final instalment of this period's trilogy on consistent valuation of pension fund assets and liabilities. They once again started from the basis that the assumed value for the valuation rate of interest should be 'the average rate of interest that is assumed for future investment over a long future term'.[27]

We saw above that Gilley and Funnell suggested that equities should be valued as their market value scaled by the ratio of the market consol yield to the valuation interest rate. Day and McKelvey suggested a variation on this theme: they argued equities should be valued as their market value scaled by the ratio of the *equity dividend yield* to the valuation interest rate. As the product of the market value and the dividend yield is the current dividend payable, it can be seen that this formula simply values the current dividends as a perpetuity at the valuation interest rate i. In what was now becoming an established actuarial asset valuation tradition, they also allowed an additional scalar parameter as 'an arbitrary factor'.

Clearly, this approach to equity valuation ignored future dividend growth and they suggested this valuation formula should only be used when inflation is ignored in the liability valuation. They then suggested that the discount rate could explicitly be set as a real rate that could be applied to both sides of the balance sheet, and suggested a real rate of 3.5–4 % would be appropriate. This was fundamentally the same result as Heywood and Lander: it was, in essence, a dividend discount method where the market's required return on equities

[24] Gilley, in Discussion, Heywood and Lander (1961), p. 342.
[25] Day, in Discussion, Heywood and Lander (1961), p. 364.
[26] Day and McKelvey (1963).
[27] Day and McKelvey (1963), p. 108.

had been replaced by the pension fund's valuation interest rate (which, in turn, was to be set to reflect the long-term expectation for the pension fund's asset return). Under these approaches, changes in equity market values that arose from changes in the dividend yield rather than changes in dividend payouts would have no impact on pension fund valuations of equity assets.

There was a broad consistency in the thinking of this collection of papers. They all eschewed market asset values on the grounds of consistency with a liability valuation that was to be made using stable assumptions. But some, like S.G. Warner many decades earlier, felt like this was tackling the consistent valuation problem from the wrong end. In the Staple Inn discussion of Day and McKelvey's paper, Plymen commented:

> They were assuming a certain figure for the rate of interest for valuing the liabilities, and twisting the valuation of the assets round to be consistent with that basis. Why not start off with the market value of the assets and try to deduce from that basis a consistent system for valuing the liabilities?[28]

Speaking at a Staple Inn meeting over 30 years after the publication of the paper, McKelvey's son, K.J. KcKelvey, also an actuary, commented on the explanation his father had given him for tackling the consistency problem this way round:

> The sole objective of that 1964 paper was to find a consistent basis for valuing assets, given an existing methodology for valuing liabilities. The liability valuation basis was off-market, by convention at that time. Therefore the asset valuation inevitably became off-market. The main aim of the authors was to move away from the valuation of assets by book value, which was still common. They simply did not think about market values since … there were no formal discontinuance tests.[29]

We will see below how funding objectives such as discontinuance testing influenced valuation thinking in the latter decades of the twentieth century. But first, we will review how actuarial thinking on investment strategy for pension funds developed alongside the above ideas on asset and liability valuation.

[28] Plymen, in Discussion, Day and McKelvey (1963), p. 134.

[29] K.J. McKelvey, in Discussion, Exley, Mehta and Smith (1997), p. 950.

Actuarial Thought on Pension Fund Investment Strategy (1957–1985)

As noted above, the exceptional UK inflation experience that accompanied the First World War and its aftermath provided a salutary lesson to pension actuaries on the vulnerability of pension fund solvency to inflation shocks. But this period of extraordinary inflation proved short-lived: between the years of 1921 and 1941, the UK CPI actually fell. However, the post-Second World War period saw the return of high inflation, and, perhaps equally importantly, high levels of inflation volatility and uncertainty. By the mid-1950s, there was a growing sense that the prevailing political and economic environment made this inflationary experience the new normal rather than merely a transient post-war adjustment. The political focus on full employment and the increasing strength of trade unions appeared to make the prospect of a permanent inflationary environment quite plausible.

Pension actuaries of the post-Second World War era recognised the inflation risk exposure that arose when investing in long-term nominal fixed interest assets to back salary-linked liabilities. At this time, there was no index-linked gilt market. The only available 'real' asset classes were equities and property. Actuarial thought in pensions started to embrace the use of equities as a form of inflation hedge in the mid-1950s. As we have seen in earlier sections, this was also a period of increased equity investment for life offices, and the 'cult of the equity' was more generally in full flight. Pension funds were not immune to this trend—between 1945 and 1954, UK pension schemes' average asset allocations to equities increased from 10 % to around 30 %.[30] The inflation hedging argument may have driven this, or may have simply provided an intellectual rationale for it.

An actuarial paper by McKelvey, 'Pension Fund Finance', published in 1957, was amongst the first to argue for substantial equity investment by pension schemes on the basis of equities' inflation hedging ability.[31] The quality of inflation hedging that equities could provide was accepted as a matter of some uncertainty by McKelvey. He constrained his advocacy of equity investment to open rather than closed pension funds, and emphasised that it was the inflation-protection of long-term dividend growth (rather than equity market values) that he believed was important for such funds. He posited that the 'propaganda value' of balance sheet asset valuations of open funds was of no great importance. In essence, McKelvey was arguing that dividends

[30] McKelvey (1957), p. 136.

[31] McKelvey (1957).

provided a form of real cashflow match for salary-linked benefits. Volatility in real equity price levels could be ignored as an inconsequential market foible for an open pension fund that would never need to sell them. He emphasised that, in real terms, long-term nominal fixed income was the risky asset class for backing salary-linked pension schemes:

> The question now is not, as it used to be, dare we put more than 10% in equities? It is, dare we leave more than 50% in fixed interest investments?[32]

J.G. Day published a paper in the *Journal of the Institute of Actuaries* in 1959 that provided further support for substantial equity investment by final salary pension funds.[33] Day emphasised how final salary pension liabilities had different characteristics to life office liabilities—specifically being real rather than nominal, and having longer duration. He argued that these characteristics made pension funds more suited to equity investment than life offices and noted the high levels of equity allocation that life offices had made in recent years. He recognised that equity investment would expose pension funds to more market value volatility, but, like McKelvey, he argued that this was not fundamentally important. He was unequivocal in his advocacy of equities as 'the basic investment' of final salary pension funds:

> A pension fund by the nature of its liabilities is so vulnerable to both inflation and any movements towards higher wages and salaries and a higher standard of living that, if one takes the view for the normal pension fund, it is income that matters and provided that income is secure then fluctuations in the value of the portfolio do not matter, then in the author's opinion one should come down on the side of equities (or property) as the basic investment for a pension fund.[34]

The classic papers on pension fund valuation of the early 1960s that were discussed above also contained further advocacy of equity investment for final salary pension schemes, primarily on the basis of inflation hedging. For example, in the discussion of Heywood and Lander, Hemsted commented:

> Rising standards derive from increased productivity which in turn is based on increasing capital backing per employee. Much of this increased capital arises from retained earnings by industrial companies and these retentions in a period of expansion produce increasing equity asset values per share. There is therefore

[32] McKelvey (1957), p. 121.
[33] Day (1959).
[34] Day (1959), p. 130.

reason to think that ordinary shares have a built-in correction not only for falling money values [price inflation] but also for increasing standards of living [real salary growth].[35]

Day and McKelvey's 1963 paper was unsurprisingly supportive of pension fund equity investment given the positions they had taken in their individual papers of the late 1950s. They wrote:

It has been widely and authoritatively suggested that, to meet the dangers of inflation, pension funds should invest in equities. It has been argued that with inflation a business's real assets increase in money terms ... so that, although inflation may initially cause difficulties and distortions, an equity investment will eventually have increased in earning power in rough proportion to the effect of inflation (or very often rather more, as prior charges lose in value).[36]

All of the above development in pension fund thinking had taken place under the premise that the raison d'être of the pension fund was to generate asset income that, together with a stable pattern of contributions from the employer and possibly also the members, would reliably fund liability outgo as it fell due. (There was also an implicit assumption that the uncertainty in long-term real dividends was much less than could be inferred from short-term equity market price volatility.) The idea of the pension fund as a vehicle whose primary purpose was to ensure the security of members' accrued benefits was recognised by some actuaries, but sat largely in the background of the development of actuarial thinking on pension funds before the 1970s. However, this perspective would occasionally be considered and could lead to substantially different conclusions. Plymen noted in the discussion of Day's 1959 paper:

With any but the strongest employers the contingency existed in a slump that the employer might be bankrupt or there might perhaps be severe staff redundancy. Under those conditions, the employer would be unable to subsidize a pension fund prejudiced by depreciation of its ordinary shares. He imagined that that would be regarded as by far the greater threat, so that fixed interest securities would tend to be preferred. He maintained that the strength and security of the employer should be the major factor in determining the equity percentage.[37]

[35] Hemsted, in Discussion, Heywood and Lander (1961), p. 365.

[36] Day and McKelvey (1963), p. 107.

[37] Plymen, in Discussion, Day (1959), p. 152.

Plymen was arguing that the nature of the asset risk that mattered to pension funds was complex: asset price falls did not matter so much if the sponsor was healthy; what really mattered was how the pension fund assets behaved in the conditions that were associated with sponsor bankruptcy—that's when the assets were really needed. Such thinking would play an increasingly influential role in actuarial thinking on pension funds, their valuation and their asset strategy over the following decades. With the benefit of hindsight, it is perhaps striking that it was never more than a footnote in the thinking of the 1950s and 1960s, especially when sophisticated thinking by leading actuaries such as Plymen was in circulation.

Equity allocations continued to grow in the UK defined benefit pension fund sector, both as a result of the strong growth of pension funds' equity holdings over the 1960s, and with the investment of new contributions. By 1975, two thirds of UK final salary pension funds were invested in equities and property.[38] The extraordinary volatility of UK equities in 1973–1975, however, provided actuaries with a reminder that equity returns could be driven by factors other than inflation. In his faculty presidential address of 1977,[39] R.E. Macdonald noted:

> The market collapse of 1974 finally provided convincing evidence that any connection between the values of shares and the rate of inflation was mainly fortuitous.[40]

The 1973–1975 period certainly opened eyes to the degree of volatility that was possible in real equity price levels (despite the exceptionally high inflation rate, UK equities fell by over 70 % in 1973–1974; they then more than doubled in 1975). But McKelvey's and Day's papers had been careful not to suggest that real equity returns were not volatile: rather, they argued it did not matter if they were. Their argument was that real dividend growth could be expected to be stable, and that was what mattered for pension funds (open ones, at least). But the experience of the early 1970s also proved that this faith in the stability of the real value of dividends was misplaced: UK dividend pay-outs fell by around 45 % in real terms between 1970 and 1974 and did not fully recover their real 1970 value until the late 1980s.[41]

---

[38] Holbrook (1977), p. 58.

[39] Macdonald (1977).

[40] Macdonald (1977), p. 9.

[41] Dimson, Marsh and Staunton (2002), p. 152.

So the UK experience of the early 1970s showed that the much-vaunted inflation-protection of equity dividend income was a myth. Nonetheless, actuarial wisdom remained unperturbed. In his *Journal* paper[42] of 1977 on pension fund investment, Holbrook was sanguine on this episode of equity market volatility, noting:

> The catastrophic fall in UK equity markets in 1973 and 1974 reflected a serious crisis of confidence, which has not been wholly restored by the subsequent rise. With the benefit of hindsight it may now be said that the main importance, for a long-term investor, of the exceptionally low market level in 1974 was the opportunity it presented to buy shares cheaply.[43]

Thus, equities and property became the dominant asset class choices of defined benefit pension funds. This was doubtless in part influenced by the wider 'cult of the equity' that developed from the second quarter of the twentieth century onwards, and, from a specific actuarial perspective, because they were the only viable real asset class choice of the era with which to back undoubtedly real liabilities. The question of how much protection from short-term inflation shocks or long-term inflationary eras can be provided by equities (relative to bonds) is a question that has been much contested by actuaries and economists for decades. Whilst it is intuitive that the real characteristics of an equity claim implies some form of inflation protection, the question remains of how material a driver of returns (and dividends) inflation is relative to the many other factors that drive the behaviour of equities (over the short and long term).

A quantitative framework for analysing liability-orientated investment strategy was proposed by A.J. Wise in his 1985 paper 'The Matching of Assets to Liabilities'.[44] Wise defined a matching portfolio in a very particular way, which encompassed a much wider potential set of portfolios than the 'absolute matching' of cashflows that was first defined by Haynes and Kirton back in 1952: Wise's matching portfolio was defined as the asset portfolio that minimised the volatility of the 'ultimate surplus' of the book of business, i.e. the residual (positive or negative) asset value after the final liability cashflow had been paid. Importantly, he added the stipulation that the matching portfolio could not be re-balanced over the projection horizon.

---

[42] Holbrook (1977).

[43] Holbrook (1977), p. 17.

[44] Wise (1985).

Wise's matching portfolio could differ from a cashflow matching portfolio for a number of reasons: the universe of investable assets may not be granular enough to exactly cashflow match (i.e. there may not be a bond with a maturity date exactly coinciding with the liability cashflow); the liability cashflow could be exposed to factors such as salary inflation that may not be directly obtained in available assets; the liability cashflows may have non-linear features that are not present in available assets (and that could not be produced by a dynamic rebalancing strategy for the asset as such strategies had been outlawed by Wise's matching portfolio definition).

Wise's stipulation that the matching portfolio must be a static one came more than a decade after Black-Scholes-Merton had shown how fundamental the idea of dynamic rebalancing was to liability matching (which in Black-Scholes-Merton was just a means to arbitrage-free pricing by replication). The static assumption also ran counter to the actuarial idea of immunisation introduced by Redington in 1952 which focused on what might be called instantaneous matching and recognised the need to rebalance through time to maintain the match. But Wise believed that there was something inherently imprudent about including dynamic rebalancing assumptions in a risk modelling exercise. In introducing his paper at the Staple Inn discussion, he considered practices in the life field where modelling was comparatively more developed than in pensions, and asked:

> When assessing the solvency of a life office is it safe to assume that the office will always be able to take advantage of investment conditions in the future so as to keep itself in an immunised position? Would it not be safer to identify a minimum risk matching portfolio for solvency purposes on the assumption of no future investment activity except as required by new money?[45]

I.C. Lumsden, a senior life actuary, responded:

> It does seem to me unrealistic to consider matching portfolios only on the assumption of passive investment policies, and a pity in particular to ignore the one dynamic strategy – immunisation – which has gained our acceptance in the past.[46]

Wise went on to propose that the market value of the matching portfolio as defined above for pensions liabilities could be considered as a form of liability valuation. On the surface this seemed quite natural, and not dissimilar to the

[45] Wise (1985), p. 65.

[46] Lumsden, in Discussion, Wise (1985), p. 71.

economic idea of arbitrage-free pricing by replication. But Wise's definition of a matching portfolio was profoundly different from an economist's notion of a replicating portfolio. Wise's matching portfolio could leave the matcher significantly exposed to market risks—so whilst the portfolio might generate an *expected* surplus of zero, it may still have a significant *economic* cost. The identification and cost of Wise's matching portfolio could also be a function of subjective assumptions about long-term asset risk premia—again a feature that ran contrary to the properties of market-based valuation techniques.

It could be said that much of the confusion that has historically arisen between actuaries and financial economists has been due to a lack of clarity between the ideas of expected funding costs (based on expected long-term asset returns) and replication costs (based on the current market prices of replicating assets). Wise's matching portfolio definition and its use in liability valuation could only add to this confusion. Wise compared pension fund liability valuations produced by his matching portfolio concept with the more straightforward market-consistent approach of discounting liability cashflows with the observed market yield curves. He expressed the traditional actuarial discomfort with such direct use of market prices:

> This [market-consistent value] result might be considered realistic on the grounds that current redemption yields on secure fixed interest stocks are the best available guide to future economic conditions. Whether this is so is a somewhat philosophical point; it is certainly not generally accepted.[47]

For pension actuaries, uncoupling expected funding costs and economic values appeared interminably challenging. But the distinction would only become more relevant.

Funding for What? (1972–1997)

Actuarial thought on pension fund valuation prior to the 1970s was almost exclusively focused on valuation with the objective of determining the long-term contribution rate, expressed as a constant percentage of salary roll, that would be required to fund liability cashflows as they ultimately fell due. Similarly, thinking on investment strategy was centred on the objective of generating long-term asset income that would be used together with these contributions to fund those liability cashflows. Pension funds were managed

[47] Wise (1985), p. 55.

to provide a stable funding plan for the pension liabilities on a going concern basis. The pension fund's function as a source of security for employees' accrued benefits had, of course, been recognised by actuaries since the nineteenth century. But it did not figure prominently in the first 100 years of actuarial thought on how to manage defined benefit pension funds.

Funding with the objective of providing security for current accrued benefits started to command greater prominence in British actuarial thought in the 1970s. This was partly driven by an increased incidence of corporate mergers and takeovers, and the tendency these had to lead to wind-ups of some of the involved companies' final salary pension funds (in which circumstance, the sponsor was not obliged to make any further contributions to the fund). In occasional high-profile cases the pension fund wind-up exposed the inadequacy of the asset fund to secure the pension benefits that members expected.

D.F. Gilley, whose work we already encountered above in his 1958 paper with Funnell, published an important paper on this subject in 1972.[48] He discussed what occurred on the wind-up or dissolution of a defined benefit pension fund. There were several aspects of the dissolution process that could place considerable responsibility with the scheme actuary. The trust deed of the pension scheme would, in theory, set out what the members' benefits would be in dissolution. But these rules would generally be quite vague. Members might only be entitled to the benefits they would have obtained if they had withdrawn from the fund on the date of the dissolution of the fund. If any surplus or, more pertinently, deficit existed relative to the cost of securing those benefits from an insurance company with the pension fund's assets, the actuary's advice on how it should be equitably distributed amongst the members would usually be decisive.

The cost of securing the accrued dissolution benefits could diverge significantly from the long-term actuarial valuations undertaken to set contribution rates. The cost of the dissolution benefits was driven by the then-prevailing market gilt yield (which determined life companies' pricing terms for the deferred and immediate annuities that would need to be bought to secure the benefits). As we have seen above, this market-based rate could vary substantially from the valuation interest rate used in the pension valuation (which was based on the actuary's estimate of the expected long-term return on the pension fund assets and hence could include the actuary's estimates of the equity risk premium and the long-term expected interest rate). Although market gilt yields would typically be lower than a pension fund's valuation interest rate, the cost of the dissolution benefits would not necessarily be greater than the

[48] Gilley (1972).

long-term ongoing valuation—the withdrawal benefit that would typically be viewed as the minimum accrued benefit would break the link with real salary growth, thereby significantly reducing active members' liability values (the degree to which this impacted on the valuation would therefore be a function of the maturity of the pension fund). Overall, contribution rates and asset strategies were not set to target dissolution benefit funding, and the funding position on this basis could therefore vary markedly across pension funds.

Gilley called for the dissolution benefit funding level to be calculated as part of the regular actuarial management of defined benefit pension funds. Moreover, he proposed this funding result should be published and communicated to members. This was viewed as a fairly radical notion at the time. Indeed, the entire idea of taking into account the dissolution benefit funding level when setting contribution rates was provocative to traditional actuarial thinking. However, the economic turmoil of mid-1970s Britain prompted actuarial pensions thinking to consider some even more radical questions: should private sector defined benefit pension liabilities be funded at all?

This far-reaching question was raised in no less august and authoritative a setting than a Faculty of Actuaries presidential address—that of R.E. Macdonald in 1977.[49] The economic and political environment for defined benefit pension funds was extremely challenging in the early to mid-1970s. Annual price inflation reached a peak in excess of 24 % in 1974/1975. This high-inflation environment naturally fed through to salaries and drove pension liability values higher. Real investment returns over the first half of the 1970s were negative. There was political pressure to inflation-protect pensions in-payment, and to maintain (in real terms) the accrued benefits of withdrawing members. Taken together, and with the possibility that the high-inflation/negative real return environment could be a long-term feature of the economy, it was perhaps unsurprising that radical solutions to defined benefit pension scheme funding were being put on the actuarial table.

In this context, Macdonald noted the outspoken presence of 'those who regard funding [of defined benefit pension liabilities] as an outdated shibboleth of the actuarial profession'.[50] Macdonald expressed some sympathy with this position and, in particular, questioned the reasonableness of advance funding pension liabilities in a high-inflation, negative-real-interest rate economic environment. His point was two-fold: in such economic conditions, the funding of inflation-linked liabilities would be very demanding for sponsors, especially if pensioner inflation increases were included, and this

[49] Macdonald (1977).

[50] Macdonald (1977), p. 13.

would likely be a time when employers could least afford to contribute the additional cash; and secondly, he found it unconscionable to invest in assets that offered a negative real interest rate. In his own words, 'the acquisition of new assets during such periods [of negative real interest rates] is really impossible to justify'.[51] Macdonald suggested that, in such circumstances, pension contributions should be limited to those required to fund immediate benefit outgo, and no contributions for the funding of the accrual of longer-term liabilities should be sought.

Macdonald's ideas were quite controversial, especially coming from a leader of the profession. After all, he was proposing that contribution rates ought to be reduced at the very time when an economic view of the pension fund would suggest contributions were required more than ever—both because the funding level would be at its weakest and the credit risk of the sponsor would be at its greatest. An economist working from the premise that the primary rationale for advance funding was to ensure a high level of pension fund member security might be forgiven for thinking the actuarial dark arts were becoming too clever by half.

But President Macdonald was not the only actuary expressing reservations about advance funding of pension liabilities during this period. The same year saw the publication of a paper by J.R. Trowbridge in the *Journal of the Institute of Actuaries* which covered similar ground.[52] Trowbridge looked overseas for examples of how unfunded private sector defined benefit pension provision could function. He found that the French had been running unfunded occupational pension schemes with some success for several decades. Under this system, which was translated as *assessmentism*, the employee security concern of a non-funded approach was mitigated by having industry-level pension schemes, potentially with hundreds of different employers as sponsors collectively contributing (together with members) whatever contribution amount was required to meet each year's pension outgo as it arose. Thus the single-name credit exposure that was usually borne by British pension schemes was diversified (though an industry-level credit exposure would remain). These ideas naturally met with some suspicion amongst the UK actuarial profession. There was a concern that much of the French system was based on a voluntary cross-generational transfer that relied on notions of socialist solidarity that did not naturally fit with the British culture of self-sufficiency. Above all, there was a hope and faith that the economic norm of positive real returns in excess

[51] Macdonald (1977), p. 14.

[52] Trowbridge (1977).

of salary growth would return to Britain's pension funds, bringing with it a return of financial viability for traditional British pension schemes.

The actuarial debate of this fundamental aspect of defined benefit pension provision—to fund or not to fund—continued in the following years. Boden and Kingston followed their President's lead and further discussed the relative merits of advance funding and pay-as-you-go financing of final salary schemes in a paper published in the *Transactions of the Faculty of Actuaries* in 1979.[53] They agreed with Macdonald's central point that in a negative real interest rate environment the case for pay-as-you-go was greatly strengthened. However, Boden and Kingston went on to argue that the 1974–1975 inflation experience and the poor real equity return performance of 1968–1978 were aberrations that did not alter their forward-looking expectation of long-term positive real returns on pension fund assets. They also reviewed the French *assessmentism* approach discussed by Trowbridge and concluded: 'Once we are reasonably happy that it is possible to accumulate interest on capital at least as fast as inflation, there seems no compelling argument for going through the trauma of adopting the French system.'[54]

Boden and Kingston were even sufficiently sanguine about future real interest rates to argue that explicit allowance for post-retirement inflation increases could and should be made in pension fund valuations and contributions. This position was supported by influential actuaries such as A.D. Wilkie, who made the same argument in another Staple Inn discussion a couple of years later:

> I see no reason to assume a long-term real rate of return at present [1982] market levels of less than 4% in excess of prices. At that rate of interest, how many pension funds are in surplus, and how many cannot afford index-linked pensions?[55]

The degree of uncertainty and variation in long-term economic expectations amongst senior actuaries over this period of 1977–1982 is striking: are long-term expectations for real asset returns negative or 4 %? There was little doubt that, whether using advance funding or pay-as-you-go, the economic cost of defined benefit pension liabilities would be unsustainably high in a permanent negative real interest rate environment. A 4 % real interest rate, on the other hand, would make generous inflation-linked increases beyond the

[53] Boden and Kingston (1979).
[54] Boden and Kingston (1979), p. 412.
[55] Wilkie, in Discussion, Colbran (1982)), p. 410.

level of guaranteed benefits quite affordable. This highlighted the fundamental challenge of setting a funding objective based on subjective expectations for the behaviour of economic variables over exceptionally long time horizons.

But the optimists prevailed. British actuarial orthodoxy on the advance funding of final salary pension funds survived the economic turmoil of 1970s stagflation and its consequent financial market volatility. Moving from the late 1970s to the 1980s, actuarial debate shifted from whether or not to advance fund pension liabilities and on to funding methods and what the objectives of funding should be—back to the debate that Gilley started in 1972 which had lain largely dormant for a decade. Rather like the contemporaneous experiences of their life office brethren, influences and demands from outside the profession were the main catalysts for pension actuaries to revise and evolve their traditional approaches. In particular, and again with parallels with the life office experience, the accounting profession and government regulators had a telling influence (the Accounting Standards Committee was working on an exposure draft on treatment of pensions in company accounts; the Occupational Pensions Board was considering solvency disclosure and inflation protection for early leavers and pensions in payment).

The accounting profession was increasingly active in setting new standards for the reporting of the pension expenses in financial statements. Historically, the pension expense may have been accounted for simply as the amount the sponsor paid in contributions over the accounting period. As contribution payments became more volatile, accountants became more focused on a measure of the cost of pension benefit accrued over the accounting period. In the UK, actuaries' preference was generally to keep funding and accounting expense methods as similar as possible. There was therefore a general preference to base this measure of cost on the stable long-term percentage of salary roll that the actuary assessed would be required to meet the current members' future benefits net of the existing asset fund (the 'aggregate method'). However, accountants were increasingly aware that this contribution rate did not necessarily equate closely with the cost of the liabilities that accrued over an annual reporting period. In particular, the cost accrued might be expected to increase later in the life of a pension fund as its membership aged. So methods such as the projected unit credit method, which focused on the contribution specifically required to meet the increase in the pension liability that resulted from a year's further accrual, were preferred by accountants.

These unit credit methods were in increasingly common actuarial use in the 1970s but were initially somewhat unpopular with the British actuarial profession as it challenged the traditional view that the assessed contribution rate should be a stable long-term percentage of salary roll rather than a figure that would be expected to increase as the scheme matured. Colbran's

1982 paper reviewed this situation and called for more professional guidance to be provided by the Institute of Actuaries on pension funding methods.[56] He suggested a list of acceptable methods should be created with appropriate terminology and definitions. This suggestion was acted upon and a 'Report on Terminology of Pension Funding Methods' was jointly published by the Institute and Faculty of Actuaries in 1984.[57]

Colbran also argued in favour of a market value approach to asset valuation rather than the discounted income approaches that dominated actuarial pensions practice at this time. The Staple Inn discussion of the paper reopened the pension asset valuation debate that had lain largely dormant since Day and McKelvey's 1963 paper. In the discussion, the established actuarial preference for discounted income values over market values was clear. But this view was not unanimous, and there was a wide recognition that the first priority of the valuation method should be to ensure consistency between asset and liability values. This reopened the door to the idea of market valuation of assets and market-consistent liability valuation, as espoused as long ago as 1921 by S.G. Warner. In opening the discussion of the Colbran paper, Paul Thornton (who would go on to become Institute President in 1998) stated:

> It is possible to arrive at a set of assumptions which, when used to discount the future interest and capital proceeds from the investments, would reproduce market values. There is no reason why a valuation should not be made incorporating assets at market value if the liabilities are valued on consistent assumptions derived in this way.[58]

However, some senior pension actuaries in the discussion did indeed find reasons why this approach to consistent asset-liability valuation should not be adopted. D.E. Fellows noted:

> To rely wholly or largely on market values could lead to much volatility in the rates of interest used from valuation to valuation.[59]

And C.W.F. Low similarly stated:

> This [market-consistent liability valuation] would mean constantly changing valuation bases to reflect variations in the initial rate of interest obtainable …

[56] Colbran (1982)).
[57] Turner et al. (1984).
[58] Thornton, in Discussion, Colbran (1982), p. 387.
[59] Fellows, in Discussion, Colbran (1982), p. 390.

this approach is dangerous, as it would concentrate employers and trustees' minds on short-term investment performance which is not appropriate to the real investment needs of pension funds.[60]

1970s stagflation faded into distant memory as final salary pension funds returned to financial health in some style in the first half of the 1980s. UK price inflation appeared to have been tamed: in the mid-1980s, the UK annual inflation rate moved within the range of 3–5 %. Equities performed exceptionally strongly—the FTSE 100 increased by more than 125 % between the start of 1984 and the October 1987 crash. Real dividend growth was around 5 % per annum. Meanwhile, large-scale industrial redundancies further contributed to pension fund surpluses (redundancies shifted active members' benefits into deferred benefits that were no longer linked to real salary growth and hence reduced their value).

This turn of events may have afforded actuaries some latitude to think more broadly about funding objectives and targets. In any event, the fundamental objective of funding became a hot topic again. D.J.D. McLeish and C.M. Stewart were particularly instrumental in reopening the arguments presented by Gilley back in 1972 for focusing funding on ensuring that the cost of discontinuance benefits could be adequately covered by the pension fund at the time of the valuation. McLeish restarted this debate with a Faculty paper in 1983 which emphasised member security as the paramount rationale for advance funding of pension liabilities.[61] He expressly rejected the Faculty President's suggestion that funding should be deliberately suspended in times of negative real interest rates:

In private occupational schemes I believe that funding is necessary to provide security. The higher the yield on assets the better but there is no particular significance in the point at which the difference [between investment yield and salary inflation] changes from positive to negative.[62]

McLeish's funding-for-security arguments were further marshalled in a paper[63] co-written with C.M. Stewart in 1987 where they argued that:

The prime purpose of funding an occupational pension scheme must be to secure the accrued benefits, whatever they might be, in the event of the employer

[60] Low, in Discussion, Colbran (1982), p. 406.
[61] McLeish (1983).
[62] McLeish (1983), p. 275.
[63] McLeish and Stewart (1987).

> being unable or unwilling to continue to pay at some time in the future. To that end, the contributions would have to be sufficient both to pay the benefits as they fell due for as long as the scheme continued, and also to establish and maintain a fund which would be sufficient to secure the accrued benefits in the event of contributions ceasing and the scheme being discontinued, whenever that might occur.[64]

A practical complexity that arose in consideration of this type of funding target was the ambiguity that might exist in the pension scheme's trust deed and rules as to what members were entitled to on wind-up of the scheme. However, by 1987, statutory minimum benefits for early leavers existed that linked accrued benefits to the lower of future national average earnings inflation and 5 %. This provided an unambiguous minimum benchmark for wind-up benefits.

The approach of funding for discontinuance benefits was still generally unpopular with British actuaries at this time. Colbran's 1982 paper that discussed the need for greater professional guidance on funding methods had even called for this particular funding method to be outlawed by the profession on the grounds that it could produce a substantially lower contribution rate than other methods. But McLeish and Stewart argued that this would only be the case for immature schemes. They rejected the relevance of the level contribution rate that would ultimately fund all liability cashflows and instead focused on ensuring that accumulated assets were always sufficient to meet the accrued liabilities. They further argued that the level contribution rate approach would tend to result in the accumulation of unnecessary surpluses relative to the accrued liabilities.

The responses to the paper in its Staple Inn discussion were generally negative. Traditional actuaries were concerned that the specific targeting of the security of the accrued benefits would actually result in less security by slowing the pace of funding. But some recognition of McLeish and Stewart's arguments was won. The influential Paul Thornton agreed that the funding of accrued benefits should be part of the funding objective:

> I too believe that it is necessary for the funding policy to be such that the assets are normally sufficient to cover in full the winding-up priority levels. From the point of view of security for the members' benefits, this defines a *minimum* funding level.[65]

[64] McLeish and Stewart (1987), p. 155.
[65] McLeish and Stewart (1987), p. 221.

Like other developments in actuarial thought that emerged in a climate of traditionalist scepticism, once the genie was out of the bottle, it would never go back in. The next notable contribution to British actuarial thinking on the funding of final salary pension schemes arrived in a paper, 'A Realistic Approach to Pension Funding', by Paul Thornton and A.F. Wilson in 1992.[66]

This paper was ostensibly traditionalist in outlook, continuing to advocate a discounted income approach to asset valuation and funding methods that were focused on producing long-term contribution rates rather than funding current discontinuance benefits. But, rather like McLeish and Stewart, they argued that typical funding calculations could lead to over-funding. They noted the 'embarrassingly large surpluses' that prevailed in many pension funds at the time.[67] Their central argument was that these surpluses did not arise because of any inherent limitation in the traditional pension funding methods, but because the parameterisation of these methods generally included numerous prudent margins that were cumulatively unnecessary and sometimes accidental. They argued that the typical valuation assumption of a real return of 4–4.5 % was lower than the expected real return on equities that most pension fund assets were invested in. They believed a 5–6 % expected real return on equities was reasonable based on historical analysis and forward-looking economic arguments. They suggested that salary inflation tended to be lower than the national average for the age group of over-40s, and it was this group that was most important to the valuation of active members' benefits. They estimated that this over-stated salary inflation by around 0.5 %. According to Thornton and Wilson, these and other margins were generating unintended surpluses.

In a similar vein to Thornton's comments in the McLeish and Stewart discussion, they regarded the discontinuance position as an important reference point, though not necessarily the central driver of the contribution rate. They even went a step further, arguing that a margin of 20 % over the discontinuance cost should be maintained in light of 'the speed with which surpluses and deficits can build up'.[68] To paraphrase, they were arguing that if the purpose of the assets was to meet a market value of liabilities in the near future, it was not sufficient to have assets that met that value of liabilities today—an additional buffer would be required to the extent that the asset and liability values could diverge between now and the possible dissolution event. This logic is essentially that of the value-at-risk concept that emerged from the banking sector and, in the twenty-first century, would become the basis for insurance solvency regulation.

[66] Thornton and Wilson (1992).

[67] Thornton and Wilson (1992), p. 263.

[68] Thornton and Wilson (1992), p. 264.

A number of speakers in the Staple Inn discussion of the paper noted that the cost of transferring discontinuance benefits to insurance companies had increased considerably in recent years (partly due to improvements in early leaver rights and partly due to lower interest rates). For the first time, the cost of the non-profit immediate and deferred annuities that would meet the discontinuance benefits might typically exceed the liability values assessed by the traditional ongoing funding assessments, and indeed the market value of assets in the pension fund. Between the publication of the paper and its discussion at Staple Inn, the Robert Maxwell affair occurred, placing discontinuance security more squarely in the public eye than ever before. Thus, actuaries' 'embarrassing surpluses' had to be simultaneously reconciled with increasingly transparent discontinuance shortfalls. This was a significant challenge to the actuarial profession's credibility and communication skills. It was a challenge they struggled to meet. Ultimately, it was taken partly out of the profession's hands when the government legislated the introduction of the Minimum Funding Requirement (MFR) for final salary pension schemes in 1995.

The MFR was a controversial and contentious requirement from its inception. It was a result of a committee, the Goode Committee, which was established by the government in the wake of the Robert Maxwell scandal. However, the recession of the early 1990s was another driving factor for the interest in legislation to protect pension fund members' accrued benefits. A number of pension scheme wind-ups occurred during this recession where members did not receive their expected benefits, and this experience further contributed to the sense that the security of accrued benefits was not being adequately protected by the prevalent actuarial funding methods. So the MFR was first intended as a robust measure of the adequacy of the value of assets in the pension fund to meet the market cost of accrued benefits. However, once the financial implications of using the insurance annuity basis in the pension valuation were realised, a political compromise was made and the MFR valuation basis was fudged: the expected return on equities would be used as the discount rate for accrued benefits (except pensions in payment), even though this would undoubtedly understate the cost of securing the accrued benefits in the event of dissolution of the scheme.[69] Consequently, being 100 % funded on the MFR basis did not mean the pension fund necessarily had close to sufficient assets to meet the cost of the accrued benefits in the event of dissolution of the pension fund. The result was unsatisfactory for all concerned: actuaries lost more of their professional independence and ability to

[69] See Greenwood and Keogh (1997) for a detailed description and discussion of the actuarial bases of the MFR.

self-regulate; government had created a costly regulatory framework that did not meet its intended purpose; members may have been led to believe their benefits were secure when in fact they were not. As with their life office brethren, the 1990s was proving to be a tough decade for British pension actuaries.

Modern Thought on Pension Scheme Valuation and Investing (1990–1997)

The clash of new and traditional thinking on the funding, valuation and investment of defined benefit pension funds grew in intensity over the 1990s, representing one of the more turbulent episodes in British actuarial thought. The 1990 *Journal* paper, 'Actuaries, Pension Funds and Investment', by T.G. Arthur and P.A. Randall can be viewed as the start of this period.[70] The paper was not as radical or provocative as those that would follow over the next few years, but it unmistakably advocated a further move away from traditional actuarial valuation approaches to those that attached greater legitimacy to market values and fully embraced their use in pension fund valuation.

The central idea set out in the paper was that changes in the market values of assets that did not match liabilities should be fully reflected in the change in the actuarial assessment of surplus in the pension fund. They argued that to do otherwise was 'implicitly taking an investment view that the market was wrong' and that 'this is not our [actuaries'] job when constructing [pension fund] valuation models'.[71] They further elaborated 'there is no *actuarial* reason to suppose that any relative changes in market values that have taken place in the past, whether recent or otherwise, will be reversed in the future'.[72]

This perspective implied a significant narrowing of the pension actuary's professional input. It was a departure from the traditional actuarial tendency to attempt to 'look through' irrational short-term market volatility to the underlying intrinsic worth of the pension fund investments on a going concern basis. For the previous 40 years, pension actuaries had felt quite comfortable placing values on pension fund investments that could differ quite substantially from their market values. This was partly justified on the basis of consistency with an off-market and stable liability basis, but underlying these choices was a philosophical view that short-term market value movements were of no consequence to an open pension fund's long-term funding:

[70] Arthur and Randall (1990).

[71] Arthur and Randall (1990), p. 5.

[72] Arthur and Randall (1990), p. 6.

as a going concern, it would never have to sell its assets, and asset income was more stable than was implied by market value movements. More widely, actuarial asset models such as Wilkie's implied that future risky asset returns were significantly influenced by past returns (mean-reversion in equity returns, for example). This gave a technical manifestation to the actuarial convention that market values varied more than could be justified by changes in assets' expected future cashflows, and a justification for removing this variation from actuarial valuations. Arthur and Randall argued that all of this had no place in actuarial methods for pension fund valuation.

A natural corollary of their logic was that liabilities should be valued at the market value of those assets that matched them. In the terminology of financial economics, this was the fundamental concept of the replicating portfolio. Irrespective of what pension actuaries chose to call it, it was essentially what S.G. Warner had called for in 1921. Whilst the Thornton and Wilson 1992 paper discussed above still advocated the use of the traditional discounted income approach to asset valuation using parameters that need not be market-based, the actuarial proponents of the market value approach were undoubtedly growing in number and influence.

The final implication of the Arthur and Randall perspective was that pension fund investment strategy should be set at a scheme-specific level that reflected the specific liability cashflow profile of the pension fund. This was a fairly novel idea in 1990 that only became well established in the 2000s under the jargon of liability-driven investing. Their vision of the actuarial role in pension funds removed subjective investment views from actuarial valuations and made the valuation independent of the actual asset strategy of the pension fund. But it suggested a greater role for actuaries in pension funds' strategic asset allocation—it made actuarial estimates of liabilities central to pension fund investment strategy.

In the Staple Inn discussion there were inevitably voices who were uncomfortable with Arthur and Randall's modern perspective and its implied diminution of the role of actuarial judgement in pension fund valuation. There remained amongst some senior actuaries a strong view that the pension fund valuation should be an estimate of the amount of assets they expected to be required to meet the liability cashflows given the actual asset strategy of the fund, rather than be based on the market cost of a theoretical risk-free matching strategy. In Dr Sisson's speech to open the discussion, he argued:

> The valuation ought increasingly to reflect the realities of life. Included in this would be a recognition of the investments expected to be held by the trustees in the future and the assumed returns thereon.[73]

This encapsulated the fundamental difference of opinion as to the *purpose* of the pension fund valuation: was it to assess the current market cost of matching the liabilities (which should reflect the cost of transferring the liabilities to a third-party immediately), or to make a best estimate of the average amount of assets required to meet the ultimate payment of the liability cashflows if a particular set of investment risks (and corresponding expected rewards) were borne by the fund until its extinction? Both of these quantities were, to borrow Dr Sisson's phrase, 'realities of life'. Neither was wrong, they were merely answers to different questions. But this clarity was generally missing from the debate.

Arthur and Randall's call for a market-based valuation approach to pension fund assets and liabilities received further support in a paper by Dyson and Exley that was published in the *British Actuarial Journal* in 1995.[74] Dyson and Exley emphasised that, by this time, pension fund valuations had a wider set of applications than their traditional use in the assessment of the long-term contribution rate required to fund expected liability outgo. And whilst the assessment of the contribution rate expected to fund long-term expected liability outgo necessarily required assumptions about asset risk premia, a number of other valuation purposes had emerged that did not. In particular, the assessment of whether the fund had sufficient assets to fund the transfer of the accrued liabilities to a third-party—the discontinuance solvency assessment—was fundamentally an assessment of a market cost. It did not require subjective actuarial estimates of long-term asset risk premia. Such applications were naturally more suited to market-based approaches to asset and liability value. The traditional approach of using a discounted income asset valuation coupled with a valuation interest rate based on a long-term expectation for investment returns that was developed in the early 1960s did not answer the discontinuance solvency question.

Dyson and Exley also expressed misgivings about the use of the discounted income approach even in the context of long-term contribution rate assessment. In particular, they highlighted that the equity valuation produced by the discounted income approach was extremely sensitive to the actuary's long-term dividend growth assumption, and that this was a parameter around

[73] Sisson, in Discussion, Arthur and Randall (1990), p. 28.

[74] Dyson and Exley (1995).

which there was significant uncertainty—'the level of confidence in any estimate of *m* [the long-term rate of real dividend growth] is likely to be very low indeed'.[75] But there was no escaping this—the estimation of the expected amount of contributions required to meet ultimate liability outgo in the presence of a risky asset strategy was an inherently subjective, difficult-to-estimate judgement.

Dyson and Exley also emphasised that the existence of index-linked gilts (from the early 1980s) permitted a market-consistent approach to the valuation of inflation-linked liabilities that was not available to actuaries in the 1950s and 1960s era when discounted income methods were pioneered. They argued that pension liabilities should be valued using the nominal and real term structures of interest rates implied by gilt and index-linked gilt prices at the given valuation date. Predictably, these proposals led to discomfort for some actuaries who saw their valuation expertise being reduced to mechanically plugging in the market's latest bond prices. In opening the Staple Inn discussion of the meeting, G.J. Clark suggested:

> We must not allow ourselves to become so mesmerised by market information that we exclude one very valuable source of information – the professional judgement of the actuary.[76]

Interestingly, whilst market-based approaches to valuation had increasingly been advocated in actuarial thinking in recent years, their proposals were still viewed as quite drastic. R.C. Urwin commented:

> It [the paper] has quite a radical conclusion which is iconoclastic to current actuarial practice.[77]

Concerns were also raised as to the wisdom of relying on index-linked bond prices when the volume of issuance was low relative to the scale of pension fund assets. C.D. Daykin, the Institute President of the time, stated:

> It has been indicated by several people that the real return on index-linked gilts is not necessarily a very good indicator [of the market expectation of the real rate of return] … the low liquidity and the way in which they are influenced by a

[75] Dyson and Exley (1995), p. 489.
[76] Clark, in Discussion, Dyson and Exley, p. 543.
[77] Urwin, in Discussion, Dyson and Exley, p. 544.

particular sector of the investment market means they are of less value than they might be in determining the market value of liabilities.[78]

But the biggest protest against the extensive use of market prices was a philosophical one. Many actuaries retained the view that market prices were irrationally volatile, and that actuarial judgement should be able to look through this. S.J. Green rhetorically asked:

> What new piece of information became known on 19th October 1987 which was not known three days before? It must have been very significant to lead to falls of more than 25% in a few days, and that is the answer to the author's point where they say that the market provides rational prospective expectations.[79]

The influential Professor Wilkie struck a similar chord in defence of discounted income valuation approaches:

> Using this methodology [discounted income approach], the actuary is, indeed, saying that the market has temporarily got it wrong, but that, in due course, it will get it right, especially by the time the liabilities fall due.[80]

Wilkie's view of the role and breadth of actuarial judgement in asset valuation was the polar opposite of that advocated by Arthur and Randall five years earlier. But, as noted above, these differences were perhaps not as irreconcilable as they appeared: different assumptions were required to answer different questions. A rigorous discontinuance solvency assessment would be market-based; an assessment of the amount of contributions required to, on average, meet liability cashflows as they fell due would inevitably require someone somewhere to make an assumption about the prospective long-term risk premia offered by the pension fund's asset strategy. The two were not theoretically inconsistent or mutually exclusive, they were merely distinct.

Throughout their paper, Dyson and Exley were careful to position their arguments in as inoffensive a manner as possible to actuaries schooled in traditional methods. They were quite prepared to concede that the discounted income methods were appropriate or even impressive at the time they were conceived. Their position was that recent circumstances dictated it was time for the actuarial profession to continue to improve and modernise its methods. An important paper by Exley, Mehta and Smith that followed in 1997

[78] Daykin, in Discussion, Dyson and Exley, p. 552.
[79] Green, in Discussion, Dyson and Exley (1995), p. 546.
[80] Wilkie, in Discussion, Dyson and Exley (1995), p. 549.

adopted a markedly different tenor.[81] By contrast, its tone barely concealed a relish in smashing all actuarial shibboleths in sight.

Like Dyson and Exley before them, Exley, Mehta and Smith again highlighted the broader range of contemporary applications of pension fund valuations and how these applications were better served by market-based valuation approaches. This paper developed some further sophistication around the market-consistent valuation method for pension liabilities, tackling complexities such as the valuation of the complex non-linear inflation exposures of deferred member liabilities that arose in the form of limited price indexation. The paper also presented some empirical evidence from UK economic data to argue that national real salary growth was fairly stable. This supported the argument that index-linked bonds could be regarded as a good hedge for active liabilities (and hence helpfully avoiding the need to consider equities or other risky assets as a hedging asset within liability matching and valuation).

The main contribution of the paper was, however, in the analysis of the investment strategy of a defined benefit pension scheme from the perspective of the corporate sponsor. Little had been written in actuarial journals on the management of pension funds from the specific perspective of the sponsor. Since the 1980s, however, consulting pension actuaries often formally advised sponsors as well as pension fund trustees. Exley, Mehta and Smith pointed out that from an economic perspective, the sponsor of a defined benefit pension scheme should, in theory, be indifferent to the pension fund's investment decisions. The assets of the scheme could be considered as off-balance sheet assets of the corporation and, from a shareholder value perspective, bonds with a market value of £1 had the same economic worth as a portfolio of equities that had a market value of £1. They argued that shareholder wealth therefore could not be created or destroyed by merely switching from one to the other, in much the same way that Modigliani-Miller showed that shareholder wealth could not be created or destroyed by changing the corporate financial structure of the firm.

However, this conclusion came with some caveats. Firstly, it only strictly applied in the case of a 'pure' defined benefit scheme. If, for example, the members participated in discretionary benefits that would be increased in the event of a pension scheme surplus arising but which could not be reduced beyond a floor in the event of a deficit arising, then in economic terms the shareholder was writing a call option to the member. The greater the volatility of the pension fund's assets relative to liabilities, the greater this option would be worth. In such circumstances, the shareholders' interest would be served

[81] Exley, Mehta and Smith (1997).

by minimising this volatility, which implied matching the pension liabilities as well as possible.

The second complication was that their argument ignored the pension fund members' exposure to sponsor default—if the sponsor went bust when there was a deficit, the shareholder would never make good on the pension promises. In economic terms, this could be viewed as a put option that the pension fund member had written to the shareholder. It was a complicated option because its value was a function of the sponsor default probability and the correlation between the default event and the size of the pension fund surplus/deficit (as well as the volatility of the pension fund surplus). Nonetheless, it was an option that would increase in value (to the shareholder) as the volatility of the pension fund surplus increased. An increasingly *mismatched* asset-liability position in the pension fund would therefore increase this option's value to the shareholder. These two effects were therefore impacted in opposite directions by changes to the pension fund's degree of asset-liability matching. The specific circumstances of the pension fund and its sponsor would determine which effect dominated at a particular point in time.

The Staple Inn discussion of the paper unsurprisingly included much defence of the traditional actuarial approaches to the management of final salary schemes. A.J. Wise argued that discretionary benefit features such as early retirement pensions remained an important plank of scheme benefits, and more important than suggested by the authors. The ability to reduce such benefits in times of poor equity performance meant that 'for shareholders … the downside to equity investment is less onerous when that equity is in the pension fund'[82] (though presumably the shareholder's exposure to the upside would be similarly dampened).

Thornton pointed out that the shareholder value analysis was not a priority for trustees who were responsible for managing the pension fund assets. He also could not accept the removal of the equity risk premium from the active members' liability discount rate, though he could not articulate a rational reason why not:

> I am suspicious of their arguments that pension fund trustees should not be investing so heavily in equities, and that valuation bases should factor out any excess return expected. This conflicts with economic reality![83]

[82] Wise, in Discussion, Exley, Mehta and Smith (1997), p. 942.

[83] Thornton, in Discussion, Exley, Mehta and Smith (1997), p. 952.

As ever, 'economic reality' meant different things to traditional pension actuaries and the modernisers. The traditionalist view was that a realistic measure of liability cost meant the expected amount of (risky) assets required to fund the cashflows as they fell due; to the moderniser, economic reality was the current market cost of transferring the liabilities to a third party. This basic difference in perspective got lost in a debate of increasing passion that could verge on animosity. Nonetheless, it appeared that the use of market values for asset valuation had now reached close to universal acceptance amongst the actuarial profession. As L.P. Tomlinson noted in closing the discussion:

> I was astounded by the uniform acceptance of the use of market values. Speaker after speaker endorsed the principle of using market values for assets, and then valuing the liabilities consistently. My actuarial background was in the 1970s when no actuary would endorse market values, so we have moved a very long way since that period.[84]

Final Thoughts on the Pensions Story

In the final analysis, the actuary's traditional objective in pension fund management of estimating the stable long-term contribution rate that was expected to be required to fund ultra-long-term and economically complex forms of liability cashflow under a given risky asset strategy was an extremely challenging task. Their conclusions were highly sensitive to guestimates of non-stationary economic variables such as price inflation, real salary growth and long-term asset returns. Different economic environments came and went, each time prompting fundamental reviews of actuarial practices and assumptions. In times of particular economic uncertainty, this could become publically embarrassing. We saw how the five-year period between 1977 and 1982 produced actuarial estimates of the long-term real asset return ranging from less than zero to more than 4 % per annum. The implications of these different assumptions for the health of pension funds and their ongoing funding needs was so great as arguably to be a discredit to actuarial expertise.

Given the scale of inevitable uncertainty in the above estimates, it is perhaps surprising in retrospect that no coherent risk management framework or philosophy emerged as an intrinsic part of the apparatus of actuarial theory and practice in defined benefit pensions. By the late 1950s, actuarial thinking

[84] Tomlinson, in Discussion, Exley, Mehta and Smith (1997), p. 955.

in the life sector, propelled by the papers of actuaries such as Redington, Hayes, Kirton, Anderson and Binns, was producing and applying new technical ideas about asset-liability risk management—matching, immunisation, 'portfolio insurance'-style thinking in dynamic asset allocation and the stress testing of risky asset values relative to guaranteed liabilities all received serious consideration in this rich era of development of actuarial thought on financial risk management in the life sector. Pension actuarial thinking of that era appeared oblivious to such ideas and unwilling to consider how they could be applied in the particular context of defined benefit pension funds. It is worth pausing to ask why this was the case. It is perhaps linked to the historical lack of clarity and consensus on what advance funding of pension liabilities was for in the first place. In particular, a lack of focus on member security as the overriding purpose of the pension fund and on sponsor credit risk as a potential threat to that member security meant that pension actuaries perhaps did not have the same clarity of purpose as their life brethren when it came to thinking about managing financial risk.

The actuarial focus on the estimation of the stable long-term contribution rate arose in nineteenth-century actuarial practice. It was accountants and regulators that drove actuaries to consider other relevant financial objectives and metrics such as funding for security against near-term sponsor insolvency and measuring the cost of liability accrual. The profession struggled to establish in its thinking how these different demands were distinct questions that could be consistently reconciled within a wider technical actuarial framework.

Again with the benefit of hindsight, it now seems clear that the path the actuarial profession took to ensuring consistency between asset and liability valuation—off-market liability valuations with consistently off-market asset valuations, first proposed by Puckridge in 1948 and enthusiastically refined over the following 15 years—was a choice that created unnecessary confusion both inside the profession and in its dealings with others for half a century.

The treatment of equity investments permeated much of the above thinking, and was a particular source of complexity and, at times, controversy. Prior to the 1960s, the valuation treatment of equities was relatively straightforward, if basic. Equities would be valued at the lower of market value and book value, and the liability valuation paid no heed to the pension fund's asset allocation. The papers of Heywood and Lander, and Day and McKelvey in the early 1960s strove to create an equity valuation framework that was consistent with the established off-market liability valuation basis. Things were made more complicated still when Heywood and Lander proposed that equity allocations should permit the liability discount rate to be adjusted upwards from the actuarial estimate of the long-term risk-free rate in anticipation of the long-term excess returns of equities. This contributed to the actuarial perception of a 'free lunch' in long-term equity investment that was ultimately only dispelled in 1997 by Exley, Mehta and Smith.

The UK dividend experience of the 1970s showed that the original actuarial rationale for pension fund equity investment—stability in real growth in dividends—was misplaced. But actuarial confidence in their ability to look through short-term irrational market value volatility for the benefit of their long-term clients was difficult to entirely dispel, despite the arguments of Arthur and Randall and others in the 1990s. However, this actuarial perspective was not so far away from contemporary financial economics as the actuarial modernisers would lead the profession to believe. The empirical program of financial market research of the 1980s developed by leading academic financial economists such as Jensen, Shiller and Roll could lend some serious intellectual weight to the traditional actuarial disdain for short-term market price volatility. Curiously, this point was entirely missed in the actuarial traditionalist/financial economist moderniser debate of the 1990s. That is perhaps reflective of how this debate could generate more heat than light, and could at times enflame personal passions at the expense of objectivity and intellectual openness.

Rather like with-profits in the life sector, from the late 1990s defined benefit pension funds would start on a path of terminal decline, a structure that no longer met the needs of the modern workforce, and whose untenable economic cost was becoming increasingly apparent to employers. By the twenty-first century, British actuarial debate on how best to manage open defined benefit pension funds had become increasingly moot.

7

British Actuarial Thought in General Insurance (1851–1994)

General insurance—that is, the insurance of risks other than life contingencies—pre-dates life assurance in its development as an established business activity of critical importance to Britain's economy. Marine insurance first emerged as a significant commercial activity in early thirteenth-century Italy in ports such as Genoa. The earliest historical records of British marine insurance date from sixteenth-century London. Over the following 200 years, the City of London and its insurance institutions such as Lloyds of London developed into a leading global centre of maritime insurance. These institutions delivered dependable insurance to the cargos and sailing ships whose perilous sea voyages drove Britain's maritime trading economy and the expansion of its Empire.

The actuarial role in the British general insurance industry has, however, been most notable by its absence for most of this history. It was only in the 1970s that the actuarial profession's engagement with the industry moved significantly beyond the occasional passing observation. General insurance first appeared on the syllabus of the Institute of Actuaries in 1978. Actuarial professional guidance was not proposed in Britain until 1986. This contrasted with experience elsewhere in the world, most notably in the USA. General insurance actuaries even had their own actuarial professional organisation: the Casualty Actuarial Society, which was established in 1914. This greater actuarial role in US general insurance was prompted mainly by state regulation of premium setting, which created a demand for objective analysis and professional certification.

C. Turnbull, *A History of British Actuarial Thought*,
DOI 10.1007/978-3-319-33183-6_7

In comparison with life assurance, most forms of general insurance typically have several fundamental features that, to a greater or lesser degree, complicate and confound the use of the statistical methods pioneered by the likes of Richard Price and William Morgan in mortality modelling and the valuation of life contingencies. These include:

- Defining the unit of exposure. In life assurance, the unit of exposure is straightforward: person-years. It is unambiguous, intuitive and quite easily observed and recorded. In general insurance, the unit of exposure can be much less obvious: for example, in motor insurance, it could be vehicle-years or vehicle-miles; for fire insurance, it could be building-years, or building-floors-years, or something else. And if there is a theoretically appropriate unit, it may not be easily recorded.
- The above point is related to the heterogeneity of general insurance experience. There may be a very large number of risk factors that drive general insurance experience. Identifying and observing these risk factors may be a significant practical challenge. And once they have been identified, the resultant homogenous sub-sets of data may be very small.
- The claims experience incurred in a given insurance period may take many years to 'discover'. In some forms of general insurance, especially liability business, it may only be many years after the end of the insurance period that it transpires that a claim exists. So-called incurred but not reported (IBNR) claims are inherently challenging to predict or estimate. For some classes of business, the final settlement of the claim size may only occur a further many years from when it is reported.
- Most classes of general insurance business are inherently less stable or stationary than life assurance and its mortality rates. Social, judicial, technological, economic and commercial changes can drive changes in the expected claims behaviour of general insurance that are typically faster and more material than those in life contingencies. This can place substantial limitations on the usefulness of historical experience data, even where it is homogeneous, plentiful and complete.

As a result of these challenges, general insurance pricing and reserving have historically often been determined on a case-by-case basis by underwriters and other general insurance specialist practitioners using their expert judgement and individual experience. The use of analytical methods and statistical modelling based on empirical data has often been very limited. In this setting, British actuaries have historically been uncertain of what role they can and should play in the industry. We will see below, however, that British actuaries have been, at least sporadically, considering how to apply their technical and professional skills to general insurance pricing and reserving since the mid-nineteenth century, if not before. This engagement grew in conviction from the early 1970s, and by the 1990s actuaries in general insurance, having finally established an uncontested industry role, were faced with similar challenges and questions as their life assurance and pensions peers: did their relative lack of technical training in cutting-edge stochastic modelling and the new ideas in financial economics represent an existential threat to their established role? Could they absorb some of these concepts into the mainstream of actuarial thought and practice? First, however, we go back to the 1850s and how that generation of innovative actuarial thinkers tackled the unique challenges of general insurance.

Fire Insurance (1851–1880)

In the UK, fire insurance first emerged as a significant commercial activity in London in the decades following the Great Fire of 1666. A century later, it was well-established across the country, and increasingly covered industrial and agricultural fire risks as well as residential properties and retailers. By the nineteenth century, some of the leading British fire insurance companies had become successful multinationals, writing substantial business in North America and the British colonies.

Eighteenth- and nineteenth-century fire insurance practices did not make use of the statistical methods that were being contemporaneously developed and successfully applied in life assurance. The broad challenges of applying statistical and analytical techniques to general insurance were wholly applicable to fire insurance business: the classification of historical claims experience into homogenous groups; the modelling of claim size as well as claim frequency; making allowance for social effects such as fraud and its propensity to vary with the prevailing economic climate. These challenges led to a deep scepticism of the potential usefulness of statistical analysis of historical experience as an alternative to underwriting by case-by-case estimation.

However, some of the pioneering actuaries of the mid-nineteenth century were attracted to the challenge. Samuel Brown, who would go on to become the third President of the Institute, and who wrote one of the first actuarial papers on investment, also wrote one of the first actuarial papers on general insurance.[1] The mid-nineteenth-century challenges

[1] Brown (1851).

in developing statistical analysis of fire insurance started with a lack of available data on historical claims experience. The pooling of life assurers' mortality data was a well-established practice by the mid-nineteenth century but fire insurers showed no inclination to follow that collaborative practice. This was both due to commercial sensitivities and because the insurers recognised that their data records were generally inadequate for this purpose. Brown encouraged the fire offices to cooperate on the development of statistical claims analysis by providing him with historical data that he would pool but he found 'the course of his investigations on this subject was stopped by the determination expressed by the fire offices to refuse him all information whatever'.[2]

Brown struck upon an alternative rich source of data. The fire offices had formed the London Fire Engine Establishment in the 1830s to protect its insured properties. Its superintendent, Mr Baddeley, compiled comprehensive records of every fire to which its fire engines were called. These records included the extent of damage to the property, the occupations of the premises, the cause of the fire, and what proportion of the value of the building was insured. Brown's 1851 paper presented a tabulation and analysis of Baddeley's records for the years 1833 to 1849. Baddeley's records, however, could only tell half the story. In order to deduce some useful insights for fire insurance premium-setting, some measure of the 'exposed-to-risk' was also required. For this purpose, Brown turned to the Post Office Directory, which provided an estimate of the total number of buildings that lay within the area covered by London Fire Engine Establishment and, for commercial properties, the trades undertaken at the premises.

Armed with these estimates of the number of buildings exposed and the historical fire experience, Brown then calculated the one-year probabilities of fire for commercial properties as a function of the occupation undertaken in the property. He estimated, for example, that there was a 0.34 % probability of fire in a given year at a grocers' premises, whereas the corresponding probability for Lucifer match-makers was a rather disconcerting 30 %. By observing the distribution of the damage categories recorded by Baddeley, and making some assumptions about the cost associated with each degree of damage, he calculated the premium rates required for each class of occupation. His calculations implied an average annual premium rate of around 5 % of sum insured with significant variation by use of premises.

Brown himself recognised how the inherent complexity of fire insurance business inevitably placed significant limitations on the usefulness of his premium rate estimates:

[2] Bailey in Discussion, Brown (1851), p. 34.

> The locality, the construction of the buildings, the distance from a fire engine station, and, still more, the manner in which an adjoining building may be injured by a fire breaking out in a neighbour's house, would have to be considered.[3]

Nonetheless, he had made resourceful and practical use of whatever limited data was available, and had handled it competently to provide some illuminating insights and new quantitative reference points for insurance premium-setting.

Thomas Miller wrote two short papers on the application of statistical techniques to fire insurance that appeared in the *Journal of the Institute of Actuaries* in 1857 and 1880. Unlike in life assurance, the claim size in the event of a general insurance claim is usually not a fixed amount. Assumptions are therefore required about the behaviour of claim size as well as claim frequency. In the case of fire insurance, the claim size would be based on the cost of the fire damage incurred, up to the size of the policy's sum insured. Furthermore, fire insurance policies could insure either the full value of the property or a specified sum that was only a fraction of the value of the property.

Miller's first paper considered some of the difficulties in interpretation of claim experience that resulted from these features of fire insurance.[4] He argued that fire insurance claim data would be more useful if it reported the size of loss as a proportion of the value of the property rather than the usual practice of recording it as a proportion of the sum insured. This would allow the behaviour of claim size and hence insurance cost to be better understood as a function of how the sum insured varied as a proportion of property value. He argued that the profitability of the fire office and the adequacy of its premium bases should be analysed not by considering the actual profits that had been generated by the business, but by considering what the profits would have been if the size of all its policies was standardised to the same sums insured. This would remove the impact of the variation in policy size that occurred due only to the incidental preferences of a particular group of policyholders—'accidental circumstances' that were not controlled or targeted by the insurer. This could be considered as an early example of actuarial thinking on the analysis of surplus.

Miller's second paper considered what the appropriate unit of risk exposure should be for fire insurance.[5] He suggested it should be based on the number

---

[3] Brown (1851), p. 51.

[4] Miller (1857).

[5] Miller (1880).

of building *floors* insured rather than simply the number of buildings. He then developed an algebraic framework for calculating fire probabilities based on the two possible ways of a floor catching fire: by a fire originating on that floor; and by a fire spreading to the floor from another floor. This allowed him to produce probabilities of fire for a given floor of a building as a function of the total number of floors in that building. From this he produced the result that the probability of a given building floor catching fire increased with the number of floors in the building. Miller produced some numerical examples of his algebra, but made no attempt to calibrate his model to empirical data. It is hard to believe that fire insurance practitioners would have recognised any practical value in his approach.

These early actuarial contributions had little to no impact on British fire insurance practice. In 1878, the historian and actuary Cornelius Walford noted in a *Journal* paper,[6] 'It will no doubt be thought remarkable that, after two centuries of experience, the rates of fire insurance are still deduced by a method entirely 'non-scientific' ... But if there be no scientific base for fire insurance premiums, it is not from the want of suggestions in this direction [from actuaries].'[7]

British actuarial interest in fire insurance and property insurance subsequently waned. No paper of note on fire or property insurance was published in the *Journal* or *Transactions* between Miller's 1880 paper and the end of the Second World War.

Employers' Liability Insurance (1882–1931)

A series of Acts of Parliament came into effect during the final two decades of the nineteenth century that extended British workers' rights to compensation in the event of workplace injury or death. These laws meant a step-change in the magnitude of legal liabilities that employers would be exposed to in the event of workplace accidents. In nineteenth-century industrial Britain, this could represent a substantial business risk, particularly in the accident-prone manufacturing and extraction industries. Employers' liability insurance quickly emerged as a significant new line of general insurance business. Its complexity, potentially long-term nature and exposure to mortality risks naturally attracted strong actuarial interest—on of the face of it, it certainly appeared a more 'actuarial' type of general insurance than fire insurance.

[6] Walford (1878).

[7] Walford (1878), p. 4.

In Great Britain, as in several countries across Europe, the common law accorded injured parties some general rights to compensation from those who caused the injury. These rights were vague and limited and were not specifically related to injuries caused at work. The Employers' Liability Act of 1880 gave explicit rights to workmen who were injured in the workplace due to defective machinery or the negligence of superiors. The compensation was limited in value to three years' wages and was payable in the form of a lump sum. William Whittall wrote the first British actuarial paper to tackle the specific challenges raised by insurance of this form of liability. It appeared in the *Journal of the Institute of Actuaries* in 1882.[8] Like much of early, and indeed later, actuarial work in general insurance, many analytical compromises were required as a result of the limited quality and quantity of historical claims data. Whittall set out to analyse the rates of both fatal and non-fatal accidents and how they varied by occupation, but a sparsity of data constrained him to the statistical analysis of fatal accidents only.

In a similar spirit to Samuel Brown's fire insurance analysis, Whittall had to splice together data from disparate sources, neither of which arose directly from insurance claims data. He used the 1871 census to obtain estimates of the numbers employed in different occupations. For accident numbers, he made use of the records of fatal accidents (in 1870, 1871 and 1872) that were maintained at the government's General Records Office. These records had the helpful feature of recording the occupation of the deceased (although they did not distinguish between those accidents that occurred in the workplace and those that did not). From these data sources Whittall compiled tables of fatal accident rates for a comprehensive array of occupations. These results showed, for example, that only 1.3 authors per 10,000 died annually from a fatal accident, whereas horse breakers experienced 47.9 such deaths. Clearly occupation was a material risk factor for fatal accident rates!

The actuarial audience at the Staple Inn discussion of the paper, however, proffered a great deal of scepticism towards the reliability and relevance of these statistics. The President, A.H. Bailey, cautioned against the usefulness of the statistics for setting liability insurance premiums given that fatal accidents made up a relatively small proportion of liability claims and were not necessarily strongly correlated with the rate of non-fatal accidents. Whittall himself acknowledged that the granularity of the occupations meant that the numbers of deaths in some of the estimates resulted in unreliably small sample sizes. He also acknowledged that the fast-changing nature of some occupations could quickly render some historical rates obsolete. F.G.P. Neison, who had some

[8] Whittall (1882).

experience of working with liability claims data from the coalmines and railways industries, highlighted that the records of the General Records Office may not provide a reliable record of the occupation of the deceased, and 'did not provide any satisfactory basis for the determination of the rates of fatal accidents'.[9] As in fire insurance, the application of actuarial techniques continued to be frustrated by the lack of availability of relevant, reliable historical claims experience data.

The Workmen's Compensation Act of 1897 extended the rights of employees to compensation for workplace injuries. This act entitled the employee's dependents, if he had any, to three years' wages in the event of a fatal accident, and in the event of a non-fatal accident that rendered the employee unable to work, 50 % of his weekly earnings were payable during the period of incapacity after the second week. Whereas the 1880 act only entitled the employee to compensation in the event of a form of negligence on the part of his employer, the 1897 act entitled the employee to compensation for any workplace injury, providing it was not wilfully committed by the employee. The 1897 act was also notable for establishing injured employees' right to weekly incapacity compensation rather than only a lump sum benefit. Further Workmen's Compensation Acts of 1900 and 1906 extended the industries in which the act was applicable. The 1906 act made its application almost universal. Similar forms of legislation were introduced around continental Europe and Scandinavia during the same period (though not in the USA, Canada or Australia).

A behemoth, prize-winning paper by John Nicoll, published in 1902, represented the next notable British actuarial contribution on the subject of employers' liability insurance.[10] Some 20 years of liability insurance business experience had been accumulated in Britain by this time, but no actuarial basis for premium-setting had been established, even in theory, beyond the early efforts of Neison and Whittall on accident rates. Standard industry practice was to charge a premium amount set as a percentage of the total wages of the employees of the business.

Nicoll derived formulae for the value of fatal and non-fatal accident compensation, including the new feature of weekly compensation during the period of incapacity, from actuarial first principles. For the valuation of fatal accident compensation, Nicoll's formula was a function of the rate of fatal accident of a given occupation (which was specified as a function of age); and the probability of the employee having dependants (also specified as a

[9] Neison, in Discussion, Whittall (1882), p. 212.
[10] Nicoll (1902).

function of age). For the valuation of non-fatal accident compensation, his formula used probabilities of different degrees of incapacitation (total and partial) occurring for each occupation as a function of age; mortality rates for the incapacitated as a function of age (treating partial and total incapacity separately); and the rate at which non-fatal accidents recover from their incapacity. Assumptions were also required about wage levels and how they varied as a function of age.

Whilst the actuarial community at the Staple Inn discussion of the paper voiced appreciation for the application of actuarial techniques to this complex form of insurance, a visiting non-actuarial practitioner in employee liability insurance was much more sceptical:

> Actuaries could not tell, any more than they [general insurance underwriters] could, what the rate should be, because they had no data ... the time, he thought, was not ripe for actuarial knowledge to be called in.[11]

Thus, by the start of the twentieth century, some recurring themes were already well-established in the British actuarial relationship with general insurance: a lack of data rendered statistical and analytical methods of limited use, and the actuarial role in general insurance was therefore at best unclear.

William Penman's *Journal* paper of 1911 was the first to focus specifically on actuarial approaches to *reserving* for outstanding claims rather than premium setting.[12] Penman was another example, like Samuel Brown, of a life actuary who dabbled in general insurance thinking, as well as investment thinking, and who would go on to reach a leading position in the profession (Penman was President of the Institute of Actuaries during the years 1940–1942).

The reserves required in a general insurance business can be considered to have a few high-level components whose relative size will vary with the type of business that is written. At the highest level, reserves can be required for two basic categories of claims: claims that arise due to future events that may occur during the outstanding period of insurance covered beyond the valuation date by the premiums already received; and claims for events that have already occurred but have yet to be settled.

The reserve for the first of these categories of claims is generally referred to as an *unexpired premium reserve.* It reflects the portion of the insurance premium received that is charged to cover the period of insurance beyond the valuation date. Alternatively, an *unexpired risk reserve* may be set up that

[11] Green, in Discussion, Nicoll (1902), p. 559.

[12] Penman (1911).

is intended to reflect the portion of exposed-to-risk that is still outstanding on the insurance cover (these two quantities may differ for reasons such as seasonality effects—for example, more motor accidents might occur in winter than summer).

Reserves for claims that have already occurred are required due to delays in the claim reporting and claim settlement processes. These two forms of delay each has its own form of reserve: *incurred but not reported* (IBNR) *reserves* are for claims that have already occurred but are not yet known to the insurer (that is, there is a delay between the occurrence of the event that causes the claim and the notification of the insurer of the claim); and *outstanding claims reserves* are for claims that the insurer has been notified of but which are yet to be settled. Settlement delays may arise due to the time taken to agree the claim size. This process may simply involve negotiation between the insurer and the policyholder, or may have to be settled by the courts.

Penman's paper focused on a particular form of outstanding claim reserve that arose from the 1906 act: the reserve that arose not from settlement delays, but simply because the form of settlement for incapacity compensation was in the form of an annuity for the period of incapacity rather than an immediately payable lump sum. The standard industry practice was to reserve for these outstanding claims by making individual case estimates of the required reserve based on medical opinion on the duration of the injury. To the trained actuarial mind, this approach appeared arcane, expensive, unnecessarily subjective and open to deliberate manipulation. Penman contrasted the case estimate approach with life office practice and argued it was 'surely contrary to all the principles of valuation to deal with individual cases on their merit'.[13]

Penman's paper adopted a pattern that was similar to Nicoll's a decade earlier. He developed an actuarial method for valuing the reserves based squarely on standard life office techniques, only for general insurance practitioners to tell him that there was no adequate data with which to apply his formula. Penman's approach was intuitive enough as an actuarial approach: the outstanding claim was a weekly annuity that would remain payable as long as the employee was alive and remained incapacitated. It could be valued using a mortality table for incapacitated workers together with a table for the rate of recovery from incapacity (which could be set as a function of age) and a valuation interest rate. In referring to the calibration of the recovery rates, Penman conceded:

> The only way in which these items of information can be obtained is by a very complete analysis of a large number of claims and such an analysis has not been possible for me to make.[14]

[13] Penman (1911), p. 112.

[14] Penman (1911), p. 121.

Penman did acquire some claims experience data from two companies and he used this to develop illustrative bases and valuations. He showed that the rate of recovery from incapacity tended to be lower at older ages and argued this should be explicitly allowed for in the reserving basis.

The Staple Inn discussion was broadly negative. The themes were familiar. Several speakers voiced strong doubts about the availability of sufficient relevant data to implement the actuarial approach set out by Penman. There were so many factors affecting the duration of incapacity that the separation of claims into homogenous groups would produce such small groups that the result would not differ significantly from the case estimation approach that the author had strongly criticised. The business was in a state of flux, with factors such as judicial interpretation rendering historical experience less relevant to the future. One speaker pointed out that valuing life business by case estimation may not be so absurd if the information was readily available to support it.

It was almost two decades before another paper appeared in either of the British actuarial journals on the topic of employee liability insurance (or any other form of general insurance). Then, in 1929 and 1931, two papers were published, partly motivated by the 1925 and 1926 Workmen's Compensation Acts. Hugh Brown's paper,[15] published in the *Transactions of the Faculty of Actuaries* in 1931, reviewed the implications of the 1925 and 1926 Acts of Parliament. A notable feature of these acts was that they allowed the weekly incapacity compensation payments to be commuted to an immediately payable lump sum at the behest of the employer, providing the payment was not less than 75 % of the prevailing market value of an immediate life annuity. This provided a natural benchmark for reserving for outstanding claims, though the actuary had discretion to attach a different value to the liability in the regulatory board of trade returns providing a justification was provided along with details of the basis used. However, an actuary was only required for the valuation of weekly incapacity payments that had been in duration for five years or more. For other outstanding claims, the practice remained to value those liabilities on an individual case basis using available medical opinion. Hence, 20 years after Penman's paper and almost 30 years after Nicoll's, liability insurance practices remained largely unaltered by actuarial input: outstanding claims were either valued on an individual case basis, or were essentially assumed to be life annuities. No actuarial method for the estimation of the duration of the claims based on statistical analysis of rates of recovery from incapacity was implemented in practice.

[15] Brown (1931).

Brown advocated a pragmatic compromise. Instead of valuing the outstanding claims using intricate assumptions about incapacity recovery rates and how they varied as a function of age, together with incapacity mortality assumptions, he came up with something rather counter to actuarial convention: that for claims of less than five years' duration, age could be completely disregarded from the annuity valuation. Instead, he suggested grouping the claims by type of employment and then applying an average duration factor to each employment group that could be based on the analysis of historical data. He did not provide any analysis of such data and how these factors might vary by employment category, however. His suggestion, whilst pragmatic in theory, remained somewhat enigmatic in practice.

The motivation for the second of this pair of papers, written by R.G. Maudling and published in the *Journal* in 1929,[16] was related to the author's own experience of analysing a large amount of liability insurance claims data. The data was specifically related to the mining industry and was sourced from the experience of one particular Colliery Owners' Mutual Indemnity Fund. Mutual indemnity funds were a popular approach to co-insuring liability exposure amongst the employers of some industries, and the mining industry was one of the biggest examples. He considered the data for the period from 1908 to 1923, and his data set included 11,000 claims. Maudling categorised his data by the type of injury or disease that was the cause of the claim, and by age at date of accident. He found that the rate of discontinuance of incapacity was a function of both. He used the data to provide annuity valuation tables as a function of age and type of injury.

This was, in spirit, the implementation of the actuarial frameworks advocated by Nicoll and Penman some 30 and 20 years earlier, finally brought to life with a meaningful dataset. However, Maudling also noted that the observed rates of incapacity discontinuance had varied over the data period. In particular, the discontinuance rates tended to be lower when wages were lower and vice versa. Data heterogeneity appeared inescapable in general insurance. Maudling's papers represented the final notable contributions of British actuarial thought to general insurance business until after the Second World War.

Risk Theory (1954–1971)

British actuarial thought on general insurance over the one hundred years from 1850 to 1950 was primarily focused on attempting to apply the actuarial techniques then in successful use in life assurance to fields of general

[16] Maudling (1929).

insurance. These attempts were continually stymied by the complexities found in general insurance business: in particular, by the heterogeneity in its claims experience and by the lack of stability engendered by its greater sensitivity to unpredictable social and economic change.

Risk theory offered a different analytical approach. Traditional actuarial techniques in life assurance and pensions were predicated on the idea that insurance risks were diversifiable and that insurers wrote business in such a way that these risks were indeed diversified away. Thus actuarial work in life assurance and pensions was traditionally set within a deterministic, riskless framework. Risk theory started from a different premise: it applied stochastic modelling techniques to model the random occurrence and impact of individual general insurance risks. This explicitly captured the variability that could arise from one claim to the next due to fluctuations in both claim size and frequency. It employed sophisticated mathematical models that went far beyond the quantitative techniques used in British actuarial science in mid-twentieth-century life and pensions.

The germs of risk theory—which we could loosely define as the stochastic modelling of individual insurance claim outcomes—were developed in Scandinavia and particularly in Sweden during the early decades of the twentieth century.[17] This initial work was focused on the modelling of mortality risk in life assurance, but it was found to have no practical use there as its application to the large sizes of homogeneous policies in life business left little random fluctuation—risk theory essentially produced the same results as the standard deterministic actuarial models. Following the end of the Second World War, actuarial researchers began increasingly to consider risk theory in the setting of general insurance. This largely occurred outside the UK. Researchers were again very active in Scandinavia during this period, most notably in Finland.[18] Somewhat later, technical actuaries such as Hilary Seal further developed and applied these ideas in the USA.[19]

The essence of the post-war general insurance risk theory framework can be summarised as follows. Individual claim events and individual claim sizes were each modelled as random variables. Claim events were typically assumed to be generated by a Poisson process. This could be generalised into a mixed Poisson process, where the Poisson rate was itself assumed to be a random variable. The claim size was typically assumed to have a skewed probability distribution such as the lognormal or gamma distribution. In *individual* risk

[17] See Lundberg (1909) for example.

[18] See, for example, Pentikainen (1952).

[19] See, for example, Seal (1969).

theory, each policy in the portfolio had its own specified probability distribution for claim frequency and claim size. In *collective* risk theory, the models and calibrations were specified at the portfolio level without distinguishing which policies gave rise to the claims (but individual claim events were still modelled).

This analytical framework could provide many theoretical insights. At the most basic level, it was one way of calculating the expected level of claims, and hence the pure premium for insurance. It could also be used to model claims with and without reinsurance and to set the pure premium for a particular form of reinsurance. For a given level of premium and starting capital, the evolution of the 'risk reserve' could be modelled—that is the stochastic projection over time of assets and premiums less claims and expenses. The modelling of this reserve could be used to estimate probabilities of ruin over various time horizons for a given starting reserve. Equivalently, it could be used to establish the starting level of reserve required to support a given probability of ruin.

This analysis of ruin probabilities was fundamentally very similar to the approach pioneered by Sidney Benjamin and adopted by the Maturity Guarantees Working Party for unit-linked investment guarantee reserving during the 1970s. In maturity guarantee reserving, the ruin probability modelling focused on the impact of stochastic variation in non-diversifiable (financial market) risk. Risk theory focused on the stochastic impact of what might be called under-diversification of risks that were theoretically diversifiable—that is, the impact of random fluctuations in claims processes that were usually assumed to be independent (claims may also be highly correlated due to exposure to a common event or risk factor). As we shall see below, Benjamin also played a significant role in developing British actuarial thought in general insurance.

Whilst the theoretical framework of risk theory was intuitive and powerful in the context of general insurance business, it did not necessarily provide a solution that was any less challenged than traditional actuarial techniques by the age-old practical general insurance modelling issues of data, heterogeneity and stability. Could the models of risk theory be reliably calibrated such that it could produce dependable quantitative output?

Risk theory never reached the mainstream of the British actuarial profession. Its primary British actuarial exponent during the post-war decades was Robert Beard. He liaised extensively with international actuarial colleagues from the late 1940s onwards, especially the leading Finnish thinkers in the discipline, to apply the subject to British general insurance business. Beard's interest in risk theory was mainly stimulated by his experience in quantitative operational research during the Second World War rather than by his actuarial training.[20] He appealed to the Institute of Actuaries as early as 1948 to engage more actively in general insurance, but he received little encouragement. Faced with a less-than-enthusiastic institute, Beard helped to establish ASTIN—Actuarial Studies in Non-Life Insurance—as a section of the International Actuarial Association.[21] Its journal—the *Astin Bulletin*—carried much of Beard's substantial research output of the 1950s and 1960s.

Beard did manage to get a couple of papers on risk theory and its application to general insurance published in the *Journal of the Institute of Actuaries* in 1967. He also had a greater number published in the perhaps more liberally minded *Journal of the Institute of Actuaries' Student Society*. Beard also co-wrote a book,[22] first published in 1969, on risk theory with two of the leading Finnish actuarial thinkers on risk theory, Professor Pentikainen and Dr Pesonen. The book was widely used in European actuarial university courses and subsequent editions were published in the 1970s and 1980s.

A notable Beard paper was published in the *Journal of the Institute of Actuaries Students' Society* in 1954.[23] Beard gave British actuarial students an overview of risk theory and demonstrated its application in general insurance with an example based on a large US fire insurance claims dataset. The paper showed that the distribution of claim size could be well-fitted with a log-normal distribution, except for a handful of exceptionally large claims out of a dataset of more than a quarter of a million. It also showed how an excess-of-loss reinsurance treaty impacted on the insurance portfolio's net claims probability distribution. He went on to analyse how the probability of ruin behaved over various time horizons for differing levels of starting reserves, and how the excess-of-loss treaty impacted on these results. It was an accessible and comprehensive practical case study on the application of risk theory to general insurance reserving, solvency assessment and risk management.

Beard was not a mere theorist. He was a working actuary who recognised the practical limitations of the mathematical framework and the data that was used to fuel it. For example, he noted that other US statistics showed that higher claim frequencies tended to be experienced in years of economic depression, and less in boom years, and that this could be an important source of heterogeneity and non-stationarity. He also noted that claims would tend not to be independent due to geographical exposures. Nonetheless, he held a

[20] Beard in Discussion, Plackett (1971), p. 355.

[21] Beard in Discussion, Abbott et al. (1974), p. 277.

[22] Beard et al. (1969).

[23] Beard (1954).

deep conviction that there were significant practical insights for general insurance premium-setting and reserving that could be obtained from a sensible utilisation of available data within the sophisticated mathematical modelling of risk theory.

The editors of the *Journal of the Institute of Actuaries* permitted Beard a full seven pages in its September 1967 edition.[24] He argued that the risk theory approach of modelling claim frequency and claim size as separate statistical processes could make it easier to detect changes in patterns of claims experience and thus allow for faster premium rate adjustments to be made by the insurance office. This was a topical issue for the general insurance industry at the time, especially in motor insurance, where a deterioration in experience had occurred over a number of years without premium bases adequately reacting to avoid a sequence of multiple years of loss.

Although it was not accepted into the mainstream of the profession's thinking at the time, Beard's work of the 1950s and 1960s laid the foundations and provided the inspiration for broader quantitative research in general insurance by the British actuarial profession. Meanwhile, progress started to be made in the collation of relevant claims data in some general insurance lines, most notably motor insurance. In 1967, the British Insurance Association established the Motor Risk Statistics Bureau to pool claims experience data for several member insurers. This data, whilst challenged by differences in rating structures and policy types across member firms, provided a new source for the application of statistical techniques to pricing and reserving.

The 1970s saw more actuaries become engaged in general insurance practice. As noted above, the Institute eventually added general insurance to its examination syllabus in 1978. Some new British actuarial thought-leaders emerged in the 1970s who were seeped in practical experience in general insurance business. G.B. Hey was one such example and he helped to break new ground in 1971 when he and P.D. Johnson wrote the first ever paper on motor insurance to be published in the *Journal of the Institute of Actuaries*.[25] Their paper was not as mathematical as typical risk theory works, but it was strongly influenced by Beard and his work in ASTIN.

Johnson and Hey analysed the efficacy of 'experience rating' in motor insurance—that is 'a system by which the premium of the individual risk depends upon the claims experience of this same individual risk'.[26] In British motor insurance, this was known as the No Claims Discount (NCD) system, and

[24] Beard (1967).
[25] Johnson and Hey (1971).
[26] Johnson and Hey (1971), p. 202.

it was a well-established feature of that market by the mid-1960s. The individual risk data used in an experience rating system still faced the same challenge as a risk rating approach based on broader historical claims experience: a longer period of history was required to effectively differentiate the risk heterogeneity amongst different policyholders, but older data may be of less relevance for the projection of future claims behaviour. For example, in NCD systems, the advances in the policyholder's driving skills that would typically accompany increasing age and driving experience could render earlier claims experience irrelevant to expected future claims behaviour.

A typical UK NCD system of the time allowed the maximum claims discount to be obtained after four or five consecutive claim-free years. Johnson and Hey analysed how effective this chosen length of period was in differentiating different risks (under the assumption the risks were indeed stationary). They considered a theoretical pool of policyholders where 75 % of the group had a claim frequency of 0.1 and the remaining 25 % had a higher claim frequency of 0.25. Their analysis showed that a policyholder in the low-claim-frequency group had a two in three chance of obtaining the maximum discount rating after four years, whilst the high-claim-frequency group policyholder had a corresponding probability of two in five. NCD systems made some contribution to distinguishing policyholder heterogeneity, but they were a blunt and limited tool that could only be highly effective with a long history of stationary data. This analysis highlighted that the source of the poor motor insurance industry experience in the years preceding the paper may not only have been due to high claims inflation as was generally suspected (and which was partly driven by changing judicial treatment of third-party injury claims). Another factor may have been that the average policyholder who earned the maximum NCD discount was not actually as good a risk as they had been assumed to be.

Aside from this theoretical analysis, Johnson and Hey's paper focused on analysing the data made available through the Motor Risk Statistics Bureau (though they only considered the experience of a single member firm to avoid complications with comparability and consistency). They fitted an eight-factor regression model of claims frequencies using a least squares optimisation (factors were intuitive motor insurance risk factors such as age of policyholder, car rating group and NCD category). They found that the NCD category was a significant variable in the regression, even in the presence of the other factors, and hence concluded it was still an important element in the premium rating process, despite its noted limitations.

Despite such attempts to illustrate the practical application of risk theory ideas and statistical methods, it was generally viewed through the 1950s,

1960s and even 1970s as too theoretical and quantitative to be of much practical value to the British actuary. This perspective is well-represented, though perhaps a little overstated, by the following passage in a *Journal* paper by Ryder published in 1976:

> Risk theory is a rather esoteric branch of actuarial theory which has been extensively developed by the more theoretical continental actuarial tradition. The practical actuary, however, finds that he hardly ever uses this theory.[27]

The application of risk theory, however, did become an established element of British actuarial practice in general insurance in the 1980s and 1990s—a period when the profession developed a more prominent role in the sector and the actuarial approaches employed there became increasingly technical.

Claim Reserving (1974–1996)

Between the years of 1850 and 1970, the British actuarial profession made little headway in establishing itself as an influential and important component of Britain's highly successful, global general insurance industry. It had tried to import its life assurance methods into general insurance and had learned that they were not adequate for the task of pricing and reserving for general insurance. It had seen and broadly rejected the stochastic approach of risk theory that had been developed and applied by overseas actuaries. Was there a role for British actuaries to play in general insurance? If so, what was it and what kinds of actuarial skills would it employ?

As noted above, over the course of the 1970s, there was an increase in the numbers of British actuaries working in general insurance business. A General Insurance Study Group was formed by the Institute in 1974. It produced many research papers, though most of these were merely deposited at the Staple Inn library and were not deemed worthy of publication in the *Journal* or discussion at sessional meetings. The profession recognised that there may be an opportunity for actuaries to play a more formal role in statutory reserving for general insurance business. The 'freedom with publicity' regime of actuarial life assurance reserving, which had been in place in some form or another since the 1870 Life Assurance Companies Act, provided an explicit, professional and statutory role for the actuary that gave them considerable discretion to exercise their expert professional judgement. No such equivalent

[27] Ryder (1976), p. 71.

role existed in general insurance reserving in the 1970s. General insurance firms had to calculate statutory reserves, but no actuarial role was mandatory in reserving and no detailed disclosure of the methods use in assessing the reserves was required.

The collapse in 1971 of Vehicle and General, a British motor insurance firm, provided a further opportunity for the actuarial profession to press for a more active and statutory role in general insurance reserving. An Institute working party was established in the early 1970s with a remit to consider the statutory reserving regime for general insurers 'with reference to uniformity and the verification of non-life reserves'.[28] Perhaps unsurprisingly, its report,[29] published in the *Journal* in 1974, concluded that only professional certification could address the challenges inherent in general insurance reserving. Nonetheless, even amongst actuaries there was scepticism that this could be a cure-all. In the Staple Inn discussion of the paper, J.L. Manches noted:

> It was perhaps an oversimplification to believe that actuarial techniques were the best answer to accurate claims reserving. Life valuations were based on the application of established claim probabilities (i.e. mortality tables) to homogeneous groups of policies which generated claims of known amounts. None of that applied to non-life business, and there was bound to be the utmost difficulty in reaching agreement on any standard approach to reserving.[30]

There is certainly a hint of arrogance in the profession's position that it could improve standards in general insurance reserving when it had done so little over the previous decades to encourage actuarial research and education in the field. However, it is important to note a shift in emphasis. No longer was the profession positioning itself as the leading provider of statistical or analytical methods. By the early 1970s it was clear that advanced techniques in mathematical statistics were emerging faster than the profession's appetite to apply them. On the contrary, the profession's argument was increasingly based on its ability to meet the need for sound professional judgement and broader financial acumen in the setting of prudential insurance reserves.

From the early 1970s until his death in the early 1990s, Sidney Benjamin—whom we met earlier as the frustrated pioneer of stochastic risk-based approaches to maturity guarantee reserving—relentlessly argued for a statutory role for actuaries in general insurance reserving that was equivalent to their position in life assurance. This argument was based on the inadequacies

[28] Abbott et al. (1974), p. 217.

[29] Abbott et al. (1974).

[30] Manches, in Discussion, Abbott et al. (1974), p. 268.

in general insurance statutory reserving that he perceived arose from the absence of a professional actuarial role:

> Historically, the first job of the actuary is to safeguard the interests of policyholders. The life actuary determines the amount of risk capital which should be set aside to give an acceptable level of safety to the policyholders. In non-life insurance, that does not happen. The amount of backing solvency capital for any volume of business is set vaguely, according to an informed perceived wisdom, with: no scientific justification; no explicit public justification; no published standards of consistency within any one company from year to year; and no apparent standards of consistency between companies in any one year.[31]

Benjamin's paper, 'Profit and Other Financial Concepts in Insurance',[32] appeared in the *Journal* in 1976 and presented this case with typical Benjamin gusto. He discussed the universal applicability of fundamental actuarial concepts such as the use and disclosure of actuarial bases that provided transparency and consistency to the reserving process; the difference between reserving and premium bases and its implications for new business strain and the emergence of surplus; implicit and explicit reserving margins for prudence; asset mismatching reserves; differences between provisions and reserves; prospective and retrospective reserving approaches. Benjamin argued that these actuarial concepts were as applicable to general insurance as to life assurance, and that they were the unique domain of the actuarial profession.

Benjamin's paper also offered some specific criticisms of the practices that then prevailed in general insurance reserving. Like other actuaries before him, he viewed reserving by case-by-case estimation as 'fundamentally subjective ... not inherently stable in the way a fixed basis will be' and hence 'an inadequate substitute for a reserving basis'.[33] He was also critical of the standard industry practice of not discounting projected claims cashflows when setting general insurance reserves. Whilst this provided a form of implicit reserving margin, its size was arbitrary and it distorted the stated form and timing of the release of surplus. He also argued that required solvency margins should be risk-based and should be a function of the asset mix of the business, which was again not standard practice at the time.

His most forceful argument, however, was reserved for the use in general insurance of the professional framework of 'freedom with publicity' in statutory reserving: the idea that the actuary, as the professional expert, should

---

[31] Benjamin in Discussion, Ryan and Larner (1990), p. 658.

[32] Benjamin (1976b).

[33] Benjamin (1976b), pp. 252–53.

have the freedom to use his judgement to set appropriate assumptions and methods for the business he is reserving for, providing he disclosed his assumptions sufficiently such that another actuary could reproduce his results and opine on the reasonableness of the approach taken. To Benjamin, this was the fundamental reserving discipline that actuaries could bring which was lacking in British general insurance. Whilst the Staple Inn discussion of the paper was broadly supportive of Benjamin's thinking, there were some general insurance practitioners who felt the need to highlight that the intrinsic differences in general insurance and life assurance could limit the direct applicability of some of his arguments and that 'actuarial advice [in general insurance] ... would continue to be sought ... if, and only if, actuaries showed a proper understanding and humility of approach to an industry which had operated successfully without actuaries for many years'.[34]

In the years following Benjamin's 1976 paper, a flurry of papers appeared in the *Journal* which did indeed attempt to develop 'a proper understanding ... of an approach' to claim reserving. These papers were some of the most technical and quantitative papers ever to appear in the *Journal*, and were mostly written by actuaries, and in some cases non-actuaries, with PhDs in advanced quantitative fields. They were mainly concerned with developing improvements in methods for the projection of claims from a run-off triangle (which tabulated claims paid in rows of the year of insurance, which was usually referred to as the underwriting year and sometimes as the origin year; and columns of 'development year', i.e. the number of years after the origin year when the claim was paid). The concept of a run-off table of general insurance claims had been around for a long time. For example, at the Staple Inn discussion of Penman's 1911 paper, W.R. Strong noted:

> If the claim payments in respect of the business accepted in any given year were traced separately until the claims of the year were finally disposed of, the total volume of payments year by year fell into a somewhat regular sequence of diminishing amounts ... It might perhaps be practicable when a sufficient period had elapsed to construct a table by means of which, given the claim payments up to the end of the year, an estimate might be formed of the ultimate cost of disposing of the liability in respect of the policies of the first year.[35]

The standard method of estimation of the ultimate cost from the run-off data was the ubiquitous chain ladder method. The method assumes that future

[34] Scurfield, in Discussion, Benjamin (1976b), p. 300.

[35] Strong, in Discussion, Penman (1911), pp. 137–38.

claims for each underwriting year will accumulate over their outstanding development years in proportion to how claims paid up to that point in the origin year's development differ from the average observed across all the origin years in the run-off table for which the outstanding development period has been observed. Its greatest limitation is embedded in its basic assumption: it implies that an unexpected claim size in an early year of development impacts proportionally on all future development years for that underwriting year. This sensitivity could result in noisy, unstable estimates for business in its early years of development.

The first of this series of technical papers was written by D.H. Reid and published in the *Journal* in 1978.[36] Reid began by noting that case-by-case estimation was still the prevalent industry practice for setting outstanding claims reserves in general insurance. He voiced the usual actuarial concerns with this approach, but particularly highlighted that whilst the approach might sometimes have some merit in reserving for claims that had been reported but not yet settled, it was entirely inapplicable to IBNR claims and their required reserves. The reserving approach for these claims must involve some form of statistical method as there was no case-by-case information that could be used in the reserving assessment—by definition, the claim was completely unknown to the insurer at this point in its development.

Reid developed a mathematical framework for the emergence of claims payments over time by specifying a cumulative joint probability function for claims paid and development year. He then fitted this function to the available data for the claims experience from the earliest available underwriting year of the run-off data only. His model was essentially smoothing the observed experience of a single complete sample path for the claims development. Once this function had been fitted, it was then transformed into a function for use in projected future claims via parameters for inflationary changes in claim size and changes in the rate of settlement that were assumed to be experienced between the time of the first underwriting year's claims and the projected times of future claims. These parameters could be fitted to the claims run-off data for the sequence of subsequent underwriting years for which data was available. Reid applied his framework to example claims data for a variety of lines of business such as employers' liability, fire and motor insurance. The approach was somewhat aligned to risk theory in that it provided a full probabilistic description of (aggregate) claims. But its formulation was complex, and the model contained a very large number of parameters that needed to be fitted to typically very limited data. His presentation was rather impenetrable

[36] Reid (1978).

for the typical British actuary of the time. Crucially, it was extremely difficult to ascertain from his analysis whether his complex modelling would result in a more accurate or reliable estimate of outstanding claims than a much simpler modelling approach.

D.H. Craighead produced a more accessible paper for the *Journal* in 1979.[37] Craighead was an experienced actuarial practitioner in the Lloyd's of London market. His paper gave a broad overview of its business practices and institutional arrangements. It also contained an important section with his views on how to model the run-off of claims and hence establish claim reserves at any point of time. He proposed fitting a formula for the incurred loss ratio of a given underwriting year as a function of the development year. He applied this approach to the proprietary claims data of an anonymous reinsurance company using a three-parameter exponential form of function. This produced fits of varying degrees of quality for different lines of underlying business and forms of reinsurance.

The fitted curves provided a smoothed description of how claims had historically run-off over their development period. It did not provide an explicit statistical predictive model. The fitted parameters could then be used to extrapolate the claims run-off of the underwriting years that were not yet fully developed. This was essentially a parametric form of the chain ladder method and, in similarity with that method, Craighead noted the sensitivity of the reserve estimate to unexpectedly large claims that arose early in the claim development period. In the Staple Inn discussion, this theme was expanded upon by J.P. Ryan, a general insurance actuary who made notable contributions to actuarial research in the 1980s and 1990s. Ryan highlighted that the work done in the USA by Bornhuetter and Ferguson might provide the solution to this problem.[38] The Bornhuetter-Ferguson approach reduced the sensitivity of projected ultimate claims to the claims experience data by mixing the estimate implied by the 'raw' claims data with some specified prior expectations for the claims development pattern. It was essentially Bayesian in spirit, and its inherent limitation was in the potentially arbitrary specification of the 'prior' estimates.

Three further technical papers on statistical methods for general insurance claim reserving appeared in the *Journal* in 1982 and 1983. The first of these was written by J.H. Pollard.[39] Pollard focused on the run-off behaviour of the aggregate claims of a large book of business. He invoked the Central Limit Theorem to justify assuming the aggregate claims would be normally distributed.

[37] Craighead (1979).

[38] Bornhuetter and Ferguson (1972).

[39] Pollard (1982).

Pollard moved away from the direct manipulation of the run-off triangle and instead set up a matrix algebra to describe how claims paid behaved through their development years: he specified vectors for the mean and variance of the claims paid in each development year, together with a covariance matrix to capture the correlations between claims paid in different development years (for example, the correlation between claims paid in development year 1 and development year 3). This, almost tautologically, allowed the expected claims, and indeed the whole probability distribution of future claims to be determined for a given claims development period to date.

The second useful feature of Pollard's set-up was that it provided a statistical basis for assessing whether statistically significant changes in claim settlement patterns had arisen—the multivariate normal framework allowed Chi-Square significant tests to be used to assess if the claims development was statistically different to the previously fitted distributions. Pollard's framework was mathematically elegant and intuitive, but, as ever, it relied entirely on how the model was parameterised and the reliability of the available data for that purpose.

The two claim reserving papers published in the *Journal* in 1983 were the most statistically complex. The first of these was written by de Jong and Zehnwirth.[40] Their paper applied recent technical developments in the statistical modelling literature to the claim reserving problem, particularly the times series modelling of Box and Jenkins.[41] This involved 'state-space models' and the Kalman filter, which was essentially a recursive application of Bayes' Theorem. This mathematical statistical technology allowed claim reserve estimates to be dynamically updated in a Bayesian way as new data arose. The second paper,[42] by G.C. Taylor, drew parallels with the claim projection problem and 'invariance problems' in physics and used this observation to apply variational calculus to the stochastic modelling of run-off triangles. The invariance assumption was that 'the expected amount of outstanding claims, deflated to current values, is invariant under all variations of future speed of finalizations'.[43] Taylor recognised that such an assumption would not hold if a change in rate of settlement arose, for example, due to a change in negotiating stance of the insurer. But of course, no quantitative method could readily incorporate such factors into claims projections. Neither of these papers were presented at Staple Inn and it is hard to believe they resonated strongly with either general insurance practitioners or the British actuarial profession at large.

[40] de Jong and Zehnwirth (1983).

[41] Box and Jenkins (1970).

[42] Taylor (1983).

[43] Taylor (1983), p. 211.

A working party was established in 1982 by the Institute's General Insurance Study Group to formally consider general insurance solvency and 'the methods and bases used for the valuation of assets and liabilities'.[44] This was partly inspired by a recent study of general insurance solvency by Finnish actuaries which was particularly notable for its application of simulation modelling as a solution to risk theory problems.[45] The profession had embraced simulation modelling methods in its recently-published and influential maturity guarantee research. It is certainly easy to see why it seemed like a more appealing route for the profession to pursue than further developing the work of Reid, De Jong and Zehnwirth, and Taylor.

The report of the working party was published as a *Journal* paper in 1984.[46] No significant regulatory or professional guidance applied to general insurance reserves at this time other than that valuations should be made in accordance with generally accepted accounting principles or other accepted methods. Statutory reserving required the assessment of two key quantities: the technical provisions (i.e. liability valuation) and the solvency margin. The standard industry practice was for the technical provisions assessed in solvency reporting to also be used as the provisions shown in financial statements. Conceptually, the solvency margin provided an asset buffer over the cost of meeting the liabilities. The regulator had powers to intervene in the running of the insurer in the event that assets were insufficient to cover the solvency margin as well as the provisions.

The working party argued that, for the purposes of solvency reporting, technical provisions should contain a prudent margin over the best estimate of the cost of meeting the liabilities. This margin should be set such that there was a 'relatively low risk of [the technical provisions] proving inadequate'. Meanwhile, the solvency margin would protect against 'the more remote adverse contingencies of the run-off'.[47] This approach was difficult to reconcile with accounting principles for the valuation of provisions (which required the provisions to be a true and fair estimate of the liability value). It also ran counter to the applicable European Commission regulatory framework, which assumed technical provisions were best estimates, with a solvency margin calibrated to produce a risk-of-ruin of one-in-1000 over a three-year horizon. The working party was satisfied with the solvency margin target but suggested that technical provisions should be set at a one-in-200 risk-of-ruin

[44] Daykin et al. (1984), p. 279.

[45] Pentikainen and Rantala (1982).

[46] Daykin et al. (1984).

[47] Daykin et al. (1984), p. 288.

over the three-year horizon (which was clearly substantially different to the one-in-two risk-of-ruin loosely implied by a best estimate approach).

The working party also advocated an approach similar in spirit to life assurance reserving's 'freedom with publicity' concept: insurers and their actuaries would be free to choose their own methods to determine the technical provisions and solvency margin, but disclosure of those methods and professional certification by a 'loss-reserving specialist' would be required. However, in the Staple Inn discussion of the paper, some concerns were raised that this freedom was 'fraught with danger unless the Department [of Trade and Industry, the insurance solvency regulator of the time] is able to bring to bear a stringent monitoring of the results'.[48]

The working party recommended that the above ruin probabilities should take account of asset-side risks as well as liability risks—in essence, a mismatch reserve should be included in both the technical provisions and solvency margin. It may be recalled that Sidney Benjamin first proposed that general insurers' solvency margin should include a mismatch reserve in 1976, but it remained a significant departure from prevailing general insurance practice in 1984. Such a change in reserving method could imply a significant increase in reserving levels. However, the working party attempted to water down the ramifications of this proposed change with a caveat:

> If asset values fall only temporarily, the problem may be largely presentational, and the supervisor would not need to withdraw the authorization of companies unable to meet the solvency requirements at a particular date if the position had subsequently been rectified. Only with a prolonged shift in market values would the effects be serious.[49]

Quite how the supervisor, or the certifying loss-reserving specialist, was supposed to ascertain at a particular date whether experienced asset value falls were temporary or not was not addressed by the working party.

The working party, again led by Chris Daykin, produced a further paper that was published in the *Journal* in 1987.[50] Whereas the first paper had dealt mainly with principles and a conceptual framework, the second had a greater focus on implementation and the modelling challenges involved therein. The second paper did, however, make one conceptual U-turn—after professional and industry feedback, the working party now expressed its ambiva-

[48] Hart, in Discussion, Daykin et al. (1984), p. 320.
[49] Daykin et al. (1984), p. 302.
[50] Daykin et al. (1987).

lence on the question of the statistical standard for technical provisions, arguing that it was the total solvency level that was really relevant. This effectively rescinded their earlier proposal that technical provisions should include a margin for prudence and hence accepted a best estimate definition. This allowed closer alignment of provisions with accounting principles and 'true and fair' valuation.

The 1987 working party paper proposed that the required solvency margin be calculated using a simulation model that projected asset and liability cashflows over the run-off of the existing business. It adopted a pragmatic approach that did not attempt to reach the heights of statistical ambition that had been explored by Reid, Taylor, de Jong and Zehnwirth over the previous decade. It suggested that variations in the real amounts of claims should be modelled at an aggregate level. As in Pollard's work, the Central Limit Theorem could be used to support the use of a normal distribution to describe variation in aggregate claims. They suggested that the standard deviation of the claims paid in a given year should be a function of the size of that year's expected aggregate claim. Variations from year to year in claims paid were assumed to be independent. Asset variation and inflation were modelled using a version of the Wilkie model that was recalibrated with the objective of being more appropriate to the shorter-term horizons of general insurance business than the long-term projections for which the model was originally intended.

This modelling framework could be used to determine the starting amount of assets in excess of the technical provisions that was required to support a given probability of ruin. The working party again suggested that an actuary should write a public report on the financial strength of the company that presented these findings. The paper showed that illustrative calibrations of this modelling framework could generate intuitive assessments of solvency margin—for example, the central case of their example suggested a solvency margin of around 10 % of the (best-estimate) technical provisions would be required. Of course, such results were entirely predicated on their assumptions about the scale of variability in the claims run-off. Their example calibration was developed in a very heuristic way. The paper did not offer any substantial guidance on how these parameters could be robustly calibrated to reflect the specific features of a given general insurance business.

Whilst the Solvency Working Group was developing its vision of a solvency framework and the role of the actuary within it, other actuarial researchers continued with the investigation of quantitative techniques for estimating claim reserves. The most intuitive and accessible paper of the 1980s on general insurance claim reserving methods was a *Journal* paper written by Sidney

Benjamin and Ian Eagles, published in 1986.[51] Benjamin and Eagles analysed historical claims run-off patterns in Lloyd's syndicates and found that the ultimate loss ratio tended to have a strong linear dependence on the year one paid claims ratio. This linear dependency relationship had already been alluded to by Pollard's development year correlation matrix but Benjamin and Eagle's presentation implied a very simple mathematical rendering that appeared to produce good empirical fits: ultimate cumulative claims, and hence required current reserves, could be estimated by simple linear regression of the cumulative claims of a given development year with the ultimate cumulative claims paid. The empirically observed variation around the line of best fit could also provide some heuristic indication of the extent to which ultimate cumulative claims could deviate from the extrapolated estimate. These regression relationships could be fitted to different business lines and years of development. Naturally, the later the year of development, the more confidence could be had in the ultimate loss estimate. It was classic Sydney Benjamin—avoiding unnecessary statistical niceties, it cut to the chase and delivered powerful practical actuarial insight. In a world increasingly characterised by rocket science and advanced computing technology, Benjamin had a knack of making original and informative use of the back of an envelope.

The working party's normal distributions and Benjamin's linear regressions may have led the British actuarial profession to exhale a collective sigh of relief at the potential accessibility of new general insurance reserving methods. The investigation, however, of the use of advanced statistical techniques in claim reserving was far from over. Three further papers of a high statistical ambition were published in the *Journal* in 1989 and 1990. The first two of these papers were the fruits of a seminar 'Applications of Mathematics in Insurance, Finance and Actuarial Work', jointly sponsored by the Institute of Actuaries and the Institute of Mathematics and its Applications. These two papers, by R.J. Verrall[52] and A.E. Renshaw[53] respectively, covered broadly similar statistical ground. They both observed that the ubiquitous chain ladder method could be considered as a form of two-way analysis of variance (ANOVA). From this observation, a similar recursive Bayesian estimation approach as that developed by de Jong and Zehnwirth could be developed. It could also be shown that, under the statistical assumptions of the ANOVA model and an assumed lognormal distribution for claims, the chain ladder method did not produce the maximum likelihood estimates for the expected claims. This

[51] Benjamin and Eagles (1986).

[52] Verrall (1989).

[53] Renshaw (1989).

more statistically sophisticated analytical approach could also shed some light on the parameter stability (or lack thereof) of the chain ladder method, especially for the most recent underwriting years.

T.S. Wright's paper, the third of this series of highly technical investigations of advanced statistical techniques for claim reserving, appeared in the *Journal* in 1990.[54] Like the work of de Jong and Zehnwirth, this again used the Bayesian Kalman filter statistical technology to produce stochastic projections of claims run-off. However, unlike the papers by Verrall and Renshaw, Wright's approach did not rely on the assumption that claims were log-normally distributed. This was an assumption which Wright regarded as an 'untenable' description of general insurance claims distributions. Several technical assumptions about the distributional characteristics of the claims process were still required by Wright's approach, but it allowed the statistical insights to be placed in a more general family of distributions than the previous research.

Whilst these advanced statistical methods may have had some application in insurers' internal assessments of claims experience and profitability, they were too exploratory and complex for use in 1990s statutory solvency assessment. As has so often been the case in the history of British actuarial engagement in general insurance, further inspiration was sought from overseas. A group of British actuaries authored a paper reviewing the recent US regulatory developments which appeared in the *British Actuarial Journal* in 1996.[55] The US regulatory authorities introduced a Risk-Based Capital (RBC) system in the early 1990s. The essential idea was that the capital requirement (similar to the solvency margin in the UK) would be calculated using a series of prescribed factors that were applied to a defined metric of volume of business (such as premiums earned or reserves net of reinsurance). Hence the capital requirement would be determined as some percentage of net reserves, and the percentage would be determined formulaically as a function of the mix of insurance business and asset risks on the balance sheet. The underwriting and claim reserve risk factors were set on a rolling basis to reflect the worst industry experience of the previous ten-year period. This mechanical calibration approach was open to the criticism of being too retrospective for the fast-changing world of general insurance.

The authors highlighted some of the limitations of this simplified one-size-fits-all formula application, most notably its inability to capture the aggregation of exposure to a single underlying risk event. They suggested that a *Dynamic Solvency Testing* (DST) approach, implemented within a statutory

[54] Wright (1990).

[55] Hooker et al. (1996).

professional framework, would be preferable to a formula-based approach. DST meant the insurer developing their own financial model of their business and using it to project and analyse the health of the business under a range of selected adverse stress scenarios. They also considered the use of *dynamic financial analysis* within this system, where the stress modelling is replaced with a full set of stochastic scenarios—in essence, what had been proposed by the Solvency Working Party in 1987. The application of the DST approach had since been pioneered in statutory reserving for Canadian general insurance in the early 1990s. The Canadian Institute of Actuaries had been active in developing professional guidance on the professional role actuaries could perform in a DST statutory framework. As might be expected, this call for a broad-ranging role that involved substantial actuarial judgement and freedom was generally welcomed in the Staple Inn discussion of the paper.

Yet the elephant in the room still remained—did the British actuarial profession have the skills and experience required to perform this type of role in general insurance? In the discussion, the influential general insurance actuarial practitioner D.H. Craighead tried his best to take a positive stance: 'I expressed reservations then [several years ago], but I think that we are now beginning to be ready for such a role'.[56]

Actuaries and the Financial Management of General Insurers (1981–1994)

In the 1980s and 1990s, technical actuarial research in general insurance broadened beyond the assessment of claim reserves and solvency margins to consider the wider horizons of the financial management of insurance firms. Sidney Benjamin was ahead of his time in this field as he was in maturity guarantee reserving, and his 1976 paper 'Profit and Other Financial Concepts in Insurance' was an early forerunner to this strand of thinking. But the 1981 *Journal* paper by Abbott, Clarke and Treen, 'Some Financial Aspects of a General Insurance Company', with its focus on the measurement and management of shareholder returns, arguably marked the real departure point from traditional actuarial ground.[57]

Like Benjamin in 1976, Abbott et al. advocated the use of discounted reserves, providing inflation was consistently applied to the claims projections. The paper particularly focused on how undiscounted reserves resulted in a

[56] Craighead in Discussion, Hooker et al. (1996), p. 313.

[57] Abbott et al. (1981).

misleading profit emergence pattern over the life of the business. It highlighted how material this effect could be in the high inflation environment that prevailed at the time the paper was written and how *changes* in inflation and interest rates over the run-off of the business could further distort the emergence of profits when using the undiscounted reserving approach. The paper also discussed how to assess the return on shareholder capital that had been earned by a general insurance business. It did not, however, venture to answer what return ought to be *required* by general insurance shareholders other than to suggest that an arbitrary amount in excess of the real risk-free interest rate should be a form of profit objective.

A short paper by G.C. Taylor published in the *Journal* in 1984 considered the capital that should be required to write new business and from where this capital should come.[58] In particular, in general insurance in the 1970s, the notion had developed that insurance business should be self-financing: that is, solvency margin requirements should be funded entirely from premium loadings. Taylor argued that solvency margin requirements should be considered as part of the working capital needs of a general insurance business and should therefore be funded by the shareholder rather than the policyholder. He further argued that whilst these funds could be invested and would generate investment return for shareholders, this return would be inadequate compensation for shareholders due to effects such as double-taxation. Hence premiums should be loaded to generate an acceptable expected return on that capital (which would be less than the loading for full self-financing). He then developed some algebra to show how to calculate this premium loading, though the required return on shareholder capital was taken as an arbitrary parameter that was 'dictated by the equity market'.[59] This work can be seen as a step towards a more economically coherent perspective on insurance premium setting. The concept of loading premiums to include a cost of capital has endured.

A 1990 paper by Daykin and Hey,[60] two leading general insurance actuaries that we already met earlier in our general insurance discussion, discussed how the stochastic asset-liability cashflow modelling that had been developed by the Solvency Working Group in 1987 for the purposes of solvency assessment could be extended for use in the wider financial management of a general insurance firm. The 1987 approach to solvency assessment had explicitly assumed that the insurer ceased to write new business. The 1990 paper considered how the simulation model could be extended to include the

[58] Taylor (1984).
[59] Taylor (1984), p. 178.
[60] Daykin and Hey (1990).

long-term projection of new business and its financial impacts. This naturally required a modelling framework to describe the behaviour of new business in terms of business volumes, profitability, competitiveness of pricing relative to the wider market, and how these variables behave jointly with each other and with the other stochastic variables and outputs in the model. For example, the business may price more aggressively when its balance sheet is strong relative to the required solvency margin. Daykin and Hey produced results for the ten-year stochastic projection of asset and liability cashflows, profits and balance sheets under a variety of different assumptions for new business writing and pricing strategies. Their objective was to show that technical actuarial skills could be employed as an intrinsic part of business strategy development and planning in a general insurance firm.

Daykin and Hey also considered the potential use of financial economics in general insurance business management. Once again, this topic of general insurance research had first been explored overseas—in this case in the USA. The particular stimulus for the use of financial economics in US general insurance had been in determining the return required by investors for funding general insurance business. This had relevance beyond the commercial management considerations of a given company—in the USA, premium rates were controlled by government regulation, and this required an objective and rigorous framework for establishing a 'fair' premium rate. This, in turn demanded an assumption about the level of return reasonably required by the providers of insurance capital.

The application of the Capital Asset Pricing Model to determine required returns on general insurance equity capital was initially explored in the USA in the late 1970s and early 1980s.[61] This work highlighted that where general insurance claims are uncorrelated with market returns, no risk premium should be required by shareholders for bearing this (diversifiable) risk. Daykin and Hey expressed discomfort with this implication for general insurance required returns, arguing that it 'suggests that the CAPM is missing some important aspects'.[62] A natural candidate for these missing aspects was the frictional cost of capital argument that had been presented by Taylor in 1984, but this line of thought was not pursued by Daykin and Hey.

Daykin and Hey also briefly considered the potential use of Merton-style option modelling of an insurer's capital structure, where shareholder equity is modelled as a call option on the net assets of the general insurance company.

[61] See, for example, Biger and Kahane (1978); Hill (1979).

[62] Daykin and Hey (1990), p. 197.

This had also recently been explored in the USA.[63] By explicitly capturing the leverage implied by insurance business (where policyholders are essentially debtholders), higher required shareholder returns (and hence premium rates) could be implied, which Daykin and Hey welcomed. They suggested that their simulation modelling approach could be used to analyse more realistic forms of optionality in shareholder returns than the simplistic analytical models presented in the literature.

Daykin and Hey widened the actuarial perspective on general insurance from solvency to include profitability, return of capital and new business pricing strategy. Later in 1990, a paper by Ryan and Larner was published in the *Journal* that widened the actuarial horizons of general insurance further to include company valuation.[64] This paper discussed the application to general insurance of the *appraisal value* method that actuaries had recently started to apply in merger and acquisition work in both life and general insurance business. The essence of the appraisal value was that it involved an explicit projection of the cashflow earnings of the business, using assumptions set following actuarial investigation. These cashflows would then be discounted at appropriate risk discount rates to obtain the company's appraisal value.

Ryan and Larner considered the appraisal value in three components: the adjusted net asset value (the current balance sheet net asset value, adjusted for the cost of those assets being 'locked-in' to the insurance balance sheet); other value arising from past written business (expected release of surplus from insurance reserves); and the value arising from future written business. The adjusted net asset value would be based on the market value of assets, and the deduction from the net asset value for capital 'lock-in' would be based on the shareholders' cost of capital. They explained the rationale for the cost of capital as follows:

> They [shareholders] will require a larger return on their funds [if invested in an insurance operation] than if they invested them separately. This arises, partly because the capital is being exposed to the risk of loss in the insurance business and partly because it could be used elsewhere.... We use the term "cost of capital" to mean the value of the shortfall in net earnings between the risk return required by shareholders and the actual [expected] investment return.[65]

Their cost of capital represented the incremental additional return that arose from the consequences of investing via an insurance entity rather than

[63] Doherty and Garven (1986).

[64] Ryan and Larner (1990).

[65] Ryan and Larner (1990), pp. 603–4.

directly in financial markets. Taylor had taken a similar position in his 1984 paper on loading premiums for the cost of capital, though he had argued that this additional cost arose from 'frictional' sources such as double-taxation rather than because of exposure to insurance risk.

The risk discount rate would reflect the riskiness of the shareholder cashflow streams. The authors suggested different risk discount rates should be applied to different elements of the projected earnings stream to reflect their risk characteristics (for example, a higher discount rate would be applied to earnings from future business than that used for future profits emerging from existing business). Whilst the authors were aware of financial economists' approaches to valuation, some of their suggestions were not necessarily consistent with those ideas. In particular, their suggestion that the risk discount rate should be a function of the extent to which the business was exposed to diversifiable insurance risk ran contrary to a fundamental principle of financial economics. This idea ran throughout the paper—for example, later the paper suggested that a well-diversified insurance business could command a lower risk discount rate and, hence, higher appraisal value than a less-diversified business.

A 1994 *Journal* paper by Bride and Lomax[66] provided a perspective on the financial management of general insurance business that was more closely aligned to financial economics. Their starting point was that the appraisal value implementation that had been developed by actuaries to support shareholder valuations in merger and acquisition activity 'fails to capture the operational dynamics of non-life insurance business and obscures the issues surrounding the nature of shareholders', policyholders' and other creditors' claims on the assets of the firm'.[67] However, they did not propose to reject the entire framework of discounted cashflow projections of the appraisal value—rather, they attempted to show how the risk-adjusted discount rate and cost of capital adjustments could be rigorously set.

They emphasised that the idea that shareholders did not require a reward for bearing diversifiable risk was the most fundamental and least controversial result produced by financial economics. This had implications both for the setting of the risk discount rate and the cost of capital adjustment—Ryan and Larner had argued that one of the reasons for the cost of capital adjustment was that shareholders required compensation for their exposure to the (diversifiable) risk of future insurance losses. This was categorically rejected by Bride and Lomax.

[66] Bride and Lomax (1994).

[67] Bride and Lomax (1994), p. 363.

They also highlighted what they viewed as other actuarial misconceptions in valuation. In particular, they argued that the choice of asset strategy of a general insurance company should have no impact on the value of the firm: 'an investment in gilts and an investment in equities have equal risk-adjusted total yields from the perspective of a shareholder with a diversified portfolio of assets'.[68]

In short, Bride and Lomax's 1994 paper was the general insurance equivalent of Exley, Mehta and Smith's 1997 defined benefit pension paper. It attempted to bring actuarial practice on the measurement and management of shareholder value in general insurance fully in line with core financial economics concepts. This was arguably less of a challenge in general insurance than in defined benefit pensions, with its legacy of 100 years of satisfied actuarial service. But the authors still wryly noted in their response to the Staple Inn discussion that 'the subject of financial economics continues to raise temperatures in the profession'.[69]

Actuaries in general insurance as well as pensions found it difficult to accept many of the insights and implications of financial economics. A reflex to reject them as academic theories based on unrealistic assumptions remained prevalent. In opening the discussion, Duffy commented:

> It is questionable whether the shareholders, let alone the management, of the major UK composites and other non-life firms follow the precepts of financial economic theory.[70]

The influential general insurance actuary J.P. Ryan expressed his opinion at Staple Inn in unambiguous terms: 'The authors use a number of oversimplified economic assumptions when applying their model. It is largely these oversimplified assumptions that give rise to some of the odd results in the paper.'[71]

Incorporating financial economics into the practices of any British actuarial field was no straightforward task in the 1990s. But in general insurance, as in life assurance and defined benefit pensions, the genie was out of the bottle. The process of embracing and incorporating the financial economic insights of the previous half-century into core actuarial thinking and practices was irrevocably underway.

---

[68] Bride and Lomax (1994), p. 388.

[69] Bride and Lomax (1994), p. 438.

[70] Duffy, in Discussion, Bride and Lomax (1994), p. 421.

[71] Ryan, in Discussion, Bride and Lomax (1994), p. 429.

Conclusions: Looking Back, Looking Forward

The intellectual foundations of actuarial methods—probability, statistics, mortality modelling, the valuation of life-contingent claims—were first laid in the latter stages of the Scientific Revolution by some of the most esteemed thinkers of that age such as Pascal, de Witt, Halley, Bernoulli, de Moivre, Bayes and Laplace. The history of British (and indeed global) actuarial thought, however, started in earnest in the second half of the eighteenth century with Richard Price. It was Price who developed the actuarial disciplines of reserving, assessment of surplus and distribution of bonus, as well as furthering technical methods for mortality modelling. Price's clarity of thought brought James Dodson's original conception of a new form of financial institution to life and ensured it was built on sustainable ground. His actuarial ideas on the management of with-profit business supported a product that prospered in a fundamentally similar form to his design for the next 200 years.

Relatively speaking, the economically serene nineteenth century was a period of incrementalism for actuarial thought. Mortality modelling continued to develop in areas such as graduation and the development of tables based on the pooled experience data of life assurers. Actuaries debated different approaches to life assurance liability reserving, and there was a growing recognition that valuations served two distinct purposes—solvency assessment and the equitable distribution of surplus—and that these purposes could make use of different valuation methods. British actuaries attempted to apply the techniques they had developed in life assurance to general insurance business (particularly fire and employers' liability insurance), but without any discernible impact on industry practices.

C. Turnbull, *A History of British Actuarial Thought*,
DOI 10.1007/978-3-319-33183-6

Nineteenth-century actuaries showed relatively little interest in the asset side of the balance sheet. From William Morgan's time onwards, however, life actuaries were often doubtful of the relevance of market values. At the start of the twentieth century, the emerging field of defined benefit pensions was broadly aligned to their life assurance colleagues. Ultra-long-term estimates of liability valuation parameters, including interest rates, were used alongside similarly off-market asset valuations to produce stable assessments of pension fund surplus.

In the late nineteenth century, actuarial thought began to actively consider investment strategy for the assets backing life assurance business. Most notably, in 1862 Arthur Bailey recognised the advantages that the long-term illiquid liabilities of life offices provided by enabling investment in highly illiquid assets that offered additional yields relative to equivalent more widely traded instruments. This thinking was reflected in the British life office asset allocations of the late nineteenth century—in 1890, some 80 % of life office funds were invested in illiquid asset classes such as mortgages and loans. A significant rotation of life office assets out of illiquid asset types and into traded securities followed in the subsequent decades as liability liquidity became more transparent and the availability of illiquid assets diminished.

Broader thinking on asset strategy for life business did not emerge until the First World War shattered the economic climate of Edwardian stability, resulting in unprecedented volatility in interest rates and inflation. The economic volatility of the 1920s was followed by a decade of falling long-term interest rates. By the mid-1930s, long-term gilt yields were alarmingly close and sometimes even below the 3 % rate assumed almost universally in British with-profit premium bases of the time.

This challenging environment encouraged actuaries to support greater levels of investment in equities by both life offices and pension funds. This was partly because the extraordinary inflation experience of post-First World War period increased the appetite for real assets that could offer some form of protection from future inflation shocks. Whilst life offices did not have liabilities that were guaranteed in real terms, there was a policyholder expectation that with-profit bonuses would offer some element of inflation protection to their pay-outs. Similarly, pension fund trustees often aimed to provide pensioners of defined benefit schemes with discretionary increases in their pension that offset the impact of inflation on the purchasing power of their pension. At this time, decades before the launch of index-linked bonds, the equity market was arguably the only liquid, real asset class available to these institutions. The potential inadequacy of long-term gilt yields to meet with-profit guarantees provided some offices with a further rationale to allocate to equities as part of an increase in investment risk appetite. The 1930s also represents the start of the period of the cult of equities, and actuaries and their institutions were not immune to its siren call. Leading actuaries of the era such as H.E. Raynes were heavily influenced by contemporary economists such as Edgar Lawrence Smith and John Maynard Keynes, and indeed by the indisputably excellent performance of equity markets over the preceding decades (especially relative to long-term bonds). Thus, from the 1920s onwards, the financial market risks embedded in British life office balance sheets started to materially grow.

The economic volatility of the post-First World War era also further disinclined pension actuaries to use market values in assessments of pension fund surplus and required long-term contribution rates. By the early 1920s a general actuarial consensus was reached that the historically high levels of prevailing long-term market bond yields should not be fully reflected in pension fund valuations. Assets were valued at above market value, though with perhaps a partial write-down from their book value. Liability discount rates were left unchanged from their pre-war assumptions. The prevailing market yield was treated as an aberration that could be assumed away. The consensus was not unanimous, however. Most notably, S.G. Warner, the senior actuary who was President of the Institute in the years 1916–1918, argued in 1921 that assets should be valued at market value and liabilities should be valued using a discount rate based on the market yield. This is perhaps the earliest actuarial call for a market-based approach to pension fund valuation, but it went unheeded.

The quarter-century following the Second World War was a period of strong economic growth and rapid technological change. For actuaries, developments in computer science and financial economics pointed to an increasingly quantitative environment for their work. More advanced quantitative concepts were introduced to British actuaries, or indeed developed by British actuaries, during this period and found varying degrees of interest amongst the profession. The application of risk theory to general insurance business found few supporters within the British profession, despite the unrelenting efforts of R.E. Beard during the 1950s. On the other hand, the work of Redington (1952), Haynes and Kirton (1952) and Anderson and Binns (1957) introduced a new technical sophistication to thinking on asset-liability management in life assurance. Duration and immunisation were concepts introduced and mathematically codified by Redington that would eventually permanently resonate with the wider financial world. The work of Haynes and Kirton, and Anderson and Binns tentatively pointed towards ideas such as dynamic hedging and portfolio insurance that would again become part of mainstream financial risk management in the decades to follow. Whilst British life actuaries had historically displayed an ambivalence towards market values, all of these risk management ideas were based on the objective of managing exposure to changes in market prices. Defined benefit pensions thinking, however, remained resolutely detached from market values and indeed from any explicit framework of risk management thinking.

These post-war decades also saw the development in US academia of the fundamental theories of financial economics—portfolio theory (1952), Modigliani–Miller (1958), the Capital Asset Pricing Moody (1964), the Efficient Markets Hypothesis (1970) and Black–Scholes–Merton (1973). Modigliani–Miller pioneered the use of arbitrage in the derivation of finance theories; the other theories were built to various degrees on contemporary developments in computer science and quantitative analysis: the expanding computational capabilities of computers, the mathematisation of finance and economics, and the rigorous statistical analysis of empirical market data were the underlying building blocks of this stream of research. The British actuarial profession was largely oblivious to these ideas in the years following their emergence and when it did start to read about them, its predominant reaction was to dismiss them as irrelevant theory.

Nonetheless, it is at least a notable historical curiosity that a number of the technical ideas that dominated financial economic theory (such as portfolio diversification, dynamic hedging, portfolio insurance) and financial economic empirical research (excess volatility and predictability of returns) were at least heuristically explored by twentieth-century British life actuaries ahead of their rigorous academic development by professors of financial economics. It is also notable that this actuarial research output tended to emerge from life actuaries rather than pensions actuaries. Actuarial work in pensions was ultimately an exercise in long-term financial planning and budgeting; pension actuaries generally did not focus on the proactive risk management of the financial security of members' accrued pension benefits with the same sense of discipline and purpose that life actuaries, as custodians of policyholder promises, attempted when measuring and managing financial risks of life offices.

Whilst the British profession of the 1970s displayed a general disdain for the actuarial application of the ideas of financial economics, it did, however, attempt its own application of computer science and statistical modelling to financial risk management. These efforts had natural applications given the market risk-laden balance sheets that actuaries presided over by this time—full of equities on the asset side and long-term guaranteed returns to life policyholders or pension fund members on the liability side. In the life sector, the provision of guarantees in the form of unit-linked business rather than in the opaque with-profits format made them more transparent to those outside the actuarial profession—they could not be cloaked in the mystique of actuarial management. Regulators worried and agitated for the actuarial profession to provide a reserving solution for these guarantees that was evidently prudent and objective.

The actuarial solution took ten years to deliver and generated much internal debate along the way. The 1971 sessional meeting that discussed Sidney Benjamin's guarantee reserving paper provoked such a response that the President converted the sessional meeting into a private discussion so as to avoid documenting it. The paper was never published by the Institute. What was the cause of such controversy? Benjamin proposed a risk-based reserving approach that explicitly recognised the non-diversifiable nature of the financial market risk that was embedded in life assurers' products and balance sheets. Up until this time, actuarial reserving worked in a deterministic setting that assumed the risks written by the office could be largely diversified away. Some margins for prudence were applied to reserving bases relative to premium bases, but these were of a small scale relative to Benjamin's risk-based reserve. The Maturity Guarantee Working Party's report of 1980, with its recommendation of a risk-based reserved assessed using a stochastic simulation model, was therefore a watershed moment in the history of British actuarial thought. It also resulted in the ending of the British life office practice of offering maturity guarantees to unit-linked business.

The working party's risk assessment methodology, however, rejected the insights available from financial economics' option pricing literature. Instead of reserving for guarantees on the basis that they were put options that could be transferred to a third party at a market price, the risk methodology considered the amount of assets required to fund the ultimate losses generated by the guarantee with a high level of probability, under the explicit assumption that no form of hedging could take place. This method therefore did not provide any incentive to life offices or their actuaries to consider how to *manage* the risks they had written. Moreover, the method was highly sensitive to assumptions about the very long-term behaviour of equity returns. Here again, the approach rejected ideas from financial economics: instead of using a random walk-style model for equity returns that was consistent with market efficiency, the modelling, which was essentially a prototype of the Wilkie model, assumed equity market returns had a predictable component that resulted from market prices being more volatile than was rationally implied by assumed dividend variability. Interestingly, however, the empirical research produced by financial economics academia in the 1980s by leading academics such as Shiller, Roll, Fama and French provided some intellectual support for this approach (although this strand of research was ignored in the later 'traditionalist versus modernist' debates within the British actuarial debates of the 1990s).

The stagflation of late 1970s Britain generated even more existential questions for actuaries in defined benefit pensions: should pensions be advance funded in an economic environment of negative real interest rates? The improving economic conditions of the 1980s made this question moot, but there remained a philosophical debate about the purpose of funding. Specifically, was it to produce a stable set of asset cashflows with a similar profile and character to the liability cashflows; or was the central purpose of advance funding to provide a current pot of assets with a market value that was sufficient to secure the accrued pension benefits in the event of sponsor insolvency? The profession slowly moved from the former towards the latter over a period of several decades, but never fully reached a consensus. When, in the early 1990s, the government tried to enforce a funding standard that would secure accrued benefits in the event of insolvency—the Minimum Funding Requirement (MFR)—the result was a fudge that failed to meet its stated objective and otherwise created cost and confusion.

Whilst the 1970s and early 1980s was a period of profound challenge for British life and pensions actuaries (challenges that arose from legacy balance sheets, difficult economic conditions and the circulation of new ideas that ran counter to actuarial convention), in contrast this was a period when actuaries' profile and role in British general insurance notably increased. This increased role capitalised on the accumulating volume and improving accessibility of relevant historical claims data in sectors such as motor insurance, and on actuaries' traditional skills in the analysis of claims data for the purposes of pricing and reserving. However, the claim for a greater actuarial role in general insurance was also strongly based on the argument that statutory reserving in general insurance needed the same professional judgement that was present in life assurance reserving.

The 1990s was a remarkably tumultuous and difficult decade for the British actuarial profession. Again, the economic environment played its part in exposing the limitations of the financial risk management approaches adopted in actuarial balance sheets. In particular, long-term interest rates steadily fell throughout the decade. By the middle of the decade, the fall in rates had triggered the Guaranteed Annuity Option crisis in the life sector, which claimed the scalp of Equitable Life, the storied institution of Richard Price and William Morgan. There were many institutional factors that contributed to the Equitable's downfall, but from the perspective of actuarial thinking, there was at least one fundamental lesson. The profession chose not to apply to with-profit guarantees the risk-based reserving methods it developed in the 1970s for unit-linked maturity guarantees. This was largely because actuaries were confident that the various powerful levers they had at their disposal to steer with-profit businesses—specifically, wide discretion in bonus policy and in investment strategy—would allow them to avoid engaging in the rigours of risk-based reserving, market risk management and hedging. The exposure of GAO business to low interest rates had been fully understood by actuaries for many years—Dick Gwilt, the Faculty President of 1952 to 1954, had argued as early as in 1948 against exposing life offices to this form of policyholder optionality. But the sales and marketing pressure to provide such product features held sway. And actuaries held on to the belief that with-profit funds had the means to manage the risks they created. Ultimately, this proved to be a misplaced confidence. The entire episode substantially damaged the public, government and regulatory confidence in actuarial judgement and discretion.

These turbulent episodes left the profession embattled as never before. Painful lessons were learned, however, and by the late 1990s there was a perceptible change in the profession's intellectual outlook and its openness to change. Renewed impetus was given to modernising actuarial thinking that finally embraced the fundamental ideas of financial economics and attempted to incorporate them into standard actuarial methods. Across each of the three main practice areas of general insurance, defined benefit pensions and life assurance, important papers were published in the 1990s and early 2000s that, whilst at times stirring controversy and discontent amongst traditionalists, paved the way for a more open-minded and technically trained era of British actuarial thought. In life assurance in particular, albeit at the behest of an impatient regulator reacting in the aftermath of the GAO debacle, new market-based, risk-based reserving methods for with-profit business were implemented by actuaries quickly, successfully and permanently. These methods were amongst the most sophisticated implemented in any life assurance market in the world, and were arguably influential in establishing the blueprint for the European Solvency II framework that was first conceptualised in the early 2000s.

It is no easy task, from today's vantage point of hindsight, to sit in objective judgement of this story and the events and characters that shaped it over hundreds of years. Perhaps the best we can do is point to where notable progress evidently did and not occur, and where particular events and decisions were clearly consequential.

Since the early nineteenth century, the very long-term nature of life and pensions business had philosophically inclined actuaries to use stable measures of cost and value. Market values have, on the most part, been treated with suspicion, disdain or indifference. There are a couple of possible kinds of explanations for this: in the case of with-profit life business, it could be argued that this lack of use of market values merely reflected the way the product was supposed to work. That is, with-profit bonuses were not supposed to reflect market value changes; at the heart of the product was the concept of intergenerational cross-subsidy and smoothing of returns between policy-holders. There is good evidence to support this: there is much recognition in the historical actuarial literature of the need for different types of valuation for different purposes. Classically, a net premium valuation with off-market asset valuation for the equitable distribution of surplus; and a gross premium valuation with market asset values for the assessment of solvency. There is a second kind of explanation: that actuaries held the belief that market values were often 'wrong'; that they were too volatile and the changes in market value were often irrational and irrelevant to the assessment of the expected cost of funding long-term liability cashflows. Pension fund valuation methods have arguably been based on this perspective. And much of modern actuarial risk methodology (such as the Wilkie model) has also been consistent with this outlook. It also arguably was a factor in the increasing appetite for market risk on the asset side of life and pensions balance sheets over the twentieth century.

Whilst such an outlook was less inconsistent with financial economics research than actuaries were often led to believe, it nonetheless inhibited the profession's willingness to embrace new ideas and techniques that could manage market risk over shorter-term horizons. In hindsight this was unfortunate, as governments and regulators became increasingly interested in solvency measures that were based on the short-term resilience of the balance sheets of long-term business on a market value basis (in both life and pensions). When reviewing the twentieth-century actuarial literature, it is quite striking how a rich stream of market value-based risk thinking emerged from 1950s life actuaries which was not successfully built upon in the actuarial research of the 1960s and 1970s. One can only speculate whether it was that philosophical actuarial disdain for managing short-term market value solvency risks that lay behind this.

To quote the cliché, history does not repeat, it rhymes. It also never ends. This historical account closed around the start of the twenty-first century. Does it point to some important trends for the future of British actuarial thought? And how has the extraordinary decade of financial turmoil that followed the end of this historical account impacted on the future direction of actuarial thought?

The global financial crisis of 2007–2008 systemically challenged financial institutions (and their regulators) in a way that had not occurred since the 1930s. This was, first and foremost, a global banking crisis. Widening credit spreads and rising doubts about the quality of the opaque assets that played an increasingly prominent role on bank balance sheets resulted in a banking solvency and liquidity crisis that started in the USA and quickly reverberated around the globe. With the notable exception of AIG (with its notorious financial products division that was far removed from conventional insurance business), the financial institutions that relied on actuarial advice were not in the front line of the crisis and they generally weathered its immediate impact successfully. Insurance companies and pension funds had equity, credit and other risk asset investments which performed poorly over the period of the crisis, but there was no widespread immediate set of failures in these institutions that resulted from the unfolding of the crisis over 2007 and 2008. This may have been a source of some satisfaction and vindication for actuaries around the world: if they could weather a once-in-a-century financial storm such as this, clearly they were doing something right.

But the full picture is more complex than this: the long-term nature of life and pension liabilities means that they are exposed not only to the short-term gyrations of financial market corrections but also to the longer-term economic consequences that a financial crisis such as this may trigger. In the years following the financial crash of 1929, long-term UK government bond yields fell from 4.6 % to below 3 %. The years following the financial crisis of 2007–2008 produced a similar pattern. We saw in the Guaranteed Annuity Options discussion that long-term government bond yields fell from 15 % to 5 % over the three decades between 1975 and 2005. One of the economic consequences of the global financial crisis of 2007–2008 has been to further extend this trend: by 2015, British government long bond yields had fallen below 3 %, to levels not seen since the Dalton era at the end of the Second World War.

For UK life assurers, whilst unwelcome, this fall in rates did not present an existential solvency threat: the regulatory changes of the early 2000s, which were directly motivated by the GAO crisis of the late 1990s and the industry's exposure to further interest rate falls that it highlighted, had strongly encouraged UK life offices to reduce the financial risk exposures in their back books and to stop writing new business with significant financial guarantees (at least without pricing it on a market basis and managing its risk actively). The same conclusion could not, however, be applied to UK defined benefit pension funds, many of which were left with very substantial levels of funding deficit on a market value basis by the post-crisis falls in long-term rates. At the time of writing it remains to be seen how the cost of meeting these deficits would be shared amongst corporate sponsors (via additional deficit-funding contributions), the pension fund members (through reductions in future pension payments) and the government (to the extent that the Pension Protection Fund cannot be adequately financed via pension fund levies). And the possibility remains that a substantial rise in long-term nominal and real interest rates could yet reverse these deficits.

These falls in long-term interest rates between 2008 and 2015 have not been a UK-specific phenomenon, but have been the experience of most developed economies around the world. Indeed, long-term interest rates in Europe have fallen even more precipitously than in the UK: by 2015 German long-term bond yields had fallen below 2 %. Unlike their UK counterparts, regulators in northern European countries such as Germany did not put a market-based, risk-based regulatory solvency system in place for insurers prior to the global financial crisis and the insurers generally did not put financial risk management strategies in place that would protect them from the asset-liability consequences of falls in the long-term interest rate to the levels of 2015. Some of these institutions may struggle to meet their long-term contractual liabilities from their existing reserves if interest rates do not revert to something like their pre-crisis levels in the coming years.

Since around the start of the twenty-first century, the European Union has been working on the implementation of a harmonised regulatory solvency system for insurers that is market-based, risk-based and principle-based (and which has similar fundamental features to the system implemented very quickly by the UK regulator in the early 2000s). This regulatory system, known as Solvency II, was finally implemented at the start of 2016 after more than a decade of political horse-trading. The result is arguably rather reminiscent of the British attempt at improved pension fund supervision in the 1990s. The Minimum Funding Requirement was a regulatory system driven by an ambition to provide a high standard of protection to the beneficiary based on short-term market-based solvency measures and the cost of transferring liabilities to a third party; but it was fudged in implementation so as to avoid the economic reality of the scale of capital that that protection implied. The MFR ultimately produced, at significant cost, numbers that were not particularly useful or meaningful, and that certainly did not have the meaning they were originally intended to have. This is arguably what Solvency II looks like in parts of the life assurance sector of northern Europe.

One might argue that it is unfair to ask actuaries to have anticipated these unprecedented economic conditions. Moreover, since the global financial crisis, government and central bank policy across the developed world has been to regard the impact of extraordinary low interest rates on long-term savings institutions such as life assurers and pension funds as mere collateral damage that is necessarily incurred in efforts to protect banks and the wider economy. This policy places insurance and pension institutions in a strong headwind that is not of their own making. But one might also look at the interminable decline in long-term interest rates since the mid-1970s and ask at what point it would be reasonable to view further significant falls as something that ought to be considered and proactively managed against. The UK life offices showed that this could be done (albeit at the impatient behest of their regulator). Their proactive management of interest rate risk at the start of the twenty-first century probably preserved the future solvency of several financial institutions and could perhaps be regarded as a highlight of post-war British actuarial achievement.

What might this all mean for the future of actuarial thought? Thought-leadership will continue to be driven to a large extent by the role that actuaries play in financial institutions. In the UK, the gradual but irreversible demise of with-profits life assurance and defined benefit pensions makes this future role more uncertain than it has been at perhaps any point in the last 200 years. Looking ahead, there are many possibilities and few inevitabilities. Whilst traditional with-profit business is no longer written in any meaningful volume in Britain and increasingly less is written in continental Europe, it has been selling successfully over the last decade in fast-growing life assurance markets such as China and India. Traditional life actuarial skills may be in ongoing demand in these increasingly consequential markets. The British profession is enthused by the prospect of actuaries contributing in new fields such as climate change that may make use of actuaries' quantitative skills and their experience in applying them to long-term complex socio-economic problems. This is quite plausible but it still remains to be seen if professional actuarial skills can make an impactful contribution in such fields.

Perhaps there is a further positive possibility: British life actuaries' success in implementing a principle-based reserving system in the early 2000s prompts echoes of Sidney Benjamin's historical call for the application of actuarial professional judgement to provide an independent, objective, statutory and expert role in assessing solvency and setting prudential reserving levels. In the 1970s Benjamin tirelessly argued that solvency assessment ought to go beyond the application of exploitable regulatory rules and should work outside the direct influence of the institution's managers and shareholders. He was speaking of extending the classical 'freedom with publicity' role of British life actuaries to include general insurance as well as life business. This has now largely happened. The experience of the global financial crisis perhaps highlights that such a role could also be valuable across the wider financial sector.

As any good actuary will tell you, trying to forecast the future is even harder than trying to explain the past. Future long-term socio-economic problems are unlikely to be any less complex and consequential than historical ones. Demographics, pensions, healthcare and insurance all have the potential to pose profound long-term societal challenges. Perhaps, like the work of James Dodson in the mid-eighteenth century, new conceptions of risk-sharing can be developed by actuaries that create new products, new solutions, and even new industries to tackle these challenges. For new ideas to have longevity, they require a robust implementation. The best actuarial work—such as that of Richard Price and Frank Redington—has often embraced more theory and technical complexity than the broader profession has been comfortable with at the time, but has been presented in a way that demonstrates it can be applied in accessible, practical, enlightening and highly consequential ways. This is likely to remain true of the most important actuarial thought-leadership of the next century. I look forward to seeing it.

Bibliography

Abbott, W.M. et al (1974), 'Some Thoughts on Technical Reserves and Statutory Returns in General Insurance'. *Journal of the Institute of Actuaries*, Vol. 101, pp. 217–283.

Abbott, W.M. et al (1981), 'Some Financial Aspects of a General Insurance Company'. *Journal of the Institute of Actuaries*, Vol. 108, pp. 119–209.

Anderson, J.L. (1944), 'Notes on the Effect of Changes in Rates of Interest on the Bonus-Earning Power of an Office Paying a Uniform Compound Reversionary Bonus', *Transactions of the Faculty of Actuaries*, Vol. 17, No. 162 (1938–1945), pp. 137–173.

Anderson, J.L. and J.D. Binns (1957), 'The Actuarial Management of a Life Office', *Journal of the Institute of Actuaries*, Vol. 83, No. 2, pp. 112–152.

Andras, H.W. (1896), 'On the system of bonus distribution to policyholders as a percentage per annum for the Valuation period on the sum assured, or on the sum assured and existing bonuses, considered in relation to some recent influences on Life Assurance Finance', *Journal of the Institute of Actuaries*, Vol. 32, No. 5 (April 1896), pp. 320–371.

Ansell, C. et al (1843), *Tables Exhibiting the Law of Mortality Deduced from the Combined Experience of Seventeen Life Assurance Offices*.

Arthur, T.G. and P.A. Randall (1990), 'Actuaries, Pension Funds and Investment', *Journal of the Institute of Actuaries*, Vol. 117, No. 1, pp. 1–49.

Bachelier, L. (1900), 'Theorie de la speculation', *Annales Scientifiques de l'École Normale Supérieure* **3** (17), pp. 21–86.

Bailey, A.H. (1862), 'On the Principles on which the Funds of Life Assurance Societies should be Invested', *Journal of the Institute of Actuaries*, Vol. 10, pp. 142–147.

C. Turnbull, *A History of British Actuarial Thought*,
DOI 10.1007/978-3-319-33183-6

Bailey, A.H. (1878), 'The Pure Premium Method of Valuation', *Journal of the Institute of Actuaries*, Vol. 21, No. 2 (July 1878), pp. 115–136.

Barnwell, R.G. and F. Hendricks (1852), *A Sketch of the Life and Times of John De Witt, Grand Pensionary of Holland, To which is added, His Treatise on Life Annuities.* (Note De Witt's original treatise was published in 1671.) 2012 re-print published by General Books LLC.

Bayes, T. (1764), 'An Essay Towards Solving a Problem in the Doctrine of Chances', *Philosophical Transactions of the Royal Society of London*, 53.

Bayley, G.V. and W. Perks (1953), 'A Consistent System of Investment and Bonus Distribution for a Life Office', *Journal of the Institute of Actuaries*, 79, pp. 14–73.

Beard, R.E. (1954), 'Some Statistical Aspects of Non-Life Insurance', *Journal of the Institute of Actuaries Students' Society*, Vol. 13, pp. 139–157.

Beard, R.E. (1967), 'On the Compilation of Non-Life Insurance Statistics', *Journal of the Institute of Actuaries*, Vol. 93, No. 2, pp. 271–277.

Beard, R.E., T. Pentikainen and E. Pesonen (1969), 'Risk Theory: The Stochastic Basis of Insurance', Chapman and Hall.

Benjamin, S. (1976), 'Maturity Quarantees for Equity-Linked Policies', *20th International Congress of Actuaries.*

Benjamin, S. (1976), 'Profit and Other Financial Concepts in Insurance', *Journal of the Institute of Actuaries*, Vol. 103, No. 3, pp. 233–305.

Benjamin, S. and Eagles (1986), 'Reserves in Lloyd's and the London Market', *Journal of the Institute of Actuaries*, Vol. 113, No. 2, pp. 197–256.

Benz, N. (1960), 'Some Notes on Bonus Distributions by Life Offices', *Journal of the Institute of Actuaries*, Vol. 86, No. 1, pp. 1–29.

Bernoulli, Daniel (1738), 'Specimen theoriae novae de mensura sortis', *Commentarii academiae scientarum imperialis Petropolitanae 5.*

Bernoulli, Jacob (1713), *Ars Conjectandi*, Basle.

Biger, N. and Kahane, Y. (1978), 'Risk Considerations in Insurance Ratemaking', *Journal of Risk and Insurance*, Vol. 45, pp. 121–132.

Black, F., T. Derman and W. Toy (1990), 'A One-Factor Model of Interest Rates and its Application to Treasury Bond Options', *Financial Analysts Journal*, January/February 1990, pp. 24–32.

Black, F. and P. Karasinski (1991), 'Bond and Option Pricing when Short Rates are Lognormal', *Financial Analysts Journal* (July/August 1991).

Black, F. and M. Scholes (1973), 'The Pricing of Options and Corporate Liabilities', *Journal of Political Economy*, Vol. 81, No. 3 (May–June 1973), pp. 637–659.

Boden, D.E. and T.D. Kingston (1979), 'The Effect of Inflation on Pension Schemes and Their Funding', *Transactions of the Faculty of Actuaries*, Vol. 36, No. 256, pp. 399–468.

Bolton, M.J. et al (1997), 'Reserving for Maturity Quarantees', *Report of the Annuity Guarantees Working Party.*

Boole, George (1854), *An Investigation of the Laws of Thought.* Watchmaker Publishing.

Bornhuetter, R.L. and R.E. Ferguson (1972), 'The Actuary and IBNR', *Proceedings of the Casualty Actuarial Society*, Vol. 59, pp. 181–195.

Box, G.E.P. and Jenkins, G.M. (1970), *Time Series Analysis: Forecasting and Control.* Holden-Day.

Boyle, P.P. (1978), 'Immunization Under Stochastic Models of the Term Structure', *Journal of the Institute of Actuaries*, Vol. 105, No.2, pp. 177–187.

Boyle, P.P. and Schwartz, E.S. (1977), 'Equilibrium Prices of Guarantees Under Equity-Linked Contracts' , *The Journal of Risk and Insurance*, XLIV, 4, pp. 639–660.

Boyle, P.P. and M. Hardy (1978), 'Guaranteed Annuity Options', *Astin Bulletin*, Vol. 33, No. 2, pp. 125–152.

Brennan, M.J. and E.S. Schwartz (1976), 'The Pricing of Equity-Linked Life Insurance Policies with an Asset Value Guarantee', *Journal of Financial Economics*, Vol. 3, pp. 195–213.

Brennan, M.J. and E.S. Schwartz (1979), 'A Continuous Time Approach to the Pricing of Bonds', *Journal of Banking and Finance*, Vol. 3 (1979), pp. 133–155.

Brennan, M.J., and E.S. Schwartz (1979; 2), 'Alternative Investment Strategies for the Issuers of Equity Linked Life Insurance Policies with an Asset Value Guarantee', *Journal of Business*, Vol. 52, pp. 63–93.

Brennan, M.J. and E.S. Schwartz (1982), 'An Equilibrium Model of Bond Pricing and a Test of Market Efficiency', *Journal of Financial and Quantitative Analysis*, Vol. 17, pp. 301–329.

Bride, M., and Lomax, M.W. (1994), 'Valuation and Corporate Management in a Non-Life Insurance Company', *Journal of the Institute of Actuaries*, Vol. 121, No. 2, pp. 363–440.

Brown, Hugh (1931), 'Employers' Liability Insurance', *Transactions of the Faculty of Actuaries*, Vol. 13, No. 115, pp. 1–66.

Brown, Samuel (1851), 'On the Fires in London During the 17 Years from 1833 to 1849 Inclusive, Showing the Numbers which Occurred in Different Trades, and the Principle Causes by which they were Occasioned', *The Assurance Magazine*, Vol. 1, No. 2, pp. 31–62.

Brown, Samuel (1858), 'On the Investments of the Funds of Assurance Companies', *Journal of the Institute of Actuaries*, Vol. 7, No. 5 (April 1858), pp. 241–254.

Cardano, Gerolamo (1663), *Liber de ludo alae.* In *Opera Omnia*, Stuttgart-Bad Cannstatt, 1966.

Clarkson, R.S. (1997), 'An Actuarial Theory of Option Pricing', *British Actuarial Journal*, Vol. 3, No. 2, pp. 321–409.

Clayton, George and W.T. Osborn (1965), *Insurance Company Investment.* George Allen & Unwin.

Cochrane, John (2005), *Asset Pricing, Revised Edition.* Princeton University Press.

Colbran, R.B. (1982), 'Valuation of Final Salary Pension Schemes', *Journal of the Institute of Actuaries*, Vol. 109, No. 3, pp. 359–416.

Collins, T.P. (1982), 'An Exploration of the Immunization Approach to Provision for Unit-Linked Policies with Guarantees', *Journal of the Institute of Actuaries*, Vol. 109, No. 2, pp. 241–284.

Corby, F.B. (1977), 'Reserves for Maturity Guarantees Under Unit-Linked Policies', *Journal of Institute of Actuaries*, Vol. 104, pp. 259–296.

Corley, R.D. et al (2001), *Report of the Corley Committee of Inquiry regarding the Equitable Life Assurance Society*.

Cowles, A. (1933) 'Can Stock Markets Forecasters Forecast?' *Econometrica*, Vol. 1, No.3 (July 1933), pp. 309–324.

Cox, J.C., J.E. Ingersoll and S.A. Ross (1985), 'An Intertemporal General Equilibrium Model of Asset Prices', *Econometrica*, Vol. 53, No. 2, pp. 363–384.

Cox, J.C., J.E. Ingersoll and S.A. Ross (1985), 'A Theory of the Term Structure of Interest Rates', *Econometrica*, Vol. 53, No. 2, pp. 385–408.

Cox, J.C. and S.A. Ross (1976), 'The Valuation of Options for Alternative Stochastic Processes', *Journal of Financial Economics*, Vol. 3, pp. 145–166.

Cox, J.C., S.A. Ross and M. Rubinstein (1979), 'Option Pricing: A Simplified Approach', *Journal of Financial Economics*, Vol. 7, No. 3.

Cox, P.R. and R.H. Storr-Best (1962a), 'Surplus: Two Hundred Years of Actuarial Advance', *Transactions of the Faculty of Actuaries*, 28 (1962–1964), pp. 19–60.

Cox, P.R. and R.H. Storr-Best (1962b), *Surplus in British Life Assurance: Actuarial Control over its Emergence and Distribution during 200 years*. Cambridge University Press.

Coutts, C.R.V. (1908), 'Bonus Reserve Valuations', *Journal of the Institute of Actuaries*, Vol. 42, No. 2 (April 1908), pp. 161–177.

Crabbe, R.J.W. and C.A. Poyser (1953), *Pension and Widows' and Orphans' Funds*. Cambridge University Press.

Craighead, D.H. (1979, 'Some Aspects of the London Reinsurance Market in World-Wide Short-Term Business', *Journal of the Institute of Actuaries*, Vol. 106, No. 3, pp. 227–287.

Daston, Lorraine (1988), *Classical Probability in the Enlightenment*. Princeton University Press.

Davis, Mark and Alison Etheridge (2006), *Louis Bachelier's Theory of Speculation; The Origins of Modern Finance*. Princeton University Press.

Day, J.G. (1959), 'Developments in Investment Policy During the Last Decade', *Journal of the Institute of Actuaries*, Vol. 85, No. 2, pp. 123–164.

Day, J.G. and K.M. McKelvey (1963), 'The Treatment of Assets in the Actuarial Valuations of a Pension Fund', *Journal of the Institute of Actuaries*, Vol. 90, pp. 104–147.

Daykin, C.D. et al., (1984), 'The Solvency of General Insurance Companies', *Journal of the Institute of Actuaries*, Vol. 111, pp. 279–336.

Daykin, C.D. et al., (1987), 'Assessing the Solvency and Financial Strength of a General Insurance Company', *Journal of the Institute of Actuaries*, Vol. 114, pp. 227–325.

Daykin, C.D. and G.B. Hey (1990), 'Managing Uncertainty in a General Insurance Company', *Journal of the Institute of Actuaries*, Vol. 117, pp. 173–277.

Deane, Phyllis and W.A. Cole (1967), *British Economic Growth, 1688–1959*, Second Edition. Cambridge University Press.

De Jong, P. and B. Zehnwirth (1983), 'Claim Reserving, State-Space Models and the Kalman Filter', *Journal of the Institute of Actuaries*, Vol. 110, No. 1, pp. 157–181.

De Moivre, Abraham (1718), *Doctrine of Chances*.

De Moivre, Abraham (1724), *Annuities on Lives*.

De Morgan, Augustus (1838), *An Essay on Probabilities and On Their Applications to Life Contingencies and Insurance Offices*. Longman, Orme, Brown, Green & Longmans.

De Morgan, Augustus (1839), 'Mortality', Penny Cylopedia.

De Wittt, Johan (1671), 'In A Series of Letters to the States-General', In R.G. Barnwell and F. Hendricks, *A Sketch of the Life and Times of John De Witt, Grand Pensionary of Holland*. New York, 1856.

Deuchar, David (1890), 'The Progress of Life Assurance Business in the United Kingdom during the last Fifty Years', *Journal of the Institute of Actuaries*, Vol. 28, No. 6 (October 1890), pp. 442–463.

Dimson, E., P.R. Marsh and M. Staunton (2002), *Triumph of the Optimists: 101 Years of Global Investment Returns*. Princeton University Press.

Dodds, J.C. (1979), *The Investment Behaviour of British Life Insurance Companies*. Croom Helm.

Dodson, James (1756), *First Lectures in Insurance*.

Doherty, N.A. and J.R. Garven (1986), 'Price Regulation in Property-Liability Insurance: A Contingent Claims Approach', *Journal of Finance*, Vol. 41, No. 5, pp. 1031–1050.

Dullaway, D. and P. Needleman (2004), 'Realistic Liabilities and Risk Capital Margins for With-Profits Business', *British Actuarial Journal*, Vol. 10, No. 2, pp. 185–222.

Dyson, A.C.L. and C.J. Exley (1995), 'Pension Fund Asset Valuation and Investment', *British Actuarial Journal*, Vol. 1, No. 3, pp. 471–557.

Edmonds, T.R. (1832), *Life Tables, Founded upon the Discovery of a Numerical Law Regulating the Existence of Every Human Being*. James Duncan.

Edmonds, T.R. (1837–38). 'On the Influence of Age and Selection on the Mortality of the Members of the Equitable Life Insurance Society During a Period of Sixty-Seven Years, Ending in 1829', *The Lancet*, Vol. 1, No. 739 (1837–38), pp. 154–162.

Elderton, W. Palin (1934), 'An Approximate Law of Survivorship and Other Notes on the use of Frequency Curves in Actuarial Statistics', *Journal of the Institute of Actuaries*, Vol. 65, No. 1 (March 1934), pp. 1–36.

Epps, G.S.W. (1921), 'Superannuation Funds. Notes on Some Post-War Problems, Together with an Account of a Pensioners' Mortality Experience (Civil Service Pensioners, 1904–1914)', *Journal of the Institute of Actuaries*, Vol. 52, No. 4, pp. 405–453.

Euler, Leonhard (1749), *Recherches sur la question des inegalites du movement de Saturne et Jupiter, sujet propose pour le prix de l'annee 1748*, par l'Academie royal des sciences de Paris.

Exley, C.J., S.J.B. Mehta and A.D. Smith (1997), 'The Financial Theory of Defined Benefit Pension Schemes', *British Actuarial Journal*, Vol. 3, No. 4, pp. 835–966.

Fama, E. (1965), 'The Behaviour of Stock Market Prices', *Journal of Business*, Vol. 38, No. 1 (January 1965), pp. 34–105.

Fama, E. (1970), 'Efficient Capital Markets: A Review of Theory and Empirical Work', *Journal of Finance*, Vol. 25, No. 2 (May 1970), pp. 383–417.

Fama, E. and M. Blume (1966), 'Filter Rules and Stock Market Trading Profits', *Journal of Business*, Vol. 30 (Special Supplement, January 1966), pp. 226–241.

Fama, E. and K. French (1988a), 'Permanent and Temporary Components of Stock Prices', *Journal of Political Economy*, Vol. 96, pp. 246–273.

Fama, E. and K. French (1988b), 'Dividend Yields and Expected Stock Returns', *Journal of Financial Economics*, Vol. 22, pp. 3–25.

Financial Services Authority (2002), Consultation Paper 143.

Financial Services Authority (2003), Consultation Paper 195.

Finlaison, John (1829), *Report of John Finlaison, Actuary of the National Debt, on the Evidence and Elementary Facts on which the Tables of Life Annuities are founded.*

Fisher, Irving (1896), 'Appreciation and Interest', *Publications of the American Economic Association* (1896).

Fisher, Irving (1930), *The Theory of Interest.* Macmillan.

Fisher, Sir Ronald (1956), *Statistical Methods and Scientific Inference.* Oliver and Boyd.

Ford, A. et al (1980), 'Report of the Maturity Guarantees Working Party', *Journal of the Institute of Actuaries*, Vol. 107, No. 2, pp. 103–112.

Francis, John (1853), *Annals, Anecdotes and Legends; A Chronicle of Life Assurance.* Longman, Brown, Green & Longmans.

Galloway, Thomas (1841), *Tables of Mortality Deduced from the Experience of the Amicable Society for a Perpetual Insurance Office.*

Gauss, Carl Friedrich (1809), *Theoria motus corporum celestium.*

Geoghegan, T.J. et al (1992), 'Report on the Wilkie Stochastic Investment Model', *Journal of the Institute of Actuaries*, Vol. 119, pp. 173–228.

Geske, R. (1979), 'The Valuation of Compound Options', *Journal of Financial Economics*, Vol. 7, No. 1 (March 1979), pp. 63–81.

Gilley, D.F. (1972), 'The Dissolution of a Pension Fund', *Journal of the Institute of Actuaries*, Vol. 98, No. 3, pp. 179–232.

Gilley, D.F. and D. Funnell (1958), 'Valuation of Pension Fund Assets', *Journal of the Institute of Actuaries Students' Society*, Vol. 15, No. 1, pp. 43–68.

Gompertz, Benjamin (1825), *On the Nature of the Function Expressive of the Law of Human Mortality, and on a New Mode of Determining the Value of Life Contingencies.*

Gompertz, Benjamin (1871), 'On one Uniform Law of Mortality from Birth to extreme Old Age, and on the Law of Sickness', *Journal of the Institute of Actuaries,* Vol. 16, No. 5 (October 1871), pp. 329–344.

Graunt, John (1662), *Natural and Political Observations Mentioned in a Follow Index, and Made Upon the Bills of Mortality by John Graunt, Citizen of London.*

Greenwood, P.M. and T.W. Keogh (1997), 'Pension Funding and Expensing in the Minimum Funding Requirement Environment', *British Actuarial Journal,* Vol. 3, No. 3, pp. 497–582.

Hacking, Ian (1965), *Logic of Statistical Inference.* Cambridge University Press.

Hacking, Ian (1975), *The Emergence of Probability.* Cambridge University Press.

Hairs, C.J. et al (2002), 'Fair Valuation of Liabilities', *British Actuarial Journal,* Vol. 8, No. 2, pp. 203–340.

Halley, Edmond (1693), *An Estimate of the Degrees of the Mortality of Mankind, drawn from curious Tables of the Births and Funerals at the City of Breslau, with an Attempt to ascertain the Price of Annuities upon Lives.*

Hare, D.J.P. et al (2000), 'A Market-Based Approach to Pricing With-Profit Guarantees', *British Actuarial Journal,* Vol. 6, I, pp. 143–213.

Harrison, M. and D.M. Kreps (1979), 'Martingales and the Arbitrage in Multiperiod Securities Markets', *Journal of Economic Theory,* Vol. 20, pp. 381–408.

Harrison, M. and S.R. Pliska (1981), 'Martingales and Stochastic Integrals in the Theory of Continuous Trading', *Stochastic Processes and Their Applications,* Vol. 11, pp. 215–260.

Haynes, A.T. and R.J. Kirton (1952), 'The Financial Structure of a Life Office', *Transactions of the Faculty of Actuaries,* Vol. 21, No. 178, pp. 141–218.

Heath, D., R.A. Jarrow and A. Morton (1992), 'Bond Pricing and the Term Structure of Interest Rates: A New Methodology for Contingent Claims Valuation', *Econometrica,* Vol. 60, No. 1 (January 1992), pp. 77–105.

Heywood, G. and M. Lander (1961), 'Pension Fund Valuations in Modern Conditions', *Journal of the Institute of Actuaries,* Vol. 87, No. 3, pp. 314–370.

Hibbert, A.J. and C.J. Turnbull (2003), 'Measuring and Managing the Economic Risks and Costs of With-Profit Business', *British Actuarial Journal,* Vol. 9, No. 4, pp. 725–786.

Hicks, J. (1939), *Value and Capital.* Oxford University Press.

Hill, R.D. (1979), 'Profit Regulation in Property-Liability Insurance', *Bell Journal of Economics,* Vol. 10, pp. 172–191.

Holbrook, J.P. (1977), 'Investment Performance of Pension Funds', *Journal of the Institute of Actuaries,* Vol. 104, No. 1, pp. 15–91.

Homans, Sheppard (1863), 'On the Equitable Distribution of Surplus', *Journal of the Institute of Actuaries,* Vol. 11, No. 3 (October 1863), pp. 121–129.

Homer, Sidney and Richard Sylla (1996), *A History of Interest Rates, Third Edition Revised.* Rutgers University Press.

Hooker, N.D. et al (1996), 'Risk-Based Capital in General Insurance', *British Actuarial Journal,* Vol. 2, No. 2, pp. 265–323.

Hull, J.C. and A.D. White (1990), 'Pricing Interest Rate Derivatives', *Review of Financial Studies*, Vol. 3, No. 4, pp. 573–592.
Hume, David (1739), *A Treatise of Human Nature, Being an Attempt to Introduce the Experimental Method of Reasoning into Moral Subjects.*
Huygens, C. (1657), *Ratiociniis in aleae ludo.*
Institute of Actuaries (1869), *The Mortality Experience of Life Assurance Companies.* Charles and Edwin Layton.
Jellicoe, C. (1851), 'On the Determination and Division of Surplus, and on the Modes of Returning it to the Contributors', *The Assurance Magazine*, Vol. 1, No. 2, pp. 22–28.
Jellicoe, C. (1852), 'On the Conditions which give rise to Surplus in Life Assurance Companies, and on the Amount of the Return, or "Bonus", which such Conditions justify', *The Assurance Magazine,* Vol. 2, No. 4, pp. 333–341.
Jensen, M. (1978), 'Some Anomalous Evidence Regarding Market Efficiency', *Journal of Financial Economics,* Vol. 6, No. 2 (1978), pp. 95–101.
Jensen, M. and W.H. Meckling (1976), 'Theory of the Firm: Managerial Behaviour, Agency Costs and Ownership Structure', *Journal of Financial Economics,* Vol. 3, No. 4, pp. 305–360.
Johnson, P.D. and G.B. Hey (1971), 'Statistical Studies in Motor Insurance', *Journal of the Institute of Actuaries,* Vol. 97, No. 2/3, pp. 199–249.
Kendall, M. (1953) 'The Analysis of Economic Time-Series, Part I: Prices', *Journal of the Royal Statistical Society* (Series A, General), Vol. 116, No. 1 (1953), pp. 11–34.
Kennedy, S.P.L. et al (1976), 'Bonus Distribution with High Equity Backing', *Journal of the Institute of Actuaries,* Vol. 103, pp. 11–58.
Keynes, J.M. (1936), *The General Theory of Employment, Interest and Money.* Palgrave Macmillan.
Keynes, J.M. (1983), edited by D. Moggridge. *The collected writings of John Maynard Keynes, Volume XII.* Cambridge University Press.
King, G. (1905), 'On Staff Pension Funds', *Journal of the Institute of Actuaries,* Vol. 39, No. 2, pp. 129–206.
Knight, Roger (2013), *Britain Against Napoleon.* Allen Lane.
Laplace, Pierre Simon (1774), *Memoire sur la probabilite des causes par les evenemens.*
Laplace, Pierre Simon (1781), *Memoire sur les probabilities.*
Laplace, Pierre Simon (1810), *Memoire sur les approximations des formules qui sont fonctions de tres grand nombres et sur leur application aux probabilities.*
Leeson, Francis (1968), *A Guide to the Records of the British State Tontines and Life Annuities of the 17th and 18th Centuries.* Pinhorns.
Legendre, Adrien Marie (1805), *Nouvelles methods pour la determination des orbites des cometes.*
Lintner, J.V. (1965), 'The Valuation of Risk Assets and the Selection of Risky Investments in Stock Portfolios and Capital Budgets', *Review of Economics and Statistics,* Vol. 47, No. 1 (February 1965), pp. 13–37.
Lochhead, R.K. (1932), *Valuation and Surplus.* Cambridge University Press.

Lundberg, F. (1909), 'Zur Theorie der Ruckversicherung', *Transactions of the International Congress of Actuaries.*

Macaulay, Frederick R. (1938), *The Movements of Interest Rates, Bond Yields and Stock Prices in the United States Since 1856.* Reprinted by Risk Books.

Macdonald, R.E. (1977), 'Presidential Address', *Transactions of the Faculty of Actuaries*, Vol. 36, No. 253, pp. 1–16.

Mackenzie, A.G. (1891), 'On the Practice and Powers of Assurance Companies in regard to the Investment of their Life Assurance Funds', *Journal of Institute of Actuaries*, Vol. 29, No. 3, pp. 185–232.

Maclean, J.B. (1948), 'Some Recent Actuarial Developments in the United States of America', *Transactions of the Faculty of Actuaries*, Vol. 18, No. 169, pp. 281–321.

Makeham, William Matthew (1859), 'On the Law of Mortality and the Construction of Annuity Tables', *Journal of the Institute of Actuaries*, Vol. 8, pp. 301–310.

Makeham, William Matthew (1867), 'On the Law of Mortality', *Journal of the Institute of Actuaries*, 13, pp. 325–358.

Makeham, William Matthew (1870), 'On the Objections to the Net-Premium Mode of Valuation', *Journal of the Institute of Actuaries*, Vol. 15, No. 6 (July 1870), pp. 449–452.

Makeham, William Matthew (1889), 'On the Further Development of Gompertz's Law', *Journal of the Institute of Actuaries*, Vol. 28, No. 2 (October 1889), pp. 152–159.

Manly, H.W. (1911), 'On Staff Pension Funds: The Progress of the Accumulation of the Funds; The Identity of a Valuation with the Future Progress of a Fund; The Manner of dealing with the Funds which are insolvent; and Sundry Observations', *Journal of the Institute of Actuaries*, Vol. 45, No. 2, pp. 149–231.

Markowitz, Harry (1952), 'Portfolio Selection', *Journal of Finance*, Vol. 7, No. 1 (March 1952).

Marsh, T. and R.C. Merton (1986), 'Dividend Variability and Variance Bound Tests for the Rationality of Stock Market Prices', *American Economic Review*, Vol. 76, No. 3, pp. 483–398.

Maudling, R.G. (1929), 'On the Classification and Duration of Compensation Claims in the Mining Industry', Vol. 60, No. 3, pp. 251–296.

May, G.E. (1912), 'The Investment of Life Assurance Funds', *Journal of the Institute of Actuaries*, Vol. 46, No. 2 (April 1912), pp. 134–168.

McKelvey, K.M. (1957), 'Pension Fund Finance', *Transactions of the Faculty of Actuaries*, Vol. 25, No. 193, pp. 113–165.

McLeish, D.J.D. (1983), 'A Financial Framework for Pension Funds', *Transactions of the Faculty of Actuaries*, Vol. 38, pp. 267–314.

McLeish, D.J.D. and C.M. Stewart (1987), 'Objectives and Methods of Funding Defined Benefit Pension Schemes', *Journal of the Institute of Actuaries*, Vol. 114, No. 2, pp. 155–225.

Meech, Levi W. (1881), *System and Tables of Life Insurance: A Treatise Developed from the Experience and Records of Thirty American Life Offices Under the Direction of a Committee of Actuaries.*
Merton, R.C. (1973), 'Theory of Rational Option Pricing', *Bell Journal of Economics and Management*, Vol. 4, No. 1 (Spring 1973), pp. 141–183.
Merton, R.C. (1974), 'On the Pricing of Corporate Debt: The Risk Structure of Interest Rates', *Journal of Finance*, Vol. 29, No. 2 (May 1974), pp. 449–470.
Merton, R.C. (1990), *Continuous-Time Finance*. Blackwell.
Miller, M. (1977), 'Debt and Taxes', *Journal of Finance*, May 1977, pp. 261–275.
Miller, Merton and Franco Modigliani (1961) 'Dividend Policy, Growth, and the Valuation of Shares', *Journal of Business*, Vol. 34, No. 4, pp. 411–433.
Miller, T. (1857), 'Some Suggestions respecting Fire Insurance Statistics', *Journal of the Institute of Actuaries*, Vol. 6, No. 6, pp. 333–344.
Miller, T. (1880), 'Fire Insurance. A Theory of Statistics', *Journal of the Institute of Actuaries*, Vol. 22, No. 2, pp. 103–117.
Milne, Joshua (1815), *A Treatise on the Valuation of Annuities and Assurances on Lives and Survivorships*. Longman, Hurst, Rees, Orme and Brown. Reprinted by Elibron Classics.
Mitchell, B.R. (1988), *British Historical Statistics*. Cambridge University Press
Modigliani, Franco and Merton Miller (1958) 'The Cost of Capital, Corporation Finance and the Theory of Investment', *American Economic Review*, Vol. 48, No. 3 (June 1958), pp. 261–297.
Modigliani, F. and R. Sutch (1966), 'Innovations in Interest Rate Policy', *American Economic Review*, (May 1966), pp. 178–197.
Moody, P.E. (1964), 'Life Funds and Equity Investment', *Journal of the Institute of Actuaries*, Vol. 90, No. 2 (September 1964), pp. 175–210.
Morgan, William (1779), *The Principles and Doctrines of Assurances, Annuities on Lives, and Contingent Reversions*. [Note the first edition was published in 1779, the second edition was published over forty years later, but the exact year of publication is unclear.]
Morgan, William (1829), *A View of the Rise and Progress of the Equitable Society, and of the Causes Which have Contributed to Its Success*. Longman, Rees, Orme, Brown and Green.
Murray, A.C. (1937), 'The Investment Policy of Life Assurance Offices'. *Transactions of the Faculty of Actuaries*, Vol. 16, No. 152, pp. 247–284.
Nicoll, J. (1902), 'The Actuarial Aspects of Recent Legislation, in the United Kingdom and other Countries, on the subject of Compensation to Workmen for Accidents', *Journal of the Institute of Actuaries*, Vol. 36, No. 5/6, pp. 417–564.
Ogborn, Maurice Edward (1962), *Equitable Assurances*. George Allen and Unwin.
Ogborn, M.E. (1953), 'On the Nature of the Function Expressive of the Law of Human Mortality'. *Journal of the Institute of Actuaries*, Vol. 79, pp. 170–212.
Osborne, M.F.M. (1959), 'Brownian Motion in the Stock Market', *Operational Research*, Vol. 7, No. 2, pp. 145–173.

Parliamentary Papers (1825), *Report of the Select Committee to Consider the Laws Respecting the Friendly Societies.*

Parliamentary Papers (1853), *Report of the Select Committee on Assurance Associations.*

Pegler, J. B. H (1948), 'The Actuarial Principles of Investment'. *Journal of the Institute of Actuaries*, Vol. 74, pp. 179–211.

Penman, William (1911), 'On the Valuation of the Liabilities of an Insurance Company under its Employers' Liability Contracts', *Journal of the Institute of Actuaries*, Vol. 45, No. 2, pp. 101–149.

Penman, William (1933), 'A Review of Investment Principles and Practices'. *Journal of the Institute of Actuaries*, Vol. 64, No. 3, pp. 384–427.

Pentikainen, T. (1952), 'On the Net Retention and Solvency of Insurance companies', *Scandinavian Actuarial Journal.*

Pentikainen, T. and Rantala (1982), *Solvency of Insurers and Equalisation Reserves.* Helsinki.

Plackett, R.L. (1971), 'Risk Theory', *Transactions of the Faculty of Actuaries*, Vol. 32, No. 237, pp. 337–362.

Pollard, J.H. (1982), 'Outstanding Claims Provisions: A Distribution-Free Statistical Approach', *Journal of the Institute of Actuaries*, Vol. 109, No. 3, pp. 417–433.

Porteous, D.A. (1936), *Pension and Widows' and Orphans' Funds.* Cambridge University Press.

Price, Richard (1772), *Observations on Reversionary Payments; on Schemes for Providing Annuities for Widows, and for Persons in Old Age; To Which Are Added Four Essays.* T. Cadfll.

Puckridge, C.E. (1948), 'The Rate of Interest which Should be Employed in the Valuation of Pension Fund and the Values which Should be Placed on Existing Investments'. *Journal of Institute of Actuaries*, Vol. 74, No. 1, pp. 1–30.

Raynes, H.E. (1928), 'The Place of Ordinary Stocks and Shares (as distinct from Fixed Interest bearing Securities) in the Investment of Life Assurance Funds'. *Journal of the Institute of Actuaries*, Vol. 59, No. 1, pp. 21–50.

Raynes, H.E. (1937), 'Equities and Fixed Interest Stocks During Twenty-Five Years'. *Journal of the Institute of Actuaries*, Vol. 68, No. 4, pp. 483–507.

Redington, F.M. (1952), 'Review of the Principles of Life-Office Valuations'. *Journal of the Institute of Actuaries*, Vol. 78, No. 3, pp. 286–340.

Redington, F.M. (1981), 'The Flock and the Sheep and Other Essays'. *Journal of the Institute of Actuaries*, Vol. 108, pp. 361–404.

Redington, F.M. (1982), 'The Phase of Transition – An Historical Essay'. *Journal of the Institute of Actuaries*, Vol. 109, No. 1, pp. 83–96.

Reid, D.H. (1978), 'Claim Reserving in General Insurance', *Journal of the Institute of Actuaries*, Vol. 105, No. 3, pp. 211–315.

Renshaw, A.E. (1989), 'Chain Ladder and Interactive Modelling (Claims Reserving and GLIM)', *Journal of the Institute of Actuaries*, Vol. 116, No. 3, pp. 559–587.

Roll, Richard (1988), 'R^2', *Journal of Finance*, Vol. 43, No. 3 (July 1988), pp. 541–566.

Ross, S.A. (1976), 'The Arbitrage Theory of Capital Asset Pricing', *Journal of Economic Theory*, Vol. 13, No. 3, pp. 341–360.

Ross, S.A. (1977), 'The Determination of Financial Structure: The Incentive Signalling Approach', *Bell Journal of Economics*, Spring 1977, pp. 23–40.

Rubinstein, M. (2006), *A History of the Theory of Investments.* Wiley.

Ryan, Gerald H. (1903), 'Methods of Valuation and Distribution of Profits in the United Kingdom'. *Journal of the Institute of Actuaries*, Vol. 38, No. 1, pp. 69–97.

Ryan, J.P. and K.P.W. Larner (1990), 'The Valuation of General Insurance Companies', *Journal of the Institute of Actuaries*, Vol. 117, No. 3, pp. 597–669.

Ryder, J.M. (1976), 'Subjectivism – A Reply in Defence of Classical Actuarial Methods', *Journal of the Institute of Actuaries*, Vol. 103, No. 1, pp. 59–112.

Samuelson, P.A. (1960), 'The St. Petersburg Paradox as a Divergent Double Limit'. *International Economic Review*, Vol. 1, No. 1.

Samuelson, P.A. (1965), 'Rational Theory of Warrant Pricing', *Industrial Management Review*, Vol. 6, pp. 13–39.

Samuelson, P.A. (1965), 'Proof That Properly Anticipated Prices Fluctuate Randomly', *Industrial Management Review*, Vol. 6, No. 1 (Spring 1965), pp. 41–49.

Samuelson, P.A. and R.C. Merton (1969), 'A Complete Model of Warrant Pricing that Maximises Utility', *Sloan Management Review*, Winter 1969, pp. 17–46.

Scott, Peter (2002), *Towards the 'Cult of Equity'? Insurance Companies and the Interwar Capital Market.*

Scott, P.G. et al (1996), 'An Alternative to the Net Premium Valuation Method for Statutory Reporting', *British Actuarial Journal*, Vol. 2, No. 3, pp. 527–621.

Scott, W.F. (1977), 'A Reserve Basis for Maturity Guarantees in Unit-Linked Life Assurance', *Transactions of the Faculty of Actuaries*, Vol. 35, pp. 365–415.

Seal, H. (1969), 'Simulation of the Ruin Potential of Non-Life Insurance Companies', *Transactions of the Society of Actuaries*, Vol. 21.

Sharpe, W.F. (1963), 'A Simplified Model for Portfolio Analysis', *Management Science*, Vol. 9, No. 2 (January 1963), pp. 277–293.

Sharpe, W.F. (1964), 'Capital Asset Prices: A Theory of Market Equilibrium under Conditions of Risk', *Journal of Finance*, Vol. 19, No. 3 (September 1964), pp. 425–442.

Sheldon, T.J. and A.D. Smith (2004), 'Market-Consistent Valuation of a Life Assurance Business', *British Actuarial Journal*, Vol. 10, No. 3, pp. 543–626.

Shiller, R.J. (1981), 'Do Stock Prices Move Too Much to Be Justified by Subsequent Changes in Dividends?' *American Economic Review*, Vol. 71, No. 3 (June 1981), pp. 421–436.

Skelman, R.S. (1968), 'The Assessment and Distribution of Profits from Life Business', *Journal of the Institute of Actuaries*, Vol. 94, No. 1, pp. 53–100.

Skidelsky, Robert (2003), *John Maynard Keynes 1883 – 1946: Economist, Philosopher, Statesman.* Penguin Books.

Smith, Edgar Lawrence (1924), *Common Stocks as Long Term Investments.* Macmillan.

Sprague, Thomas Bond (1863), 'On Certain Methods for the Valuation of Liabilities of a Life Assurance Company'. *Journal of the Institute of Actuaries*, Vol. 11, No. 2, pp. 90–108.

Sprague, Thomas B. (1870), 'On the Proper Method of Estimating the Liability of a Life Insurance Company under its Policies'. *Journal of the Institute of Actuaries*, Vol. 15, No. 6, pp. 411–432.

Stigler, Stephen M. (1986), *The History of Statistics*. Belknap Harvard.

Stiglitz, J.E. (1972), 'Some Aspects of the Pure Theory of Corporate Finance: Bankruptcies and Takeovers', *Bell Journal of Economics and Management Science*, Autumn 1972, pp. 458–482.

Supple, Barry (1970), *The Royal Exchange Assurance: A History of British Insurance 1720–1970*. Cambridge University Press.

Suttie, T.R. (1944), 'The Treatment of Appreciation or Depreciation in the Assets of a Life Assurance Fund'. *Journal of the Institute of Actuaries*, Vol. 72, pp. 203–228.

Taylor, G.C. (1983), 'An Invariance Principle for the Analysis of Non-Life Insurance Claims', *Journal of the Institute of Actuaries*, Vol. 110, No. 1, pp. 205–242.

Taylor, G.C. (1984), 'Solvency Margin Funding For General Insurance Companies', *Journal of the Institute of Actuaries*, Vol. 111, No. 1, pp. 173–179.

Thiele, T.N. (1871), *En mathematisk Formel for Dodeligheden*. Reitzel, Kjobenhaven.

Thornton, P.N. and A.F. Wilson (1992), 'A Realistic Approach to Pension Funding', *Journal of the Institute of Actuaries*, Vol. 119, No. 2, pp. 229–312.

Tobin, J. (1958), 'Liquidity Preference as Behaviour Towards Risk', *Review of Economic Studies*, Vol. 25, No. 2 (February 1958), pp. 65–86.

Todhunter, Isaac (1865), *History of the Mathematical Theory of Probability From the Time of Pascal to That of Laplace*. London and Cambridge.

Trowbridge, J.R. (1977), 'Assessmentism – An Alternative to Pensions Funding?' *Journal of the Institute of Actuaries*, Vol. 104, No. 2, pp. 173–220.

Turner et al (1984), 'Terminology of pension funding methods'. Working Party Report, Institute of Actuaries.

Vasicek, O. (1977), 'An Equilibrium Characterization of the Term Structure', *Journal of Financial Economics*, Vol. 5, pp. 177–188.

Verrall, R.J. (1989), 'A State Space Representation of the Chain Ladder Linear Model', *Journal of the Institute of Actuaries*, Vol. 116, No. 3, pp. 589–609.

Walford, Cornelius (1868), *The Insurance Guide and Hand-Book*. Charles and Edwin Layton.

Walford, Cornelius (1878), 'On the Scientific Application of Data to the Purpose of Deducing Rates of Premium for Fire Insurance', *Journal of the Institute of Actuaries and Assurance Magazine*, Vol. 21, No. 1, pp. 1–37.

Warner, S.G. (1918), 'Opening Address by the President', *Journal of the Institute of Actuaries*, Vol. 51, No.1, pp. 1–24.

Watson, Alfred W. (1903), *An Account of an Investigation of the Sickness and Mortality Experience of the I.O.O.F. Manchester Unity During the Five Years 1893–1897*. Cambridge University Press (re-printed 1953).

Whittall, William (1882), 'On the Rates of Fatal Accidents in Various Occupations', *Journal of the Institute of Actuaries and Assurance Magazine*, Vol. 23, No. 3, pp. 188–220.
Whyte, Lewis G. (1947), 'Life Office Investments in a Planned Economy'. *Transactions of the Faculty of Actuaries*, Vol. 18, No. 168, pp. 215–259.
Wilkie, A.D. (1977), 'Maturity (and other) Guarantees Under Unit Linked Policies', *Transactions of the Faculty of Actuaries*, Vol. 36, pp. 27–41.
Wilkie, A.D. (1986), 'A Stochastic Investment Model for Actuarial Use', *Transactions of the Faculty of Actuaries*, Vol. 39, p. 341–403.
Wilkie, A.D. (1986; 2), 'Some Applications of Stochastic Investment Models', *Journal of the Institute of Actuaries Students' Society*, Vol. 29, pp. 25–51.
Wilkie, A.D. (1987), 'An Option Pricing Approach to Bonus Policy', *Journal of the Institute of Actuaries*, Vol. 114, pp. 21–90.
Wilkie, A.D. (1995), 'More on a Stochastic Asset Model for Actuarial Use', *British Actuarial Journal*, Vol. 1, p. 777–964.
Wilkie, A.D., H.R. Waters and S. Yang (2003), 'Reserving, Pricing and Hedging For Policies with Guaranteed Annuity Options', *British Actuarial Journal*, Vol. 9, pp. 263–391.
Williams, John Burr (1938), *The Theory of Investment Value.* BN Publishing.
Wise, A.J. (1985), 'The Matching of Assets to Liabilities', *Transactions of the Faculty of Actuaries*, Vol. 40, pp. 18–81.
Woolhouse, W.S.B. (1839), *Investigation of Mortality in the Indian Army.*
Wright, P.W. et al (1998), 'A Review of the Statutory Valuation of Long-Term Insurance Business in the United Kingdom', *British Actuarial Journal*, Vol. 4, No. 4, pp. 803–864 and *British Actuarial Journal*, Vol. 4, No. 5, pp. 1029–1058.
Wright, T.S. (1990), 'A Stochastic Method for Claim Reserving in General Insurance', *Journal of the Institute of Actuaries*, Vol. 117, No. 3, pp. 677–731.
Zillmer, A. (1863), *Beitrage zur Theorie der Pramien-Reserve bei Lebens-Versicherungs-Anstalten.*

Index

C. Turnbull, *A History of British Actuarial Thought*,
DOI 10.1007/978-3-319-33183-6

Printed by Printforce, the Netherlands